Topological Degree Approach to Bifurcation Problems

Topological Fixed Point Theory and Its Applications

VOLUME 5

For other titles published in this series, go to
www.springer.com/series/6622

Michal Fečkan

Topological Degree Approach
to Bifurcation Problems

 Springer

Michal Fečkan
Department of Mathematical Analysis
and Numerical Mathematics
Faculty of Mathematics, Physics
and Informatics
Comenius University
Mlynská dolina
842 48 Bratislava
Slovakia

ISBN 978-90-481-7969-5   e-ISBN 978-1-4020-8724-0

To my beloved family

Contents

Chapter 1

Introduction

1.1 Preface

Many phenomena from physics, biology, chemistry and economics are modeled by differential equations with parameters. When a nonlinear equation is established, its behavior/dynamics should be understood. In general, it is impossible to find a complete dynamics of a nonlinear differential equation. Hence at least, either periodic or irregular/chaotic solutions are tried to be shown. So a property of a desired solution of a nonlinear equation is given as a parameterized boundary value problem. Consequently, the task is transformed to a solvability of an abstract nonlinear equation with parameters on a certain functional space. When a family of solutions of the abstract equation is known for some parameters, the persistence or bifurcations of solutions from that family is studied as parameters are changing. There are several approaches to handle such nonlinear bifurcation problems. One of them is a topological degree method, which is rather powerful in cases when nonlinearities are not enough smooth. The aim of this book is to present several original bifurcation results achieved by the author using the topological degree theory. The scope of the results is rather broad from showing periodic and chaotic behavior of non-smooth mechanical systems through the existence of traveling waves for ordinary differential equations on infinite lattices up to study periodic oscillations of undamped abstract wave equations on Hilbert spaces with applications to nonlinear beam and string partial differential equations.

1.2 An Illustrative Perturbed Problem

For solving parameterized problems, we often apply the perturbation method, which is one of the most powerful method used in nonlinear smooth mechanics. This perturbation approach is by now known as the Melnikov method for the persistence/bifurcation of either periodics or homoclinics/heteroclinics [108]. To illustrate this, let us consider a periodically forced nonlinear oscillator like the

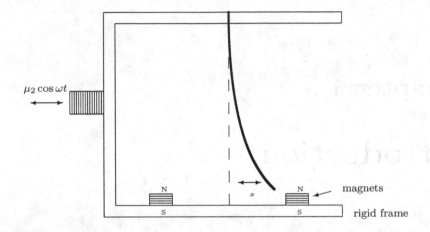

$\mu_2 \cos \omega t$

N   N   magnets

S   S   rigid frame

Figure 1.1: The magneto-elastic beam

following perturbed Duffing equation

$$\dot{x} = y, \quad \dot{y} - x + 2x^3 + \mu_1 y = \mu_2 \cos \omega t \qquad (1.2.1)$$

with $\mu_{1,2}$ small. Note

$$\ddot{x} + \mu_1 \dot{x} - x + 2x^3 = \mu_2 \cos \omega t$$

describes dynamics of a buckled beam, when only one mode of vibration is considered [115] (see also Section 8.10). In particular, an experimental apparatus in [108, pp. 83–84] is a slender steel beam clamped to a rigid framework which supports two magnets, when x is the beam's tip displacement. The apparatus is periodically forced using electromagnetic vibration generator (see Fig. 1.1).

Next, the phase portrait of

$$\dot{x} = y, \quad \dot{y} - x + 2x^3 = 0 \qquad (1.2.2)$$

is simply to find (see Fig. 1.2). There are three equilibria: $(0,0)$ is hyperbolic and $(\pm\sqrt{2}/2, 0)$ are centers. There is also a symmetric homoclinic cycle $\pm\tilde{\gamma}(t)$ with $\tilde{\gamma}(t) = (\gamma(t), \dot{\gamma}(t))$ and $\gamma(t) = \operatorname{sech} t$. The rest are all periodic solutions.

These results are consistent with the above experimental model without damping and external forcing as follows: When attractive forces of the magnets overcome the elastic force of the beam then the beam settles with its tip close to one or the other of the magnets: these are centers of (1.2.2). There is also an unstable central equilibrium position of the beam at which the magnetic forces cancel: this is the unstable equilibrium of (1.2.2).

When $\mu_{1,2}$ are small and not identically zero, then in spite of the fact that (1.2.1) is simple looking, its dynamics is very difficult. So as the first step, we try to show at least the persistence of either periodic or homoclinic solutions.

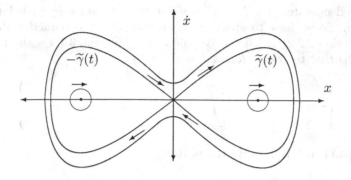

Figure 1.2: The phase portrait of $\ddot{x} - x + 2x^3 = 0$

Here we concentrate on the homoclinic case since the periodic one is similar. Since we use in this book functional-analytical methods, we explain it on this example and we refer the reader to [108, p. 184] for a geometrical approach. To find a solution near $(\gamma(t), \dot{\gamma}(t))$, we use the perturbation method, i.e. we first make the change of variables and parameters

$$x(t + \alpha) \leftrightarrow x(t) + \gamma(t), \quad y(t + \alpha) \leftrightarrow y(t) + \dot{\gamma}(t), \quad \mu_{1;2} \leftrightarrow \varepsilon\mu_{1,0;2,0} \quad (1.2.3)$$

in (1.2.1) to get

$$\dot{x} = y,$$
$$\dot{y} + (6\gamma^2 - 1)x + 2x^3 + 6x^2\gamma + \varepsilon\mu_{1,0}(y + \dot{\gamma}) = \varepsilon\mu_{2,0}\cos\omega(t + \alpha). \quad (1.2.4)$$

Here ε is small while $\mu_{1,0}^2 + \mu_{2,0}^2 = 1$ are fixed. To put (1.2.4) in a general functional framework, we take the Banach spaces $V := C_b(\mathbb{R}, \mathbb{R}^2)$ and $Z := C_b^1(\mathbb{R}, \mathbb{R}^2)$ – the spaces of bounded (together with the first derivatives) functions $z : \mathbb{R} \to \mathbb{R}^2$ with the usual supremum norms $\|z\|_0 := \sup_{\mathbb{R}} |z(t)|$ and $\|z\|_1 := \sup_{\mathbb{R}} |z(t)| + \sup_{\mathbb{R}} |\dot{z}(t)|$, respectively. $\mathbb{R}$ is the field of real numbers. We note that we are looking for homoclinic solutions, i.e. which belong to $C_b^1(\mathbb{R}, \mathbb{R}^2)$. Next by putting

$$Lz := \left(\dot{x} - y, \dot{y} + (6\gamma^2 - 1)x\right),$$
$$N(z, \alpha, \varepsilon) := \left(0, -2x^3 - 6x^2\gamma + \varepsilon\mu_{2,0}\cos\omega(t + \alpha) - \varepsilon\mu_{1,0}(y + \dot{\gamma})\right)$$

with $z = (x, y)$, (1.2.4) has the form

$$Lz = N(z, \alpha, \varepsilon). \quad (1.2.5)$$

Note $N(0, \alpha, 0) = 0$, so $z = 0$ is a solution of (1.2.5) with $\varepsilon = 0$ and any $\alpha \in \mathbb{R}$. So for $\varepsilon = 0$, there is a trivial branch of solutions $(0, \alpha)$. We intend to find conditions that (1.2.5) has a small solution for $\varepsilon \neq 0$ small. Next, the linear bounded operator $L : Z \to V$ is not invertible, since $(\dot{\gamma}, \ddot{\gamma})$ belongs to its kernel $\mathcal{N}L$. So we have to apply the Lyapunov-Schmidt reduction method (cf. [56] or Section 2.2.2) as follows: We know (see Theorem 3.1.4 with $m \to \infty$, or [56, p. 380]) that the range $\mathcal{R}L$ of L is given by

$$\mathcal{R}L = \left\{ v = (v_1, v_2) \in V \mid \int_{-\infty}^{\infty} (v_2(s)\dot{\gamma}(s) - v_1(s)\ddot{\gamma}(s)) \, ds = 0 \right\}$$

and $\mathcal{N}L = \mathrm{span}\,\{(\dot{\gamma}, \ddot{\gamma})\}$. We take the projection

$$Qv := \frac{\int\limits_{-\infty}^{\infty} (v_2(s)\dot{\gamma}(s) - v_1(s)\ddot{\gamma}(s)) \, ds}{\int\limits_{-\infty}^{\infty} (\dot{\gamma}(s)^2 + \ddot{\gamma}(s)^2) \, ds} (-\ddot{\gamma}, \dot{\gamma}) \,.$$

So $\mathcal{R}L = \mathcal{N}Q$. Then we decompose (1.2.5) as follows

$$Lz - (\mathbb{I} - Q)N(z, \alpha, \varepsilon) = 0 \tag{1.2.6}$$

and

$$QN(z, \alpha, \varepsilon) = 0 \,. \tag{1.2.7}$$

Now
$$L : Z_1 := \{z \in Z \mid x(0)\dot{\gamma}(0) + y(0)\ddot{\gamma}(0) = 0\} \to \mathcal{R}L$$

is injective and surjective. By the Banach inverse mapping theorem, it is also continuously invertible. Note $\{(x, y) \in \mathbb{R}^2 \mid (x - \gamma(0))\dot{\gamma}(0) + (y - \dot{\gamma}(0))\ddot{\gamma}(0) = 0\}$ is the transversal section to the homoclinic curve $(\gamma(t), \dot{\gamma}(t))$ at $t = 0$. Since $N(0, \alpha, 0) = 0$ and $D_z N(0, \alpha, 0) = 0$, applying the implicit function theorem, we can uniquely solve (1.2.6) in $z = z(\alpha, \varepsilon) \in Z_1$ for ε small with $z(\alpha, 0) = 0$. Inserting this solution into (1.2.7) we get the bifurcation equation

$$B(\alpha, \varepsilon) := \int_{-\infty}^{\infty} \Big(- 2x(\alpha, \varepsilon)^3(s) - 6x(\alpha, \varepsilon)^2(s)\gamma(s)$$

$$+ \varepsilon\mu_{2,0} \cos \omega(s + \alpha) - \varepsilon\mu_{1,0}(y(\alpha, \varepsilon)(s) + \dot{\gamma}(s)) \Big) \dot{\gamma}(s) \, ds = 0 \,. \tag{1.2.8}$$

Since $z(\alpha, \varepsilon) = O(\varepsilon)$, we see that $B(\alpha, \varepsilon) = O(\varepsilon)$ as well. So instead of solving (1.2.8), we put

$$\tilde{B}(\alpha, \varepsilon) = \left\{ \begin{array}{ll} B(\alpha, \varepsilon)/\varepsilon & \text{for } \varepsilon \neq 0, \\ D_\varepsilon B(\alpha, 0) & \text{for } \varepsilon = 0 \end{array} \right.$$

and solve

$$\tilde{B}(\alpha, \varepsilon) = 0 \,. \tag{1.2.9}$$

We derive

$$M(\alpha) := D_\varepsilon B(\alpha, 0) = \int\limits_{-\infty}^{\infty} \Big(\mu_{2,0} \cos \omega(s + \alpha) - \mu_{1,0} \dot\gamma(s) \Big) \dot\gamma(s) \, ds$$

$$= \mu_{2,0} \pi\omega \operatorname{sech} \frac{\pi\omega}{2} \sin \alpha\omega - \mu_{1,0} \frac{2}{3} .$$

When $|\mu_{1,0}| < |\mu_{2,0}| \frac{3\pi\omega}{2} \operatorname{sech} \frac{\pi\omega}{2}$, then clearly there is a simple zero α_0 of M, i.e. $M(\alpha_0) = 0$ and $M'(\alpha_0) \neq 0$. So we can again apply the implicit function theorem to solve (1.2.9) in $\alpha = \alpha(\varepsilon)$ for ε small with $\alpha(0) = \alpha_0$. This gives the existence of a bounded solution of (1.2.1) close to $(\gamma(t - \alpha_0), \dot\gamma(t - \alpha_0))$ (see (1.2.3)) for any $\mu_{1,2}$ small satisfying

$$|\mu_1| < |\mu_2| \frac{3\pi\omega}{2} \operatorname{sech} \frac{\pi\omega}{2} . \tag{1.2.10}$$

Next using the same approach, we can show the existence of a unique small periodic solution of (1.2.1) for any $\mu_{1,2}$ small. With a little bit more effort we can prove that this periodic solution is hyperbolic and the above bounded solution accumulates on it as $t \to \pm\infty$ [108, pp. 184–212], [157]. Moreover under conditions (1.2.10), (1.2.1) is chaotic (see Sections 2.5.3 and 4.2.1 for more details). These chaotic vibrations are also observed in the experimental apparatus of Fig. 1.1 as it is shown in [108, p. 84].

1.3 A Brief Summary of the Book

Summarizing we see that in order to find a bounded solution of (1.2.1) for $\mu_{1,2}$ small, we use the following strategy

1. First we rewrite it as an abstract equation (1.2.5) in appropriate Banach spaces.

2. Then we use the Lyapunov-Schmidt decomposition method (1.2.6–1.2.7).

3. Next we derive the bifurcation equation (1.2.8) using the implicit function theorem.

4. Finally we find conditions for the solvability of the bifurcation equation (see (1.2.9) and the analysis below it). Usually we get the corresponding Melnikov function M.

We roughly follow this way in this book for various problems. Of course the above approach is well-known [56, 157–159]. But we intend to solve problems which are not enough smooth. So our first aim is to extend this Melnikov method for nonsmooth/discontinuous mechanical systems like

$$\dot x = y, \quad \dot y - x + 2x^3 + \mu_1 \operatorname{sgn} \dot x = \mu_2 \cos \omega t \tag{1.3.1}$$

for $\mu_{1,2}$ small (see (3.1.36)). Non-smooth differential equations occur in various situations like in mechanical systems with dry frictions or with impacts. They appear also in control theory, electronics, economics, medicine and biology [45, 57, 129–131] (see also Chapters 3, 4 and 8 of this book for additional references and examples).

The plan of this book is as follows. In Chapter 2 we briefly review some known mathematical results which we use in our proofs. In Chapter 3 we study bifurcations of periodics and subharmonics from either periodics or homoclinics for systems like (1.3.1). This is the first step to show chaos for discontinuous systems. There we also study systems with small hysteresis and weakly coupled nonlinear oscillators as well. In Chapter 4 we show desired chaotic solutions for discontinuous differential equations by extended the method of Chapter 3. Then we proceed in Chapter 5 with the study of chaos for diffeomorphisms when intersections of stable and unstable manifolds of hyperbolic fixed points are only topologically transversal. To handle this problem, again topological degree methods are necessary to use. There we also deals with accumulation of periodic points of reversible diffeomorphisms on homoclinic points with extensions of this phenomenon to chains of reversible oscillators. We continue in Chapter 6 with investigation of equations on lattices which are spatially discretized partial differential equations (p.d.eqns). There we apply the known center manifold method. We investigate the persistence of kink traveling waves of p.d.eqns under discretization. Chapter 7 is devoted to the existence of periodics and subharmonics of undamped abstract wave equations using methods to avoid resonant terms for perturbed problems. Then, in the final Chapter 8, we study discontinuous wave equations with infinitely many resonances. There we develop a degree for such problems with applications to local bifurcation problems.

We note that we use the known topological degree methods as a tool for solving concrete nonlinear problems. Only in the final chapter, we also establish a suitable theoretical topological degree background for discontinuous wave equations. Next, most results of this book are based on the author published papers, but we have improved and modified the original papers to simplify and clarify final results.

Bifurcation results solved in this book are local which means that only local branches of bifurcations are studied with small parameter changes. Global bifurcations like Krasnoselski-Rabinowitz theorem are given in [47, 124]. Next, we use a topological tool based on the Leray-Schauder degree theory and its generalizations. More sophisticated topological methods based on the Nielsen fixed point theory are presented in [5, 89].

The author is indebted to coauthors of results mentioned in this book: J. Awrejcewicz, F. Battelli, M. Franca, J. Gruendler, R. Ma, P. Olejnik, V.M. Rothos and B. Thompson. He also thanks to M. Medved' for many stimulating discussions on mathematics and to L. Górniewicz for initiation to write this book. Partial supports of Grant VEGA-MS 1/0098/08, Grant VEGA-SAV 2/7140/27 and an award from Literárny fond are also appreciated.

Bratislava, April, 2008   Michal Fečkan

Chapter 2

Theoretical Background

In this chapter, we recall some know mathematical notations, notions and results used later to help the reader with reading this book. All these results are presented in any textbooks of linear and nonlinear functional analysis, differential topology, and dynamical systems, which are quoted in the text, and we refer the reader for more details to these textbooks.

2.1   Linear Functional Analysis

Let X be a *Banach space* with a norm $|\cdot|$. By $\mathbb{N}$ we denote the set of natural numbers. A sequence $\{x_n\}_{n \in \mathbb{N}} \subset X$ *(strongly) converges* to $x_0 \in X$ if $|x_n - x_0| \to 0$ as $n \to \infty$, for short $x_n \to x_0$. The *dual space* of X is denoted by X^*. It is the linear space of all bounded linear functionals on X. It generates the *weak topology* on X as follows: For any $x \in X$, we define its *weak neighborhood* as a set $\cap_{i=1}^{j} \{y \in X \mid |x_i^*(y - x)| < r_i\}$ for some $x_i^* \in X^*$, $r_i > 0$, $i = 1, \ldots, j$. Then a subset S of X is *weakly open* if any its point has a weak neighborhood laying in S. A set is *weakly compact* if from every its covering with weakly open sets it is possible to select a finite covering. If the norm $|\cdot|$ is generated by a scalar product $(\cdot, \cdot)$, i.e. $|x| = \sqrt{(x, x)}$ for any x, then X is a *Hilbert space*.

Theorem 2.1.1. *Any convex closed bounded subset of a Hilbert space is weakly compact.*

A sequence $\{x_n\}_{n \in \mathbb{N}} \subset X$ *weakly converges* to $x_0 \in X$ if $x^*(x_n) \to x^*(x_0)$ as $n \to \infty$ for any $x^* \in X^*$, for short $x_n \rightharpoonup x_0$. We have the following *Mazur's theorem.*

Theorem 2.1.2. *If $x_n \rightharpoonup x_0$ then there is a sequence $\{y_n\}_{n \in \mathbb{N}} \subset X$ such that $y_n \in \mathrm{con}\,[\{x_m\}_{m \geq n}]$ and $y_n \to x_0$ in X.*

Here $\mathrm{con}\,[S]$ is the *convex hull* of a subset $S \subset X$, i.e. the intersection of all convex subsets of X containing S.

M. Fečkan, *Topological Degree Approach to Bifurcation Problems*, 7–21.
© Springer Science + Business Media B.V., 2008

Let X and Y be Banach spaces. The set of all linear bounded mappings $A : X \to Y$ is denoted by $L(X, Y)$, while we put $L(X) := L(X, X)$. $A \in L(X, Y)$ is *compact* if $\overline{A(B_1)}$ is compact in Y when $B_1 := \{x \in X \mid |x| \leq 1\}$ is the unit ball in X. Then $x_n \rightharpoonup x_0 \Rightarrow Ax_n \to Ax_0$. If $X \subset Y$ and the inclusion $X \hookrightarrow Y$ is bounded (compact) then we say that X *is continuously (compactly) embedded into* Y.

In using the Lyapunov-Schmidt method, we first need the following *Banach inverse mapping theorem*.

Theorem 2.1.3. *If $A \in L(X, Y)$ is surjective and injective then its inverse* $A^{-1} \in L(Y, X)$.

Then this lemma.

Lemma 2.1.4. *Let $Z \subset X$ be a linear subspace with either* $\dim Z < \infty$ *or Z be closed with* $\operatorname{codim} Z < \infty$. *Then there is a bounded projection $P : X \to Z$. Note* $\operatorname{codim} Z = \dim X/Z$ *and X/Z is the factor space of X with respect to Z.*

More details and proofs of the above results can be found in [60, 199].

2.2   Nonlinear Functional Analysis

2.2.1   Implicit Function Theorem

Let X, Y be Banach spaces and let $\Omega \subset X$ be open. A map $F : \Omega \to Y$ is said to be *(Fréchet) differentiable* at $x_0 \in \Omega$ if there is an $DF(x_0) \in L(X, Y)$ such that

$$\lim_{h \to 0} \frac{|F(x_0 + h) - F(x_0) - DF(x_0)h|}{|h|} = 0 \, .$$

If F is differentiable at each $x \in \Omega$ and $DF : \Omega \to L(X, Y)$ is continuous then F is said to be continuously differentiable on Ω and we write $F \in C^1(\Omega, Y)$. Higher derivatives are defined in the usual way by induction. Similarly, the partial derivatives are defined standardly [60, p. 46]. Now we state the *implicit function theorem* [56, p. 26].

Theorem 2.2.1. *Let X, Y, Z be Banach spaces, $U \subset X$, $V \subset Y$ are open subsets and $(x_0, y_0) \in U \times V$. Consider $F \in C^1(U \times V, Z)$ such that $F(x_0, y_0) = 0$ and $D_x F(x_0, y_0) : X \to Z$ has a bounded inverse. Then there is a neighborhood $U_1 \times V_1 \subset U \times V$ of (x_0, y_0) and a function $f \in C^1(V_1, X)$ such that $f(y_0) = x_0$ and $F(x, y) = 0$ for $U_1 \times V_1$ if and only if $x = f(y)$. Moreover, if $F \in C^k(U \times V, Z)$, $k \geq 1$ then $f \in C^k(V_1, X)$.*

We refer the reader to [31, 127] for more applications and generalizations of the implicit function theorem.

2.2.2 Lyapunov-Schmidt Method

Now we recall the well-known *Lyapunov-Schmidt method* for solving locally nonlinear equations when the implicit function theorem fails. So let X, Y, Z be Banach spaces, $U \subset X$, $V \subset Y$ are open subsets and $(x_0, y_0) \in U \times V$. Consider $F \in C^1(U \times V, Z)$ such that $F(x_0, y_0) = 0$. If $D_x F(x_0, y_0) : X \to Z$ has a bounded inverse then the implicit function theorem can be applied to solve

$$F(x, y)=0 \qquad (2.2.1)$$

near (x_0, y_0). So we suppose that $D_x F(x_0, y_0) : X \to Z$ has no a bounded inverse. In general this situation is difficult. The simplest case is when $D_x F(x_0, y_0)$: $X \to Z$ is *Fredholm*, i.e. $\dim \mathcal{N} D_x F(x_0, y_0) < \infty$, $\mathcal{R} D_x F(x_0, y_0)$ is closed in Z and $\operatorname{codim} \mathcal{R} D_x F(x_0, y_0) < \infty$. Here $\mathcal{N} A$ and $\mathcal{R} A$ are the *kernel* and *range* of a linear mapping A. Then by Lemma 2.1.4, there are bounded projections $P : X \to \mathcal{N} D_x F(x_0, y_0)$ and $Q : Z \to \mathcal{R} D_x F(x_0, y_0)$. Hence we split any $x \in X$ as $x{=}x_0{+}u{+}v$ with $u \in \mathcal{R}(\mathbb{I} - P)$, $v \in \mathcal{R} P$, and decompose (2.2.1) as follows

$$H(u, v, y) := Q F(x_0 + u + v, y) = 0, \qquad (2.2.2)$$

$$(\mathbb{I} - Q) F(x_0 + u + v, y) = 0. \qquad (2.2.3)$$

Observe that $D_u H(0, 0, y_0) = D_x F(x_0, y_0) | \mathcal{R}(\mathbb{I} - P) \to \mathcal{R} D_x F(x_0, y_0)$. So $D_u H(0, 0, y_0)$ is injective and surjective so by the Banach inverse mapping Theorem 2.1.3, it has a bounded inverse. Since $H(0, 0, y_0) = 0$, the implicit function theorem can be applied to solve (2.2.2) in $u = u(v, y)$ with $u(0, y_0) = 0$. Inserting this solution to (2.2.3) we get the *bifurcation equation*

$$B(v, y) := (\mathbb{I} - Q) F(x_0 + u(v, y) + v, y) = 0. \qquad (2.2.4)$$

Since $B(0, y_0) = (\mathbb{I} - Q) F(x_0, y_0) = 0$ and

$$D_v B(0, y_0) = (\mathbb{I} - Q) D_x F(x_0, y_0) (D_v u(0, y_0) + \mathbb{I}) = 0,$$

the function $B(v, y)$ has a higher singularity at $(0, y_0)$, so the implicit function theorem is not applicable, and the bifurcation theory must be used [56].

2.2.3 Leray-Schauder Degree

Let X be a Banach space and let $\Omega \subset X$ be open bounded. A continuous map $G \in C(\bar{\Omega}, X)$ is *compact* if $\overline{G(\Omega)}$ is compact in X. The set of all such maps is denoted by $K(\Omega)$. A triple (F, Ω, y) is *admissible* if $F = \mathbb{I} - G$ for some $G \in K(\Omega)$ (so F is a compact perturbation of identity) and $y \in X$ with $y \notin F(\partial \Omega)$, where $\partial \Omega$ is the border of a bounded open subset $\Omega \subset X$. A mapping $F \in C([0, 1] \times \bar{\Omega}, X)$ is an *admissible homotopy* if $F(\lambda, \cdot) = \mathbb{I} - G(\lambda, \cdot)$ with $G \in C([0, 1] \times \bar{\Omega}, X)$ compact, i.e. $\overline{G([0, 1] \times \bar{\Omega})}$ is compact in X, along with $y \notin F([0, 1] \times \partial \Omega)$. Let $\mathbb{Z}$ be the set of all integer numbers. Now on these admissible triples (F, Ω, y), there is a $\mathbb{Z}$-defined function deg [60, p. 56].

Theorem 2.2.2. *There is a unique mapping* deg *defined on the set of all admissible triples* (F, Ω, y) *determined by the following properties:*

(i) If $\deg(F, \Omega, y) \neq 0$ *then there is an* $x \in \Omega$ *such that* $F(x) = y$.

(ii) $\deg(\mathbb{I}, \Omega, y) = 1$ *for any* $y \in \Omega$.

(iii) $\deg(F, \Omega, y) = \deg(F, \Omega_1, y) + \deg(F, \Omega_2, y)$ *whenever* $\Omega_{1,2}$ *are disjoint open subsets of* Ω *such that* $y \notin F\left(\bar{\Omega} \setminus (\Omega_1 \cup \Omega_2)\right)$.

(iv) $\deg(F(\lambda, \cdot), \Omega, y)$ *is constant under an admissible homotopy* $F(\lambda, \cdot)$.

The number $\deg(F, \Omega, y)$ is called the *Leray-Schauder degree* of the map F. If $X = \mathbb{R}^n$ then $\deg(F, \Omega, y)$ is the classical *Brouwer degree* and F is just $F \in C(\bar{\Omega}, \mathbb{R}^n)$ with $y \notin F(\partial \Omega)$. If x_0 is an isolated zero of F in $\Omega \subset \mathbb{R}^n$ then $I(x_0) := \deg(F, \Omega_0, 0)$ is called the *Brouwer index* of F at x_0, where $x_0 \in \Omega_0 \subset \Omega$ is an open subset such that x_0 is the only zero point of F on Ω_0 (cf. [56, p. 69]). $I(x_0)$ is independent of such Ω_0. Note, if $y \in \mathbb{R}^n$ is a regular value of $F \in C^1(\bar{\Omega}, \mathbb{R}^n)$, i.e. $\det DF(x) \neq 0$ for any $x \in \Omega$ with $F(x) = y$, and $y \notin F(\partial \Omega)$, then $F^{-1}(y)$ is finite and $\deg(F, \Omega, y) = \sum\limits_{x \in F^{-1}(y)} \operatorname{sgn} \det DF(x)$. In particular if x_0 is as *simple zero* of $F(x)$, i.e. $F(x_0) = 0$ and $\det DF(x_0) \neq 0$, then $I(x_0) = \operatorname{sgn} \det DF(x_0) = \pm 1$.

It is useful for computation of the Leray-Schauder degree the following product formula

$$\deg\left(F_1 \times F_2, \Omega_1 \times \Omega_2, (y_1, y_2)\right) = \deg(F_1, \Omega_1, y_1) \deg(F_2, \Omega_2, y_2),$$

where (F_i, Ω_i, y_i), $i = 1, 2$ are admissible triples and the mapping

$$F_1 \times F_2 : \bar{\Omega}_1 \times \bar{\Omega}_2 \to X_1 \times X_2$$

is defined by

$$(F_1 \times F_2)(x_1, x_2) := (F_1(x_1), F_2(x_2)) \quad \forall (x_1, x_2) \in \bar{\Omega}_1 \times \bar{\Omega}_2$$

for Banach spaces X_1, X_2 with $\Omega_i \subset X_i$, $i = 1, 2$. In particular, it holds

$$\deg\left(\mathbb{I} \times F_2, \Omega_1 \times \Omega_2, (y_1, y_2)\right) = \deg(F_2, \Omega_2, y_2), \tag{2.2.5}$$

where (F_2, Ω_2, y_2) is an admissible triple and $y_1 \in \Omega_1$ for a bounded open subset Ω_1 of a Banach space X_1.

Finally, we state the *Schauder fixed point theorem* [101].

Theorem 2.2.3. *Let* Ω *be a closed convex bounded subset of a Banach space* X. *If* $G \in K(\Omega)$ *and* $G : \Omega \to \Omega$ *then* G *has a fixed point in* Ω.

For finite dimensional cases it is the *Brouwer fixed point theorem*.

2.3   Differential Topology

2.3.1   Differentiable Manifolds

Let M be a subset of $\mathbb{R}^k$. We use the *induced topology* on M, that is $A \subset M$ is open if there is an open set $A' \subset \mathbb{R}^k$ such that $A = A' \cap M$. We say that $M \subset \mathbb{R}^k$ is a C^r-*manifold* $(r \in \mathbb{N})$ *of dimension* m if for each $p \in M$ there is a neighborhood $U \subset M$ of p and a homeomorphism $x : U \to U_0$, where U_0 is an open subset in $\mathbb{R}^m$, such that the inverse $x^{-1} \in C^r(U_0, \mathbb{R}^k)$ and $Dx^{-1}(u) : \mathbb{R}^m \to \mathbb{R}^k$ is injective for any $u \in U_0$. Then we say that (x, U) is a *local C^r-chart around* p and U is a *coordinate neighborhood* of p. It is clear that if $x : U \to \mathbb{R}^m$ and $y : V \to \mathbb{R}^m$ are two local C^r-charts in M with $U \cap V \neq \emptyset$ then $y \circ x^{-1} : x(U \cap V) \to y(U \cap V)$ is a C^r diffeomorphism. This family of local charts is called a C^r-*atlas* for M [1, 114, 156].

If there is a C^r-atlas for M such that $\det D(y \circ x^{-1})(z) > 0$ for any $z \in x(U \cap V)$ and any two local C^r-charts $x : U \to \mathbb{R}^m$ and $y : V \to \mathbb{R}^m$ of this atlas with $U \cap V \neq \emptyset$ then M is *oriented*.

Let $\alpha \in C^1((-\varepsilon, \varepsilon), \mathbb{R}^k)$ be a differentiable curve on M, i.e. $\alpha : (-\varepsilon, \varepsilon) \to M$ with $\alpha(0) = p$. Then $\alpha'(0)$ is a *tangent vector* to M at p. The set of all tangent vectors to M at p is the *tangent space to M at p* and it is denoted by $T_p M$. The *tangent bundle* is the set

$$TM := \left\{ (p, v) \in \mathbb{R}^k \times \mathbb{R}^k \mid p \in M, \ v \in T_p M \right\}$$

with the *natural projection* $\pi : TM \to M$ given as $\pi(p, v) = p$. If M is a C^r-manifold with $r > 1$ then TM is a C^{r-1}-manifold.

Let M and N be two C^r-manifolds. We say that $f : M \to N$ is a C^r-*mapping* if for each $p \in M$ the mapping $y \circ f \circ x^{-1} : x(U) \to y(V)$ is C^r-smooth, where $x : U \to \mathbb{R}^m$ is a local C^r-chart in M around p and $y : V \to \mathbb{R}^s$ is a local C^r-chart in N with $f(U) \subset V$. This definition is independent of the choice of charts. The set of C^r-mappings is denoted by $C^r(M, N)$. Take $f \in C^r(M, N)$. Let $\alpha : (-\varepsilon, \varepsilon) \to M$ be a differentiable curve on M with $\alpha(0) = p$ and $\alpha'(0) = v$. Then $f \circ \alpha : (-\varepsilon, \varepsilon) \to N$ is a differentiable curve on N with $(f \circ \alpha)(0) = f(p)$, so we can define $Df(p)v := D(f \circ \alpha)(0) \in T_{f(p)} N$. This is independent of curve α. The map $Df(p) : T_p M \to T_{f(p)} N$ is linear, and if $r > 1$, $Df : TM \to TN$ defined as $Df(p, v) := (f(p), Df(p)v)$ is C^{r-1}-smooth.

A set $S \subset M \subset \mathbb{R}^k$ is a C^r-*submanifold* of M of dimension s if for each $p \in S$ there are open sets $U \subset M$ containing p, $V \subset \mathbb{R}^s$ containing 0 and $W \subset \mathbb{R}^{m-s}$ containing 0 and a C^r-diffeomorphism $\phi : U \to V \times W$ such that $\phi(S \cap U) = V \times \{0\}$.

A C^r-mapping $f : M \to N$ is an *immersion* if $Df(p)$ is injective for all $p \in M$. If $f : M \to N$ is an injective immersion we say that $f(M)$ is an *immersed submanifold*. If in addition $f : M \to f(M) \subset N$ is a homeomorphism, where $f(M)$ has the induced topology, then f is an *embedding*. In this case, $f(M)$ is a submanifold of N.

2.3.2   Symplectic Surfaces

$\mathcal{M}$ is a *smooth symplectic surface* if it is a 2-dimensional C^∞-smooth manifold with $\omega \in C^\infty(T\mathcal{M} \times T\mathcal{M}, \mathbb{R})$ such that $\forall m \in \mathcal{M}$, the restriction $\omega_m :=$ $\omega/T_m\mathcal{M} \times T_m\mathcal{M}$ is bilinear, antisymmetric and nonzero, i.e.

$$\omega_m(a_1 v_1 + a_2 v_2, b_1 w_1 + b_2 w_2) = \sum_{i,j=1}^{2} a_i b_j \omega_m(v_i, w_j)$$

and $\omega_m(v_1, v_2) = -\omega_m(v_2, v_1)$ for any $a_{1,2}, b_{1,2} \in \mathbb{R}$, $v_{1,2}, w_{1,2} \in T_m\mathcal{M}$. In a local coordinate $U \subset \mathbb{R}^2$, ω has the form $\omega_{(x,y)}(v,w) = a_U(x,y)v \wedge w$, $a_U \in C^\infty(U, \mathbb{R})$, $a(x,y) \neq 0$ for any $(x,y) \in U$ and $v \wedge w := x_1 y_2 - x_2 y_1$ is the *wedge product* for $v = (x_1, y_1)$, $w = (x_2, y_2)$. Then ω is a *non-degenerate differential 2-form* or *symplectic area form* on $\mathcal{M}$.

A mapping $\alpha \in C^\infty(T\mathcal{M}, \mathbb{R})$ is a *differential 1-form* on $\mathcal{M}$ if for any $m \in \mathcal{M}$, the restriction $\alpha_m := \omega/T_m\mathcal{M} \in T_m\mathcal{M}^*$. In a local coordinate $U \subset \mathbb{R}^2$, α has the form $\alpha_{(x,y)}v = a_{U,1}(x,y)x_1 + a_{U,2}(x,y)y_1$ for $a_{U,1,2} \in C^\infty(U, \mathbb{R})$ and $v = (x_1, y_1)$. Then $d\alpha$ is defined by

$$d\alpha_{(x,y)}(v,w) := \left(\frac{\partial a_{U,2}}{\partial x} - \frac{\partial a_{U,1}}{\partial y} \right) v \wedge w.$$

For $f \in C^\infty(\mathcal{M}, \mathcal{M})$, we define a differential 1-form as

$$f^*(\alpha)_m v := \alpha_{f(m)}(Df(m)v) \quad \forall m \in \mathcal{M}, \forall v \in T_m\mathcal{M}.$$

Finally, $f \in C^\infty(\mathcal{M}, \mathcal{M})$ is *symplectic* or *area-preserving* if

$$\omega_{f(m)}(Df(m)v, Df(m)w) = \omega_m(v, w) \quad \forall m \in \mathcal{M}, \forall v, \forall w \in T_m\mathcal{M}.$$

2.3.3   Intersection Numbers of Manifolds

Let W be an oriented C^r-manifold of dimension $m+n$ and M, N be oriented C^r-submanifolds of W of dimensions m, n, respectively, while M is compact and N is closed. A point $x \in M \cap N$ is *transversal* if $T_x M \cap T_x N = \{0\}$. It is a *positive (negative) kind* [114] if the composite map $T_x M \to T_x W \to T_x W / T_x N$ preserves (reserves) the orientation. Then we write $\#_x(M, N) := 1(-1)$. If all intersection points of $M \cap N$ are transversal then there is a final number of them and we set $\#(M, N) := \sum_{x \in M \cap N} \#_x(M, N)$. A similar approach is used like in the Brouwer degree theory to extend $\#(M, N)$ for general nontransversal intersections [114]. The number $\#(M, N)$ is called the *(oriented) intersection number* of manifolds M, N in W. If U is a precompact open subset of W and $\partial U \cap M \cap N = \emptyset$, then similarly we can define a *local (oriented) intersection number* $\#(M \cap U, N \cap U)$ of the manifolds $M \cap U$ and $N \cap U$ in $U \subset W$. These intersection numbers have similar properties as the Brouwer degree in Theorem 2.2.2.

2.3.4 Brouwer Degree on Manifolds

Let M, N be oriented C^1-manifolds with $\dim M = \dim N = n$. Let $f \in C^1(M,N)$, $y \in N$ and Ω be an open precompact subset of M such that $y \notin f(\partial\Omega)$. Suppose y is regular, i.e. $\forall x \in \Omega$ such that $f(x) = y$, $Df(x)$ is injective. Then $f^{-1}(y) \cap \Omega$ is finite, so $f^{-1}(y) \cap \Omega = \{v_1, v_2, \cdots, v_k\}$. We take disjoint local coordinates (U_i, x_i) of v_i and (V, x) of y. Then there are Brouwer indices $I(x_i(v_i))$ of $x(f(x_i^{-1}))$, which are independent of local coordinates. We set $\deg(f, \Omega, y) := \sum_{i=1}^{k} I(x_i(v_i))$. Then like in the classical degree theory, this degree is extended to any continuous f and nonregural y. This is the *Brouwer degree on manifolds* [114].

Next, let M, N be oriented C^1-manifolds and $f \in C^1(M, \mathbb{R}^p)$, $g \in C^1(N, \mathbb{R}^p)$, $p = \dim M + \dim N$ be embeddings. Let U be a bounded open subset of $\mathbb{R}^p$ such that $\partial U \cap f(M) \cap g(N) = \emptyset$. Then

$$|\#(f(M) \cap U, g(N) \cap U)| = |\deg(G, U_1 \times U_2, 0)|$$

for $U_1 = f^{-1}(U \cap f(M))$, $U_2 = g^{-1}(U \cap g(N))$ and $G(x, y) := f(x) - g(y)$. If in addition M is a submanifold and N is a linear subspace of $\mathbb{R}^p$ with a projection $P : \mathbb{R}^p \to N$ then

$$|\#(M \cap U, N \cap U)| = |\deg((\mathbb{I} - P), U \cap M, 0)| .$$

The independence of P follows from the fact that if Q is another projection on N then $\mathbb{I} - Q : \mathcal{R}(\mathbb{I} - P) \to \mathcal{R}(\mathbb{I} - Q)$ is a linear isomorphism and $(\mathbb{I} - Q)(\mathbb{I} - P) = \mathbb{I} - Q$. Hence

$$|\deg((\mathbb{I} - P), U \cap M, 0)| = |\deg((\mathbb{I} - Q)(\mathbb{I} - P), U \cap M, 0)|$$
$$= |\deg((\mathbb{I} - Q), U \cap M, 0)| .$$

2.3.5 Vector Bundles

A C^r-*vector bundle of dimension* n is a triple (E, p, B) where E, B are C^r-manifolds and $p \in C^r(E, B)$ with the following properties: for each $q \in B$ there is its open neighborhood $U \subset B$ and a C^r-diffeomorphism $\phi : p^{-1}(U) \to U \times \mathbb{R}^n$ such that $p = \pi_1 \circ \phi$ on $p^{-1}(U)$ where $\pi_1 : U \times \mathbb{R}^n \to U$ is defined as $\pi_1(x, y) := x$. Moreover, each $p^{-1}(x)$ are n-dimensional vector spaces and each $\phi_x : p^{-1}(x) \to \mathbb{R}^n$ given by $\phi(y) = (x, \phi_x(y))$ for any $y \in p^{-1}(x)$ are linear isomorphisms. E is called the *total space*, B is the *base space*, p is the *projection* of the bundle, the vector space $p^{-1}(x)$ is the *fibre* and ϕ is a *local trivialization*. So the vector bundle is *locally trivial*. If $U = B$ then the bundle is *trivial*. The family $\mathcal{A} := \{(\phi, U)\}$ of these local trivializations is a C^r-*vector atlas*. The bundle is *oriented* if there is a C^r-vector atlas $\mathcal{A} := \{(\phi, U)\}$ such that for any two local trivializations (ϕ, U) and (ψ, V) with $U \cap V \neq \emptyset$ the linear mapping $\psi_x \circ \phi_x^{-1} : \mathbb{R}^n \to \mathbb{R}^n$ is orientation preserving for each $x \in U \cap V$. A C^r-smooth mapping $s : B \to E$ satisfying $p \circ s = \mathbb{I}_B$ is called a *section* of the bundle.

Typical examples of vector bundles are the tangent bundle (TM, π, M) and the *normal bundle* $(TM^{\perp}, \tilde{\pi}, M)$ defined as

$$TM^{\perp} := \left\{(q, v) \in \mathbb{R}^k \times \mathbb{R}^k \mid q \in M,\, v \in T_q M^{\perp}\right\}$$

with the projection $\tilde{\pi} : TM^{\perp} \to M$ given as $\tilde{\pi}(q, v) = q$, where $T_x M^{\perp}$ is the orthogonal complement of $T_x M$ in $\mathbb{R}^k$. A section of TM is called a *vector field* on M. When M is oriented then both TM and $TM^{\perp}$ are oriented. Here M is a C^r-manifold with $r > 1$.

2.3.6   Euler Characteristic

Let $\xi = (E, p, B)$ be an oriented C^r-vector bundle with dimension $n = \dim B$ and B be oriented. Consider its C^r-section $s : B \to E$. Let Ω be an open pre-compact subset of B such that $s \neq 0$ on $\partial\Omega$. Suppose first that s has only a finite number of zeroes in Ω, say $b_1, b_2, \cdots, b_j$. Let $\phi_i : p^{-1}(U_i) \to U_i \times \mathbb{R}^n$, $i = 1, 2, \cdots, j$ be local trivializations of the bundle ξ with $b_i \in U_i$, U_i are disjoint and (U_i, x_i) are local coordinates. Then $x_i(b_i)$ is the only zero point of $v_i(x) := \phi_{x_i^{-1}(x)}(s(x_i^{-1}(x)))$, $x \in x_i(U_i)$. Note $v_i : x_i(U_i) \to \mathbb{R}^n$ and $x_i(U_i) \subset \mathbb{R}^n$. According to Section 2.2.3 it has the Brouwer index $I(b_i)$, which is independent of ϕ_i, x_i. Then we set $\deg(s, \Omega) := \sum_{i=1}^{j} I(b_i)$. Then like in the Brouwer degree theory, $\deg(s, \Omega)$ is extended to a continuous section $s : B \to E$ with $s \neq 0$ on $\partial\Omega$. It has similar properties as the Brouwer degree in Theorem 2.2.2. When $\Omega = B$ then $\deg(s, \Omega)$ is independent of s and so $\deg(s, \Omega) = \chi(\xi)$, where $\chi(\xi)$ is the *Euler characteristic of the bundle* ξ. So if $\chi(\xi) \neq 0$ then any continuous section of ξ has a zero. Next, if $\xi = TM$ with an oriented connected compact C^r-manifold for $r > 1$, then $\chi(TM)$ is the *Euler characteristic of the manifold* M. Then we get the classical result of Hopf that $\chi(TM) \neq 0$ implies the existence of a zero of any continuous vector field of M. Moreover, if $\chi(TM) = 0$ then TM has a continuous section without zeroes on M. We refer the reader to [114] for more details and proofs of the above results.

2.4   Multivalued Mappings

2.4.1   Upper Semicontinuity

Let X, Y be Banach spaces and let $\Omega \subset X$. By 2^Y we denote the family of all subsets of Y. Any mapping $F : \Omega \to 2^Y \setminus \{\emptyset\}$ is called *multivalued or set-valued*. For such mappings we define sets

$$\text{graph}\, F := \{(x, y) \in \Omega \times Y \mid x \in \Omega,\, y \in F(x)\}, \quad F(\Omega) := \cup_{x \in \Omega} F(x),$$

$$F^{-1}(A) := \{x \in \Omega \mid F(x) \cap A \neq \emptyset\} \quad \text{for} \quad A \subset Y.$$

Definition 2.4.1. A multivalued mapping $F : \Omega \to 2^Y \setminus \{\emptyset\}$ is *upper-semicontinuous*, usc for short, if the set $F^{-1}(A)$ is closed in Ω for any closed $A \subset Y$.

This condition of usc is more transparent in terms of sequences: if $\{x_n\}_{n=1}^{\infty}$ $\subset \Omega$, $A \subset Y$ is closed, $x_n \to x_0 \in \Omega$ and $F(x_n) \cap A \neq \emptyset$ for all $n \geq 1$, then also $F(x_0) \cap A \neq \emptyset$. The following result is a part of [59, Proposition 1.2.(b)].

Theorem 2.4.2. *If graph F is closed and $\overline{F(\Omega)}$ is compact then F is usc. In particular, F is usc if it has a compact graph F.*

A typical example of an usc multivalued mapping is Sgn : $\mathbb{R} \to 2^{\mathbb{R}} \setminus \{\emptyset\}$ defined by

$$\mathrm{Sgn}\, r = \begin{cases} r/|r| & \text{for } r \neq 0, \\ [-1,1] & \text{for } r = 0. \end{cases} \qquad (2.4.1)$$

We refer the reader for more properties of usc mappings to [59, p. 3-11] and [103].

2.4.2   Measurable Selections

Let $J := [0,1]$. The *characteristic function* $\chi_A(x)$ of a subset $A \subset J$ is defined as $\chi_A(x) = 1$ for $x \in A$ and $\chi_A(x) = 0$ for $x \notin A$. Let X be a Banach space with a norm $|\cdot|$. An $f : J \to X$ is said to be a *step function* if $f = \sum\limits_{i=1}^{k} c_i \chi_{A_i}(x)$ for some Lebesgue measurable sets $A_i \subset J$, $j = 1, \cdots, k$. An $f : J \to X$ is said to be *strongly measurable* if there is a sequence $\{f_n\}_{n=1}^{\infty}$ of step functions such that $|f(x) - f_n(x)| \to 0$ as $n \to \infty$ almost everywhere (a.e.) on J. Following [59, p. 29, Problem 10] (see also [166, p. 17, Propositions 3.3–3.4], [167]), we have the following result.

Theorem 2.4.3. *Let $F : J \times X \to 2^X \setminus \{\emptyset\}$ be usc with compact values, and $v \in C(J, X)$. Then $F(\cdot, v(\cdot))$ has a strongly measurable selection, i.e. there is a strongly measurable function $f : J \to X$ such that $f(x) \in F(x, v(x))$ a.e. on J.*

The above definitions are taken from [59, pp. 21–30].

2.4.3   Degree Theory for Set-Valued Maps

Let X be a Banach space and let $\Omega \subset X$ be open and bounded. A triple (F, Ω, y) is *admissible* if $F = \mathbb{I} - G$ for some $G : \bar{\Omega} \to 2^X \setminus \{\emptyset\}$ which is usc with compact convex values and $\overline{G(\bar{\Omega})} \subset X$ is compact, and $y \in X$ with $y \notin F(\partial\Omega)$. Let M be the set of all admissible triples. Then it is possible to define (cf. [59, pp. 154–155]) a unique function deg : $M \to \mathbb{Z}$ with the properties of Theorem 2.2.2 with evident differences that in (i) is $y \in F(x)$ in place of $F(x) = y$ and the homotopy in (iv) is compact usc with compact convex values. The number $\deg(F, \Omega, y)$ is the Leray-Schauder degree of the multivalued map F. We refer the reader for more topological methods for multivalued equations to the books [5, 103].

2.5   Dynamical Systems

2.5.1   Exponential Dichotomies

Set $\mathbb{Z}_+ := \mathbb{N} \cup \{0\}$ and $\mathbb{Z}_- := -\mathbb{Z}_+$. Let $J \in \{\mathbb{Z}_+, \mathbb{Z}_-, \mathbb{Z}\}$. Let $A_n \in L(\mathbb{R}^k)$, $n \in J$ be a sequence of invertible matrices. Consider a linear difference equation

$$x_{n+1} = A_n x_n. \tag{2.5.1}$$

Its *fundamental solution* is defined as $U(n) := A_{n-1} \cdots A_0$ for $n \in \mathbb{N}$, $U(0) = \mathbb{I}$ and $U(n) := A_n^{-1} \cdots A_{-1}^{-1}$ for $-n \in \mathbb{N}$. (2.5.1) has an *exponential dichotomy* on J if there is a projection $P : \mathbb{R}^k \to \mathbb{R}^k$ and constants $L > 0$, $\delta \in (0, 1)$ such that

$$\|U(n)PU(m)^{-1}\| \le L\delta^{n-m} \text{ for any } m \le n, \, n, m \in J$$
$$\|U(n)(\mathbb{I} - P)U(m)^{-1}\| \le L\delta^{m-n} \text{ for any } n \le m, \, n, m \in J.$$

If $A_n = A$ and its spectrum $\sigma(A)$ has no intersection with the unit circle, i.e. A is *hyperbolic*, then P is the projection onto the generalized eigenspace of eigenvectors inside the unit circle and $\mathcal{N}P$ is the generalized eigenspace of eigenvectors outside the unit circle. Next we have the following *roughness of exponential dichotomies*.

Lemma 2.5.1. *Let $J \in \{\mathbb{Z}_+, \mathbb{Z}_-\}$. Let A be hyperbolic with the dichotomy projection P. Assume $\{A_n(\xi)\}_{n \in J} \in L(\mathbb{R}^k)$ are invertible matrices and $A_n(\xi) \to A$ in $L(\mathbb{R}^k)$ uniformly with respect to a parameter ξ. Then $x_{n+1} = A_n(\xi)x_n$, with the fundamental solution $U_\xi(n)$, has an exponential dichotomy on J with projection P_ξ and uniform constants $L > 0$, $\delta \in (0, 1)$. Moreover, $U_\xi(n)P_\xi U_\xi(n)^{-1} \to P$ as $n \to \pm\infty$ uniformly with respect to ξ.*

Analogical results hold for a linear differential equation $\dot{x} = A(t)x$ when $t \in J \in \{(-\infty, 0), (0, \infty), \mathbb{R}\}$ and $A(t) \in C(J, L(\mathbb{R}^k))$ is a continuous matrix function. Its *fundamental solution* is a matrix function $U(t)$ satisfying $\dot{U}(t) = A(t)U(t)$ on J. Sometimes we require that $U(0) = \mathbb{I}$ [159].

2.5.2   Chaos in Discrete Dynamical Systems

Consider a C^r-*diffeomorphism* f on $\mathbb{R}^m$ with $r \in \mathbb{N}$, i.e. a mapping $f \in C^r$ $(\mathbb{R}^m, \mathbb{R}^m)$ which is invertible and $f^{-1} \in C^r(\mathbb{R}^m, \mathbb{R}^m)$. For any $z \in \mathbb{R}^m$ we define its k-*iteration* as $f^k(z) := f(f^{k-1}(z))$. The set $\{f^n(z)\}_{n=\infty}^\infty$ is an *orbit* of f. If $x_0 = f(x_0)$ then x_0 is a *fixed point* of f. It is *hyperbolic* if the *linearization* $Df(x_0)$ of f at x_0 has no eigenvalues on the unit circle. The *global stable (unstable) manifold* $W_{x_0}^{s(u)}$ of a hyperbolic fixed point x_0 is defined by

$$W_{x_0}^{s(u)} := \{z \in \mathbb{R}^m \mid f^n(z) \to x_0 \text{ as } n \to \infty(-\infty)\},$$

respectively.

Theorem 2.5.2. $W_{x_0}^s$ *and* $W_{x_0}^u$ *are immersed* C^r-*submanifolds in* $\mathbb{R}^m$.

Furthermore, let y_0 be another hyperbolic fixed point of f. If $x \in W_{x_0}^s \cap W_{y_0}^u \setminus \{x_0, y_0\}$ then it is a *heteroclinic point* of f and then the orbit $\{f^n(x)\}_{n=\infty}^{\infty}$ is called heteroclinic. Clearly $f^n(z) \to x_0$ as $n \to \infty$ and $f^n(z) \to y_0$ as $n \to -\infty$. If $T_x W_{x_0}^s \cap T_x W_{y_0}^u = \{0\}$ then x is a *transversal heteroclinic point* of f. Note $x \in W_{x_0}^s \cap W_{y_0}^u \setminus \{x_0, y_0\}$ is a transversal heteroclinic point if and only if the linear difference equation $x_{n+1} = Df(f^n(x))x_n$ has an exponential dichotomy on $\mathbb{Z}$. When $x_0 = y_0$ then the word "heteroclinic" is replaced with *homoclinic*. We refer the reader to the book [159] for more details and proofs of the above subject.

Let $\mathcal{E} = \{0,1\}^{\mathbb{Z}}$ be a compact metric space of the set of doubly infinite sequences of 0 and 1 endowed with the metric [68]

$$d(\{e_n\}, \{e_n'\}) := \sum_{n \in \mathbb{Z}} \frac{|e_n - e_n'|}{2^{|n|+1}}.$$

On $\mathcal{E}$ it is defined the so called *Bernoulli shift map* $\sigma : \mathcal{E} \to \mathcal{E}$ by $\sigma(\{e_j\}_{j \in \mathbb{Z}}) = \{e_{j+1}\}_{j \in \mathbb{Z}}$ with extremely rich dynamics [195].

Theorem 2.5.3. σ *is a homeomorphism having*

(i) A countable infinity of periodic orbits of all possible periods

(ii) An uncountable infinity of nonperiodic orbits and

(iii) A dense orbit

Now we can state the following result about the existence of *the deterministic chaos* for diffeomorphisms, the *Smale-Birkhoff homoclinic theorem*.

Theorem 2.5.4. *Suppose* $f : \mathbb{R}^m \to \mathbb{R}^m$, $r \in \mathbb{N}$ *be a* C^r-*diffeomorphism having a transversal homoclinic point to a hyperbolic fixed point. Then there is an* $k \in \mathbb{N}$ *such that* f^k *has an invariant set* Λ, *i.e.* $f^k(\Lambda) = \Lambda$, *such that* $\varphi \circ f^k = \sigma \circ \varphi$ *for an homeomorphism* $\varphi : \Lambda \to \mathcal{E}$.

The set Λ is the *Smale horseshoe* and we say that f has *horseshoe dynamics* on Λ. Theorem 2.5.4 asserts that the following diagram is commutative

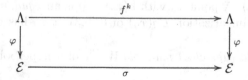

So f^k on Λ has the same dynamical properties as σ on $\mathcal{E}$, i.e. Theorem 2.5.3 gives chaos for f. Moreover, it is possible to show a *sensitive dependence on initial conditions* of f on Λ in the sense that there is an $\varepsilon_0 > 0$ such that for any $x \in \Lambda$ and any neighborhood U of x, there exists $z \in U \cap \Lambda$ and an integer $q \geq 1$ such that $|f^q(x) - f^q(z)| > \varepsilon_0$.

Remark 2.5.5. Of course the above considerations hold for a smooth diffeomorphism $f : M \to M$ on a smooth manifold.

2.5.3   Periodic O.D.Eqns

It is well-known [110], that the Cauchy problem

$$\dot{x} = g(x,t), \quad x(0) = z \in \mathbb{R}^m \qquad (2.5.2)$$

for $g \in C^r(\mathbb{R}^m \times \mathbb{R}, \mathbb{R}^m)$, $r \in \mathbb{N}$ has a unique solution $x(t) = \phi(z,t)$ defined on
a maximal interval $0 \in I_z \subset \mathbb{R}$. We suppose for simplicity that $I_z = \mathbb{R}$. This
is true for instance when g is globally Lipschitz continuous in x, i.e. there is
a constant $L > 0$ such that $|g(x,t) - g(y,t)| \le L|x - y|$ for any $x, y \in \mathbb{R}^m$,
$t \in \mathbb{R}$. Moreover, we assume that g is T-periodic in t, i.e. $g(x, t + T) = g(x,t)$
for any $x \in \mathbb{R}^m$, $t \in \mathbb{R}$. Then the dynamics of (2.5.2) is determined by the
dynamics of the diffeomorphism $f(z) = \phi(z,T)$ which is called the *time or
Poincaré map* of (2.5.2). Now we can transform the results of Section 2.5.2 to
(2.5.2). So T-periodic solutions (*periodics* for short) of (2.5.2) are fixed points of
f. Periodics of f are *subharmonic solutions* (*subharmonics* for short) of (2.5.2).
Similarly we mean a chaos of (2.5.2) as a chaos for f. To be more concrete,
we apply these results to (1.2.1). We known from introduction that for $\mu_{1,2}$
small satisfying (1.2.10), (1.2.1) has a bounded solution which tends to a small
hyperbolic periodic solution. For the time map of (1.2.1) this means that it has
a homoclinic orbit to a hyperbolic fixed point. Next after some effort [159] it is
possible to show that this homoclinic point is also transversal. So the time map
of (1.2.1) is chaotic according to Theorem 2.5.4. Consequently, (1.2.1) is chaotic
for $\mu_{1,2}$ small satisfying (1.2.10). We also refer the reader to Subsection 4.2.1
for more details.

2.5.4   Vector Fields

When (2.5.2) is *autonomous*, i.e. g is independent of t, then (2.5.2) has the form

$$\dot{x} = g(x), \quad x(0) = z \in \mathbb{R}^m . \qquad (2.5.3)$$

g is called a C^r-*vector field* on $\mathbb{R}^m$ for $g \in C^r(\mathbb{R}^m, \mathbb{R}^m)$, $r \in \mathbb{N}$. We suppose
for simplicity that the unique solution $x(t) = \phi(z,t)$ of (2.5.3) is defined on $\mathbb{R}$.
$\phi(z,t)$ is called the *orbit based at z*. Then instead of the time map of (2.5.3), we
consider the *flow* $\phi_t : \mathbb{R}^m \to \mathbb{R}^m$ defined as $\phi_t(z) := \phi(z,t)$ with the property
$\phi_t(\phi_s(z)) = \phi_{t+s}(z)$. A point x_0 with $g(x_0) = 0$ is an *equilibrium* of (2.5.3). It
is *hyperbolic* if the linearization $Dg(x_0)$ of (2.5.3) at x_0 has no eigenvalues on
imaginary axis.

The *global stable (unstable) manifold* $W_{x_0}^{s(u)}$ of a hyperbolic equilibrium x_0
is defined by

$$W_{x_0}^{s(u)} := \{z \in \mathbb{R}^m \mid \phi(z,t) \to x_0 \quad \text{as} \quad t \to \infty(-\infty)\} ,$$

respectively. These sets are immersed submanifolds of $\mathbb{R}^m$. Note for any $x \in W_{x_0}^{s(u)}$, we know that

$$T_x W_{x_0}^{s(u)} = \Big\{ v(0) \in \mathbb{R}^m \mid v(t) \text{ is a bounded solution}$$

$$\text{of } \dot{v} = Dg(\phi(x,t))v \text{ on } (0,\infty), ((-\infty,0)), \text{ respectively} \Big\} .$$

Moreover, the set

$$(T_x W^s_{x_0} + T_x W^u_{x_0})^\perp$$

is the linear space of initial values $w(0)$ of all bounded solutions $w(t)$ of the *adjoint equation* $\dot w = -Dg(\phi(x,t))^* w$ on $\mathbb{R}$ [157].

A local dynamics near a hyperbolic equilibrium x_0 of (2.5.3) is explained by the *Hartman-Grobman theorem for flows* [108].

Theorem 2.5.6. *If $x_0 = 0$ is a hyperbolic equilibrium of (2.5.3) then there is a homeomorphism h defined on a neighborhood U of 0 in $\mathbb{R}^m$ such that*

$$h(\phi(z,t)) = e^{tDg(0)} h(z)$$

for all $z \in U$ and $t \in J_z$ with $\phi(z,t) \in U$, where $0 \in J_z$ is an interval.

For nonhyperbolic equilibria we have the following *center manifold theorem for flows* [108].

Theorem 2.5.7. *Let $x_0 = 0$ be an equilibrium of a C^r-vector field (2.5.3) on $\mathbb{R}^m$. Divide the spectrum of $Dg(0)$ into three parts σ_s, σ_u, σ_c such that $\Re\lambda < 0; >
0; = 0$ if $\lambda \in \sigma_s; \sigma_u; \sigma_c$, respectively. Let the generalized eigenspaces of σ_s, σ_u, σ_c be E^s, E^u, E^c, respectively. Then there are C^r-smooth manifolds: the stable W^s_0, the unstable W^u_0, the center W^c_0 tangent at 0 to E^s, E^u, E^c, respectively. These manifolds are invariant for the flow of (2.5.3), i.e. $\phi_t(W^{s;u;c}_0) \subset W^{s;u;c}_0$ for any $t \in \mathbb{R}$. The stable and unstable ones are unique, but the center one need not be. In addition, when g is embedded into a C^r-smooth family of vector fields g_ε with $g_0 = g$, then these invariant manifolds are C^r-smooth also with respect to ε.*

Under the assumptions of Theorem 2.5.7 near $x_0 = 0$ we can write (2.5.3) in the form

$$\dot x_s = A_s x_s + g_s(x_s, x_u, x_c, \varepsilon),$$
$$\dot x_u = A_u x_u + g_u(x_s, x_u, x_c, \varepsilon), \qquad (2.5.4)$$
$$\dot x_c = A_c x_s + g_c(x_s, x_u, x_c, \varepsilon),$$

where $A_{s;u;c} := Dg(0)/E^{s;u;c}$ and $x_{s;u;c} \in U_{s;u;c}$ for open neighborhoods $U_{s;u;c}$ of 0 in $E^{s;u;c}$, respectively. Here we suppose that (2.5.3) is embedded into a C^r-smooth family. So g_j are C^r-smooth satisfying $g_j(0,0,0,0) = 0$ and $D_{x_j} g_k(0,0,0,0) = 0$ for $j, k = s, u, c$. According to Theorem 2.5.7, the *local center manifold* $W^c_{loc,\varepsilon}$ near $(0,0,0)$ of (2.5.4) is a graph

$$W^c_{loc,\varepsilon} = \{(h_s(x_c, \varepsilon), h_u(x_c, \varepsilon), x_c) \mid x_c \in U_c\}$$

for $h_{s;u} \in C^r(U_c \times V, E^{s;u})$ and V is an open neighborhood of $\varepsilon = 0$. Moreover, it holds $h_{s;u}(0,0) = 0$ and $D_{x_c} h_{s;u}(0,0) = 0$. The *reduced equation* is

$$\dot x_c = A_c x_s + g_c(h_s(x_c, \varepsilon), h_u(x_c, \varepsilon), x_c, \varepsilon), \qquad (2.5.5)$$

which locally determines the dynamics of (2.5.4), i.e. $W^c_{loc,\varepsilon}$ contains all solutions of (2.5.4) staying in $U_s \times U_u \times U_c$ for all $t \in \mathbb{R}$. In particular periodics, homoclinics and heteroclinics of (2.5.4) near $(0,0,0)$ solve (2.5.5).

Now suppose that (2.5.3) is *invariant* under a linear invertible mapping $S \in L(\mathbb{R}^m)$, i.e.

$$Sg(x) = g(Sx)$$

for all $x \in \mathbb{R}^m$. Then the uniqueness of the Cauchy problem (2.5.3) implies $S\phi(z,t) = \phi(Sz,t)$ for any $z \in \mathbb{R}^m$, $t \in \mathbb{R}$. If 0 is an equilibrium of (2.5.3) then $SDg(0) = Dg(0)S$ and $SE^{s;u;c} = E^{s;u;c}$. We have the following result [121, Theorem I.10].

Theorem 2.5.8. *Suppose $S_c := S/E^c$ is unitary, i.e. $|S_c x_c| = |x_c|$ for any $x_c \in E^c$. Then the local center manifold can be chosen invariant under S.*

More concretely, let $S_{s;u} := S/E^{s;u}$ and suppose

$$S_s g_s(x_s, x_u, x_c, \varepsilon) = g_s(S_s x_s, S_u x_u, S_c x_c, \varepsilon),$$
$$S_u g_u(x_s, x_u, x_c, \varepsilon) = g_u(S_s x_s, S_u x_u, S_c x_c, \varepsilon),$$
$$S_c g_c(x_s, x_u, x_c, \varepsilon) = g_c(S_s x_s, S_u x_u, S_c x_c, \varepsilon)$$

for any $x_{s;u;c} \in U_{s;u;c}$ and ε small. Then the functions $h_s(x_c, \varepsilon)$, $h_u(x_c, \varepsilon)$ can be chosen so that $S_s h_s(x_c, \varepsilon) = h_s(S_c x_c, \varepsilon)$ and $S_u h_u(x_c, \varepsilon) = h_u(S_c x_c, \varepsilon)$ for any $x_c \in U_c$ and ε small. Then we have

$$S_c \left(A_c x_s + g_c(h_s(x_c, \varepsilon), h_u(x_c, \varepsilon), x_c, \varepsilon) \right)$$
$$= A_c S_c x_s + g_c(h_s(S_c x_c, \varepsilon), h_u(S_c x_c, \varepsilon), S_c x_c, \varepsilon).$$

So the reduced equation is invariant with respect to S_c.

2.6 Center Manifolds for Infinite Dimensions

The center manifold theorem for flows is extended to infinite dimensional differential equations [190, 191] as follows. Let X, Y and Z be Banach spaces, with X continuously embedded in Y, and Y continuously embedded in Z. Let $A \in L(X, Z)$ and $h \in C^k(X \times \mathbb{R}^m, Y)$, $k \geq 1$, $m \geq 1$ with $h(0,0) = 0$ and $D_x h(0,0) = 0$. We consider differential equations of the form

$$\dot{x} = Ax + h(x, \varepsilon). \tag{2.6.1}$$

By a solution of (2.6.1) we mean a continuous function $x : J \to X$, where J is an open interval, such that $x : J \to Z$ is continuously differentiable and $\dot{x}(t) = Ax(t) + h(x(t), \varepsilon)$ holds on J. We need the following definition.

Definition 2.6.1. Let E and F be Banach spaces, $k \in \mathbb{N}$ and $\eta \geq 0$. Then we define

$$C_b^k(E, F) := \left\{ w \in C^k(E, F) \mid \sup_{x \in E} \|D^j w(x)\| < \infty, \, 0 \leq j \leq k \right\},$$

$$BC^\eta(\mathbb{R}, E) := \left\{ w \in C(\mathbb{R}, E) \mid \|w\|_\eta := \sup_{t \in \mathbb{R}} e^{-\eta|t|} |w(t)|_E < \infty \right\}.$$

Now concerning A we suppose the following hypothesis.

(H) There exists a continuous projection $\pi_c \in L(Z, X)$ onto a finite dimensional subspace $Z_c = X_c \subset X$ such that

$$A\pi_c x = \pi_c A x, \quad \forall x \in X,$$

and such that if we set

$$Z_h := (\mathbb{I} - \pi_c)Z, \quad X_h := (\mathbb{I} - \pi_c)X, \quad Y_h := (\mathbb{I} - \pi_c)Y,$$
$$A_c := A/X_c \in L(X_c), \quad A_h := A/X_h \in L(X_h, Z_h),$$

then the following hold

(i) $\sigma(A_c) \subset i\mathbb{R}$.

(ii) There exists a $\beta > 0$ such that for each $\eta \in [0, \beta)$ and for each $f \in BC^\eta(\mathbb{R}, Y_h)$ the linear problem $\dot{x}_h = A_h x_h + f(t)$ has a unique solution $x_h \in BC^\eta(\mathbb{R}, X_h)$ satisfying $\|x_h\|_\eta \leq \gamma(\eta)\|f\|_\eta$ for a continuous function $\gamma : [0, \beta) \to [0, \infty)$.

The next result generalize Theorem 2.5.7 to infinite dimensions.

Theorem 2.6.2. *There are open neighborhoods $\Omega \subset X$, $U \subset \mathbb{R}^m$ of origins and a mapping $\psi \in C_b^k(X_c \times \mathbb{R}^m, X_h)$ with $\psi(0,0) = 0$ and $D_{x_c}\psi(0,0) = 0$ such that for any $\varepsilon \in U$ the following properties hold:*

(i) *If $\widetilde{x}_c : J \to X_c$ is a solution of the reduced equation*

$$\dot{x}_c = A_c x_c + \pi_c h (x_c + \psi(x_c, \varepsilon), \varepsilon) \tag{2.6.2}$$

such that $\widetilde{x}(t) := \widetilde{x}_c(t) + \psi(\widetilde{x}_c(t), \varepsilon) \in \Omega$ for all $t \in J$, then $\widetilde{x} : J \to X$ is a solution of (2.6.1).

(ii) *If $\widetilde{x} : \mathbb{R} \to X$ is a solution of (2.6.1) such that $\widetilde{x}(t) \in \Omega$ for all $t \in \mathbb{R}$, then $(\mathbb{I} - \pi_c)\widetilde{x}(t) = \psi(\pi_c\widetilde{x}(t), \varepsilon)$, $\forall t \in \mathbb{R}$ and $\pi_c\widetilde{x} : \mathbb{R} \to X_c$ is a solution of (2.6.2).*

Now we generalize Theorem 2.5.8. Let $\Gamma \subset L(Z) \cap L(Y) \cap L(X)$ be a group of linear bounded operators such that

$$SA = AS, \quad Sh(x, \varepsilon) = h(Sx, \varepsilon)$$

for any $x \in X$, $\varepsilon \in \mathbb{R}^m$, $S \in \Gamma$. Then the group Γ leaves X_c invariant. Supposing that $S \in \Gamma$ are unitary, the function ψ of Theorem 2.6.2 satisfies

$$S\psi(x_c, \varepsilon) = \psi(Sx_c, \varepsilon)$$

for any $x_c \in X_c$, $\varepsilon \in U$, $S \in \Gamma$. This means that the reduced equation (2.6.2) is invariant under the action Γ on X_c.

A similar result holds when (2.6.1) is R-reversible with respect to a symmetry $R \in L(Z) \cap L(Y) \cap L(X)$:

$$RA = -AR, \quad Rh(x, \varepsilon) = -h(Rx, \varepsilon)$$

for any $x \in X$, $\varepsilon \in \mathbb{R}^m$. If R is unitary then the function ψ of Theorem 2.6.2 satisfies $R\psi(x_c, \varepsilon) = \psi(Rx_c, \varepsilon)$ for any $x_c \in X_c$, $\varepsilon \in U$, and the reduced equation (2.6.2) is R_c-reversible with R_c, the restriction of R to X_c.

Chapter 3

Bifurcation of Periodic Solutions

3.1 Bifurcation of Periodics from Homoclinics I

3.1.1 Discontinuous O.D.Eqns

We already know from Introduction and Sections 2.5.2, 2.5.3 that it is possible to show chaos for smooth systems such as

$$\ddot{x} + h(x, \dot{x}) + \mu_1 \dot{x} = \mu_2 \psi(t) \tag{3.1.1}$$

under certain conditions for h, $\psi(t)$ and small $\mu_{1,2}$. In particular, then (3.1.1) has an infinite number of subharmonics. The purpose of this section is to show that even in discontinuous perturbations there are still infinitely many subharmonics. This is our first step to show chaos for discontinuous differential equations. So in this section we study the existence of subharmonic and bounded solutions on $\mathbb{R}$ for ordinary differential equations with discontinuous perturbations. Such equations appear in nonlinear mechanical systems [45, 98] like the following problem: a dry friction force acting on a moving particle due to its contact to a wall has in certain situation the form $\mu(g_0(\dot{x}) + \operatorname{sgn} \dot{x})$, where x is a displacement from the rest, $\dot{x}$ is the velocity, μ is a positive constant, g_0 is a non–negative bounded continuous function, and $\operatorname{sgn} r = r/|r|$ for $r \in \mathbb{R} \setminus \{0\}$, see [64, 123]. So dry friction is modeled by Coulomb's friction law [98, p. 7] expressed with the discontinuous function $\operatorname{sgn} r$. Including also viscous damping, restoring and external forces, the following equation is studied (see Fig. 3.1)

$$\ddot{x} + h(x, \dot{x}) + \mu_1 \operatorname{sgn} \dot{x} = \mu_2 \psi(t), \tag{3.1.2}$$

where h, ψ are continuous and μ_1, μ_2 are parameters. We assume in this section that μ_1, μ_2 are small and ψ is periodic. Equation (3.1.2) is a discontinuous analogy of (3.1.1). Since (3.1.2) is discontinuous, by using the multivalued mapping

M. Fečkan, *Topological Degree Approach to Bifurcation Problems*, 23–119.
© Springer Science + Business Media B.V., 2008

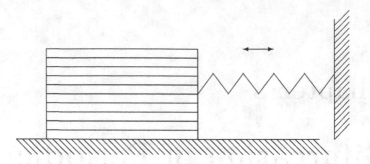

Figure 3.1: A moving block on a wall under a force

(2.4.1), (3.1.2) is rewritten as a differential inclusion

$$\ddot{x} + h(x, \dot{x}) - \mu_2 \psi(t) \in -\mu_1 \text{Sgn}\, \dot{x}.$$

To deal with much more general equations like (3.1.2), we consider differential inclusions which take the following form

$$\dot{x}(t) \in f(x(t)) + \sum_{i=1}^{k} \mu_i f_i(x(t), \mu, t) \quad \text{a.e. on} \quad \mathbb{R} \qquad (3.1.3)$$

with $x \in \mathbb{R}^n$, $\mu \in \mathbb{R}^k$, $\mu = (\mu_1, \cdots, \mu_k)$. We mean by a solution of any differential inclusion in this book a function which is absolute continuous and satisfying that differential inclusion almost everywhere. Since we are studying bifurcation from homoclinics, we set the following assumptions about (3.1.3):

(i) $f : \mathbb{R}^n \to \mathbb{R}^n$ is C^2-smooth and $f_i : \mathbb{R}^n \times \mathbb{R}^k \times \mathbb{R} \to 2^{\mathbb{R}^n} \setminus \{\emptyset\}$, $i = 1, \cdots, k$ are upper–semicontinuous with compact and convex values.

(ii) $f(0) = 0$ and the eigenvalues of $Df(0)$ lie off the imaginary axis.

(iii) The unperturbed equation has a homoclinic solution. That is, there exists a differentiable function $t \to \gamma(t)$ such that $\lim_{t \to +\infty} \gamma(t) = \lim_{t \to -\infty} \gamma(t) = 0$ and $\dot{\gamma}(t) = f(\gamma(t))$.

(iv) $f_i(x, \mu, t+2) = f_i(x, \mu, t)$ for $t \in \mathbb{R}$, $i = 1, \cdots, k$.

Under the above assumptions, we find conditions ensuring the existence of infinitely many subharmonics of (3.1.3) with the periods tending to infinity and accumulating on γ. Since our system is discontinuous, we can not apply the classical dynamical system approach based on the Smale-Birkhoff Homoclinic Theorem 2.5.4. We need a different approach. Proofs of results of this section are based on a method of Lyapunov–Schmidt decomposition, which is developed in Subsection 3.1.2, together with application of a theory of generalized Leray–Schauder degree for multivalued mappings, which is done in Subsection 3.1.3.

Let us note that periodic and almost periodic solutions to dry friction problems are also investigated in [50, 59–63]. The numerical analysis is given in [12, 164, 165] for a mechanical model of a friction oscillator with simultaneous self and external excitation. These papers [163–165] present a nice introduction to the phenomenon of dry friction as well. Finally similar equations also appear in electrical engineering (see [6, Chap. III]), related problems are studied in control systems (see [172]) as well, and dry friction problems were investigated already in [168, 169].

3.1.2 The Linearized Equation

Since we study bifurcation for (3.1.3), we begin by considering its unperturbed equation with $\mu_i = 0$, $i = 1, \cdots, k$:

$$\dot{x} = f(x). \tag{3.1.4}$$

We set $d_s = \dim W_0^s$ and $d_u = \dim W_0^u$ for the stable and unstable manifolds W_0^s and W_0^u, respectively, of the hyperbolic equilibrium $x = 0$ of (3.1.4). Clearly $\gamma \in W_0^s \cap W_0^u$. By the variational equation of (3.1.4) along γ we mean the linear differential equation

$$\dot{u}(t) = Df(\gamma(t))u(t). \tag{3.1.5}$$

Observe that as $t \to \pm\infty$, $Df(\gamma(t)) \to Df(0)$ in exponential rates [68], and $Df(0)$ is a hyperbolic matrix. Thus, the following result yields two solutions for (3.1.5) – one solution for $t \geq 0$ and one for $t \leq 0$ [106].

Lemma 3.1.1. *Let* $t \to A(t) \in L(\mathbb{R}^n)$ *be a matrix valued function continuous on* $[0, \infty)$ *and suppose there exists a constant matrix* $A_0 \in L(\mathbb{R}^n)$ *and a scalar* $a > 0$ *such that* $\sup_{t \geq 0} |A(t) - A_0| e^{at} < \infty$. *Then there exists a fundamental solution* $\widetilde{U}$ *of the differential equation* $\dot{x} = A(t)x$ *such that* $\lim_{t \to \infty} \widetilde{U}(t) e^{-tA_0} = \mathbb{I}$.

Proof. Let $P^{-1}A_0P = J$ for a regular matrix P and J be the Jordan form of A_0 with block diagonal form $J = \text{diag}(J_1, J_2, \cdots, J_r)$. The order of J_i is k_i and the corresponding eigenvalue is λ_i. We suppose that $\Re\lambda_i \leq \Re\lambda_{i+1}$. Let $y := P^{-1}x$ and $B(t) := P^{-1}A(t)P$. Then the differential equation $\dot{x} = A(t)x$ becomes

$$\dot{y} = B(t)y = Jy + (B(t) - J)y. \tag{3.1.6}$$

We construct solutions to each Jordan block. So we fix a block J_i and set $p_i := k_1 + k_2 + \cdots + k_{i-1}$. By defining q_i so that $\Re\lambda_{q_i-1} < \Re\lambda_i$ and $\Re\lambda_{q_i} = \Re\lambda_i$, we decompose e^{tJ} as follows

$$U_{1i}(t) := \text{diag}\left(e^{tJ_1}, \cdots, e^{tJ_{q_i-1}}, 0, \cdots, 0\right)$$
$$U_{2i}(t) := \text{diag}\left(0, \cdots, 0, e^{tJ_{q_i}}, \cdots, e^{tJ_r}\right).$$

We can choice $K > 0$ and b, $0 < b < a/2$ such that [68]

$$|U_{1i}(t)| \leq K e^{(\Re\lambda_i - b)t} \quad \text{for} \quad t \geq 0,$$
$$|U_{2i}(t)| \leq K e^{(\Re\lambda_i - b)t} \quad \text{for} \quad t \leq 0. \tag{3.1.7}$$

Let $t_0 \geq 0$ and consider the Banach space

$$C_i := \left\{ y \in C([t_0, \infty), \mathbb{R}^n) \mid \|y\| := \sup_{t \geq t_0} |y(t)| \, e^{-(\Re\lambda_i + b)t} < \infty \right\}.$$

Now we consider a linear operator $T_i : C_i \to C_i$ given by

$$T_i y(t) := \int_{t_0}^{t} U_{1i}(t-s)(B(s) - J)y(s)\, ds - \int_{t}^{\infty} U_{2i}(t-s)(B(s) - J)y(s)\, ds.$$

It is well defined since we easily derive $|T_i y(t)| \leq \frac{2\widetilde{K}}{a - 2b} \, e^{(\Re\lambda_i - b)t} \|y\|$ for any $t \geq t_0$ and $y \in C_i$, where $\widetilde{K} := K \sup_{t \geq 0} |B(t) - J| \, e^{at}$. Hence $\|T_i\| \leq \frac{2\widetilde{K}}{a - 2b} \, e^{-2bt_0}$. So taking $t_0 = \max\left\{ \frac{\ln[4\widetilde{K}/(a - 2b)]}{2b}, 0 \right\}$, we get $\|T_i\| \leq 1/2$. Let e_k be the kth column of the $n \times n$ identity matrix. Since for each $j \in \{1, 2, \cdots, k_i\}$, it holds $e^{tJ} e_{p_i + j} \in C_i$ we see that for each $j \in \{1, 2, \cdots, k_i\}$ and $y \in C_i$, system (3.1.6) has the form

$$y = e^{tJ} e_{p_i + j} + T_i y. \tag{3.1.8}$$

But $T_i : C_i \to C_i$ is a linear contraction with a constant $1/2$, so (3.1.8) has a unique solution $y_j \in C_i$ such that

$$\left| y_j(t) - e^{tJ} e_{p_i + j} \right| \leq \frac{2\widetilde{K}}{a - 2b} \, e^{(\Re\lambda_i - b)t} \|y_j\|.$$

By defining the matrix $Y_i(t)$ of the order $n \times k_i$ with $y_j(t)$ in column j, we obtain

$$|Y_i(t) - F_i(t)| \, e^{(-\Re\lambda_i + b)t} \leq \frac{2\widetilde{K}}{a - 2b} \sqrt{\sum_{j=1}^{k_i} \|y_j\|^2},$$

where $F_i(t)$ is the $n \times k_i$-matrix with $e^{J_i t}$ in rows $p_i + 1$ through $p_i + k_i$ and all other rows zero. Let $\mathbb{I}_{k_i \times k_i}$ be the identity matrix of order $k_i \times k_i$. Then $\lim_{t \to \infty} Y_i(t) \, e^{-J_i t} = G_i$ and G_i is the matrix of order $n \times k_i$ with $\mathbb{I}_{k_i \times k_i}$ in rows $p_i + 1$ through $p_i + k_i$ and all other rows zero. This construction is done for the block J_i. To get the result, we take the $n \times n$ matrix $Y(t)$ with $Y_i(t)$ in columns $p_i + 1$ through $p_i + k_i$ for $i = 1, 2, \ldots, r$. So $\lim_{t \to \infty} Y(t) \, e^{-Jt} = \mathbb{I}$. Finally, by putting $\widetilde{U}(t) = PY(t)P^{-1}$ we arrive at $\dot{\widetilde{U}}(t) = A(t)\widetilde{U}(t)$ satisfying

$$\widetilde{U}(t) \, e^{-A_0 t} \to \mathbb{I} \quad \text{as} \quad t \to \infty.$$

The proof is finished.   □

Using Lemma 3.1.1, the following result expresses asymptotic behavior of (3.1.5) (see [90, 106]).

Theorem 3.1.2. *Let $\mathbb{I}_s$, $\mathbb{I}_u$ denote the identity matrices of order d_s, d_u respectively. There exists a fundamental solution U for (3.1.5) along with a non-singular matrix C, constants $M > 0$, $K_0 > 0$ and four projections P_{ss}, P_{su}, P_{us}, P_{uu} such that $P_{ss} + P_{su} + P_{us} + P_{uu} = \mathbb{I}$ and that the following hold:*

(i) $|U(t)(P_{ss} + P_{us})U(s)^{-1}| \leq K_0\, e^{2M(s-t)}$ *for* $0 \leq s \leq t$

(ii) $|U(t)(P_{su} + P_{uu})U(s)^{-1}| \leq K_0\, e^{2M(t-s)}$ *for* $0 \leq t \leq s$

(iii) $|U(t)(P_{ss} + P_{su})U(s)^{-1}| \leq K_0\, e^{2M(t-s)}$ *for* $t \leq s \leq 0$

(iv) $|U(t)(P_{us} + P_{uu})U(s)^{-1}| \leq K_0\, e^{2M(s-t)}$ *for* $s \leq t \leq 0$

(v) $\lim\limits_{t \to +\infty} U(t)(P_{ss} + P_{us})U(t)^{-1} = C\begin{pmatrix} \mathbb{I}_s & 0 \\ 0 & 0 \end{pmatrix}C^{-1}$

(vi) $\lim\limits_{t \to +\infty} U(t)(P_{su} + P_{uu})U(t)^{-1} = C\begin{pmatrix} 0 & 0 \\ 0 & \mathbb{I}_u \end{pmatrix}C^{-1}$

(vii) $\lim\limits_{t \to -\infty} U(t)(P_{ss} + P_{su})U(t)^{-1} = C\begin{pmatrix} 0 & 0 \\ 0 & \mathbb{I}_u \end{pmatrix}C^{-1}$

(viii) $\lim\limits_{t \to -\infty} U(t)(P_{us} + P_{uu})U(t)^{-1} = C\begin{pmatrix} \mathbb{I}_s & 0 \\ 0 & 0 \end{pmatrix}C^{-1}$

Also, there exists an integer d with $\operatorname{rank} P_{ss} = \operatorname{rank} P_{uu} = d$.

Proof. From Lemma 3.1.1 there exist two fundamental solutions $\tilde{U}_\pm$ for (3.1.5) such that

$$\lim_{t \to \pm\infty} \tilde{U}_\pm(t)\, e^{-tDf(0)} = \mathbb{I}. \tag{3.1.9}$$

Let C be a matrix such that $J = C^{-1}Df(0)C$ is in Jordan form with $J = \begin{pmatrix} J_1 & 0 \\ 0 & J_2 \end{pmatrix}$ where the eigenvalues of J_1 satisfy $\Re\lambda_i < 0$ while those of J_2 satisfy $\Re\lambda_i > 0$. If $U_\pm(t) := \tilde{U}_\pm(t)C$, then by (3.1.9), $U_\pm(t)$ are two fundamental solutions for (3.1.5) satisfying

$$\lim_{t \to \pm\infty} U_\pm(t)\, e^{-tJ} = C. \tag{3.1.10}$$

Since these are both fundamental solutions we can write $U_+(t) = U_-(t)R$ for some constant matrix R. We now operate on R by means of elementary column operations. The objective is to obtain

$$U_+(t)Q = U_-(t)\bar{R}$$

with Q upper-triangular and $\bar{R}$ such that the first non-zero entry in each column is one with each column-leading one in a different row.

Suppose we have reached the point where the transformed R has the following property: there exist distinct integers $j_1, j_2, \ldots, j_{s-1}$ such that

$$
\begin{aligned}
r_{ij_k} &= 0 && \text{if } i < k, \\
r_{kj_k} &= 1, \\
r_{ik} &= 0 && \text{for } 1 \leq i < s - 1, \quad k \notin \{j_1, j_2, \ldots, j_{s-1}\}.
\end{aligned}
$$

In row s pick the minimum $j_s \notin \{j_1, \ldots, j_{s-1}\}$ such that $p_{sj_s} \neq 0$. Such a j_s must exist as R is non-singular. Now divide column j_s by r_{sj_s} so now $r_{sj_s} = 1$. Next, use column operations to get $r_{sj} = 0$ for $j \notin \{j_1, j_2, \ldots, j_s\}$. Notice we need operate only on columns to the right of column j_s.

Continuing this process through $s = n$ yields a non-singular, upper triangular constant matrix Q such that $U_+(t)Q = U_-(t)\bar{R}$ where $\bar{R}$ has the property that given j, $1 \leq j \leq n$, there exists $\sigma(j)$ defined by $j_{\sigma(i)} = i$ such that $\sigma(i) \neq \sigma(j)$ for $i \neq j$, $\bar{r}_{ij} = 0$ for $i < \sigma(j)$, and $\bar{r}_{\sigma(j),j} = 1$.

Define $U(t) = U_+(t)Q = U_-(t)\bar{R}$ and define four projection matrices with all zero entries except as follows:

$$
\begin{aligned}
(P_{ss})_{ii} &= 1 && \text{if } i \leq d_s \text{ and } \sigma(i) > d_s, \\
(P_{us})_{ii} &= 1 && \text{if } i \leq d_s \text{ and } \sigma(i) \leq d_s, \\
(P_{su})_{ii} &= 1 && \text{if } i > d_s \text{ and } \sigma(i) > d_s, \\
(P_{uu})_{ii} &= 1 && \text{if } i > d_s \text{ and } \sigma(i) \leq d_s.
\end{aligned}
$$

Since Q is upper triangular we can write

$$
Q = \begin{pmatrix} Q_{11} & Q_{12} \\ 0 & Q_{22} \end{pmatrix} \quad \text{and} \quad Q^{-1} = \begin{pmatrix} Q_{11}^{-1} & -Q_{11}^{-1}Q_{12}Q_{22}^{-1} \\ 0 & Q_{22}^{-1} \end{pmatrix}
$$

where Q_{11} is a $d_s \times d_s$ submatrix. We also have

$$
P_{ss} + P_{us} = \begin{pmatrix} \mathbb{I}_s & 0 \\ 0 & 0 \end{pmatrix}.
$$

For $t \geq 0$ and $s \geq 0$, these results yield

$$
U(t)(P_{ss} + P_{us})U(s)^{-1} = U_+(t)Q(P_{ss} + P_{us})Q^{-1}U_+(s)^{-1}
$$

$$
= U_+(t)\,e^{-tJ} \begin{pmatrix} e^{tJ_1} & 0 \\ 0 & e^{tJ_2} \end{pmatrix} \begin{pmatrix} \mathbb{I}_s & -Q_{12}Q_{22}^{-1} \\ 0 & 0 \end{pmatrix} \begin{pmatrix} e^{-sJ_1} & 0 \\ 0 & e^{-sJ_2} \end{pmatrix} e^{sJ}U_+(s)^{-1}
$$

$$
= [U_+(t)\,e^{-tJ}] \begin{pmatrix} e^{(t-s)J_1} & -e^{tJ_1}Q_{12}Q_{22}^{-1}\,e^{-sJ_2} \\ 0 & 0 \end{pmatrix} [e^{sJ}U_+(s)^{-1}].
$$

The expressions in square brackets are bounded for $t \geq 0$, $s \geq 0$. We can choose $K_1 > 0$ and $M > 0$ such that $\left| e^{(t-s)J_1} \right| \leq K_1\,e^{-2M(t-s)} = K_1\,e^{2M(s-t)}$ when $t - s \geq 0$. In addition, for $t \geq 0$ we can find K_2 such that $\left| e^{tJ_1} \right| \leq K_2\,e^{-2Mt}$ and $\left| e^{-tJ_2} \right| \leq K_2\,e^{-2Mt} \leq K_2\,e^{2Mt}$. This proves (i). Setting $t = s$ in the preceding equation and taking the limit yields (v). Parts (ii) and (vi) follows in a similar manner using

$$
P_{su} + P_{uu} = \begin{pmatrix} 0 & 0 \\ 0 & \mathbb{I}_u \end{pmatrix}.
$$

We now turn to (iii). If we interchange columns of $\bar{R}$ so that column j moves to column $\sigma(j)$ the result is a matrix with zeros above the diagonal. In terms of matrices there exists W such that $\hat{R} = \bar{R}W$ is lower-triangular. The matrix $P_{ss} + P_{su}$ consists of ones on the diagonal precisely when $\sigma(j) > d_s$. This means that $W^{-1}(P_{ss} + P_{su})W = \begin{pmatrix} 0 & 0 \\ 0 & \mathbb{I}_u \end{pmatrix}$.

For $t \leq 0$ and $s \leq 0$, combining these results yield

$$U(t)(P_{ss}+P_{su})U(s)^{-1} = [U_-(t)\,e^{-tJ}]\,e^{tJ}\hat{R}\begin{pmatrix} 0 & 0 \\ 0 & \mathbb{I}_u \end{pmatrix}\hat{R}^{-1}e^{-sJ}\left[e^{sJ}U_-(s)^{-1}\right].$$

Parts (iii) and (vii) follow from this; a similar argument yields (iv) and (viii). $\square$

In the language of exponential dichotomies we see that Theorem 3.1.2 provides a two-sided exponential dichotomy. For $t \to -\infty$ an exponential dichotomy is given by the fundamental solution U and the projection $P_{us} + P_{uu}$ while for $t \to +\infty$ such is given by U and $P_{ss} + P_{us}$.

Let u_j denote column j of U and assume these are numbered so that

$$P_{uu} = \begin{pmatrix} \mathbb{I}_d & 0_d & 0 \\ 0_d & 0_d & 0 \\ 0 & 0 & 0 \end{pmatrix}, \quad P_{ss} = \begin{pmatrix} 0_d & 0_d & 0 \\ 0_d & \mathbb{I}_d & 0 \\ 0 & 0 & 0 \end{pmatrix}.$$

Here, $\mathbb{I}_d$ denotes the $d \times d$ identity matrix and 0_d denotes the $d \times d$ zero matrix. For each $j = 1, \cdots, n$, let $u_j^\perp$ be the jth column of the matrix $U^\perp := U^{-1*}$, which is a fundamental solution of the adjoint linear equation $\dot{u}(t) = -Df(\gamma(t))^*u(t)$ of (3.1.5). In general we always assume $< u_{2d}^\perp, \dot{\gamma} > \neq 0$.

We use a functional-analytic method, so we fix $m \in \mathbb{N}$ and define the following Banach spaces [199]:

$$Z_m = C\left([-m, m], \mathbb{R}^n\right), \quad Z_m^p = \{z \in Z_m : z(-m) = z(m)\},$$
$$Y_m = L^\infty\left([-m, m], \mathbb{R}^n\right)$$

with the maximum norm $\|z\|_m := \max_{t \in [-m,m]} |z(t)|$ for Z_m, respectively L^∞ norm $|z|_m := \operatorname{ess\,sup}_{t\in[-m,m]}|z(t)|$, for Y_m. Integration of the inequalities in Theorem 3.1.2 yields the following result.

Theorem 3.1.3. *There exists a constant $A > 0$ such that for any $m > 0$ and any $z \in Y_m$ the following hold:*

(i) $\int_0^t |U(t)(P_{ss} + P_{us})U(s)^{-1}z(s)|\,ds \leq A|z|_m \quad$ for $t \in [0, m]$,

(ii) $\int_t^m |U(t)(P_{su} + P_{uu})U(s)^{-1}z(s)|\,ds \leq A|z|_m \quad$ for $t \in [0, m]$,

(iii) $\int_t^0 |U(t)(P_{ss} + P_{su})U(s)^{-1}z(s)|\,ds \leq A|z|_m \quad$ for $t \in [-m, 0]$,

(iv) $\displaystyle\int_{-m}^{t} |U(t)(P_{us} + P_{uu})U(s)^{-1}z(s)|\,ds \le A|z|_m \quad for\ t \in [-m, 0]\,.$

In order to apply the Lyapunov–Schmidt decomposition method, now we consider the non–homogeneous linear equation

$$\dot{z} = Df(\gamma)z + h\,, \qquad (3.1.11)$$

and we prove a Fredholm–like alternative result for (3.1.11) (see also [140]).

Theorem 3.1.4. *Let $U, P_{ss}, P_{su}, P_{us}, P_{uu}$ be as in Theorem 3.1.2. There exist $m_0 > 0$, $A > 0$, $B > 0$ such that for every $m > m_0$, $m \in \mathbb{N}$ there exists a linear function $\mathbf{L}_m : Y_m \to \mathbb{R}^n$ with $\|P_{uu}\mathbf{L}_m\| \le Ae^{-2Mm}$ and with the property that if $h \in Y_m$ satisfies*

$$\int_{-m}^{m} P_{uu}U(t)^{-1}h(t)\,dt + P_{uu}\mathbf{L}_m h = 0$$

then (3.1.11) has a unique solution in $z \in Z_m^p$ satisfying $P_{ss}U(0)^{-1}z(0) = 0$ and $\|z\|_m \le B|h|_m$. Moreover, this solution z depends linearly on h.

Proof. Given $h \in Y_m$ we use variation of constants to construct the following two solutions to (3.1.11):

$$z_1(t) = U(t)P_{su}\xi_1 + U(t)(P_{us} + P_{uu})U(-m)^{-1}\varphi_1$$

$$+ U(t)\int_0^t (P_{ss} + P_{su})U(s)^{-1}h(s)\,ds + U(t)\int_{-m}^t (P_{us} + P_{uu})U(s)^{-1}h(s)\,ds\,,$$

$$z_2(t) = U(t)P_{us}\xi_2 + U(t)(P_{su} + P_{uu})U(m)^{-1}\varphi_2$$

$$+ U(t)\int_0^t (P_{ss} + P_{us})U(s)^{-1}h(s)\,ds - U(t)\int_t^m (P_{su} + P_{uu})U(s)^{-1}h(s)\,ds$$

satisfying $P_{ss}U(0)^{-1}z(0) = 0$. Here ξ_1, ξ_2, φ_1, φ_2 are arbitrary. We consider $z_1(t)$ for $t \in [-m, 0]$ and $z_2(t)$ for $t \in [0, m]$. First we join these solutions at $t = 0$: $z_1(0) = z_2(0)$ which decomposes into the following three equations:

$$P_{su}\xi_1 - P_{su}U(m)^{-1}\varphi_2 + \int_0^m P_{su}U(s)^{-1}h(s)\,ds = 0\,, \qquad (3.1.12)$$

$$P_{us}U(-m)^{-1}\varphi_1 + \int_{-m}^0 P_{us}U(s)^{-1}h(s)\,ds - P_{us}\xi_2 = 0\,, \qquad (3.1.13)$$

$$P_{uu}U(-m)^{-1}\varphi_1 - P_{uu}U(m)^{-1}\varphi_2 + \int_{-m}^m P_{uu}U(s)^{-1}h(s)\,ds = 0\,. \qquad (3.1.14)$$

Now we join $z_1(t)$, $z_2(t)$ at the endpoints. Solving (3.1.12), (3.1.13) for ξ_1, ξ_2 respectively, then substituting these formulas for ξ_1, ξ_2 into the equation $z_1(-m) = z_2(m)$ and rearranging terms, we get the equation

$$\left[\begin{pmatrix} \mathbb{I}_s & 0 \\ 0 & 0 \end{pmatrix} + R_1(m)\right]C^{-1}\varphi_1 + \left[-\begin{pmatrix} 0 & 0 \\ 0 & \mathbb{I}_u \end{pmatrix} + R_2(m)\right]C^{-1}\varphi_2 = \Psi(m, h)\,,$$

$$(3.1.15)$$

where the matrix C is taken from Theorem 3.1.2 and

$$R_1(m) = C^{-1}\left[U(-m)(P_{us}+P_{uu})U(-m)^{-1} - C\begin{pmatrix} \mathbb{I}_s & 0 \\ 0 & 0 \end{pmatrix}C^{-1}\right.$$
$$\left. -U(m)P_{us}U(-m)^{-1}\right]C,$$

$$R_2(m) = C^{-1}\left[-U(m)(P_{su}+P_{uu})U(m)^{-1} + C\begin{pmatrix} 0 & 0 \\ 0 & \mathbb{I}_u \end{pmatrix}C^{-1}\right.$$
$$\left. +U(-m)P_{su}U(m)^{-1}\right]C,$$

$$C\Psi(m,h) = U(-m)\int_{-m}^{0}(P_{ss}+P_{su})U(s)^{-1}h(s)\,ds$$
$$+U(m)\int_{0}^{m}(P_{ss}+P_{us})U(s)^{-1}h(s)\,ds$$
$$+U(-m)\int_{0}^{m}P_{su}U(s)^{-1}h(s)\,ds + U(m)\int_{-m}^{0}P_{us}U(s)^{-1}h(s)\,ds.$$

Using Theorem 3.1.2 we see that each $|R_i(m)| \to 0$ as $m \to \infty$ and from Theorem 3.1.3 we get $|\Psi(m,h)| = |h|_m O(1)$. Writing

$$C^{-1}\varphi_1 = \begin{bmatrix} u_1 \\ 0 \end{bmatrix}, \qquad C^{-1}\varphi_2 = -\begin{bmatrix} 0 \\ u_2 \end{bmatrix}, \qquad u = \begin{bmatrix} u_1 \\ u_2 \end{bmatrix},$$

where u_1, u_2 are of order d_s, d_u respectively, (3.1.15) becomes

$$\left[\mathbb{I} + R_1(m)\begin{pmatrix} \mathbb{I}_s & 0 \\ 0 & 0 \end{pmatrix} - R_2(m)\begin{pmatrix} 0 & 0 \\ 0 & \mathbb{I}_u \end{pmatrix}\right]u = \Psi(m,h).$$

Since $|R_i(m)| \to 0$ as $m \to \infty$, there exists $m_0 > 0$ so that the coefficient matrix of u in the preceding equation is invertible whenever $m \geq m_0$. In this case the equation can be solved for u which leads to functions $\varphi_i(m,h)$ such that $|\varphi_i| = |h|_m O(1)$. Then the remaining condition (3.1.14) takes the form

$$\int_{-m}^{m}P_{uu}U(s)^{-1}h(s)\,ds + P_{uu}\mathbf{L}_m h = 0,$$

where $\mathbf{L}_m h = U(-m)^{-1}\varphi_1(m,h) - U(m)^{-1}\varphi_2(m,h)$. It follows from the properties of the φ_is and Theorem 3.1.2 that $\|P_{uu}\mathbf{L}_m h\| = |h|_m O(e^{-2Mm})$. $\qquad\square$

To formalize the preceding result we define a closed linear subspace $\widetilde{Y}_m \subset Y_m$ by

$$\widetilde{Y}_m = \left\{z \in Y_m : \int_{-m}^{m}P_{uu}U(t)^{-1}z(t)\,dt + P_{uu}\mathbf{L}_m z = 0\right\}$$

and then define a variation of constants map $K_m : \widetilde{Y}_m \to Z_m$ by taking $K_m(h)$ to be the solution in Z_m^p to (3.1.11) from Theorem 3.1.4. The norm $\|K_m\|$ is uniformly bounded with respect to m, and according to (3.1.11), we have moreover that K_m maps any bounded subset of $\widetilde{Y}_m$ into a bounded one of the Sobolev space $W^{1,2}([-m,m],\mathbb{R}^n)$ which is compactly embedded into Z_m [199]. Hence K_m is a "nice" operator, i.e. it is a compact linear operator. To use the Lyapunov-Schmidt decomposition in the next subsection, which is now based on variation of constants for (3.1.11) from Theorem 3.1.4, we need the following result.

Lemma 3.1.5. *There exist $A > 0$, $m_0 > 0$ and for each $m > m_0$, $m \in \mathbb{N}$ a projection $\Pi_m : Y_m \to Y_m$ such that*

(i) $\|\Pi_m\| < A$ *for all* $m > m_0$,

(ii) $\mathcal{R}(\mathbb{I} - \Pi_m) = \widetilde{Y}_m$.

Proof. Let u_j denote column j of U and let $\phi : \mathbb{R} \to \mathbb{R}$ be a smooth positive function such that $\int_{-\infty}^{\infty} \phi(t)\, dt = 2$ and $\sup_t |\phi(t)u_j(t)| < \tilde{A}$ for all j and some $\tilde{A} > 0$. Choose $m_0 > 0$ so that $\int_{-m}^{m} \phi(t)\, dt \geq 1$ when $m > m_0$ and then define $\phi_m(t) = \phi(t) \big/ \int_{-m}^{m} \phi(t)\, dt$. Note that we have $\int_{-m}^{m} \phi_m(t)\, dt = 1$ and $\sup_t |\phi_m(t)u_j(t)| \leq \tilde{A}$ when $m > m_0$. Now define a $d \times d$ matrix $A(m) = [a_{ij}(m)]$ by

$$a_{ij}(m) = [\mathbf{L}_m(\phi_m u_j)]_i, \qquad 1 \leq i \leq d, \quad 1 \leq j \leq d.$$

We have $|a_{ij}(m)| = O\left(e^{-2Mm}\right)$. Given $z \in Y_m$ we define $p(z) \in \mathbb{R}^d$ as

$$p_i(z) = \int_{-m}^{m} < u_i^{\perp}(t), z(t) > dt + (\mathbf{L}_m z)_i, \quad 1 \leq i \leq d.$$

So that

$$\int_{-m}^{m} P_{uu} U(t)^{-1} z(t)\, dt + P_{uu} \mathbf{L}_m z = (p_1(z), \cdots, p_d(z), 0, \cdots, 0).$$

By increasing m_0 if necessary we can assume $\|A(m)\| \leq 1/2$ for $m > m_0$ so then $\mathbb{I} + A(m)$ is invertible and we can write $\alpha = [\mathbb{I} + A(m)]^{-1} p(z)$. Let $\bar{\alpha} \in \mathbb{R}^n$ denote $(\alpha_1, \cdots, \alpha_d, 0, \cdots, 0)$ and define

$$(\Pi_m z)(t) = \phi_m(t) U(t) P_{uu} \bar{\alpha}.$$

It is straight-forward to verify that Π_m has the required properties. $\square$

3.1.3   Subharmonics for Regular Periodic Perturbations

Our aim is to find subharmonic solutions to (3.1.3) of very large periods which are close to γ. So we look for solutions to (3.1.3) in Z_m^p by rewriting (3.1.3) as an abstract operator inclusion (3.1.20). We use the Lyapunov-Schmidt method along with the Leray-Schauder degree for inclusions to handle that abstract operator inclusion. To realize this functional-analytic approach, first, we define the function $b : \mathbb{R}^{d-1} \times (0, \infty) \to \mathbb{R}^n$ by

$$b(\beta, r) = \gamma(-r) - \gamma(r) + \sum_{i=1}^{d-1} \beta_i (u_{i+d}(-r) - u_{i+d}(r)),$$

where $\beta = (\beta_1, \cdots, \beta_{d-1})$. Note that

$$|b(\beta, r)| = O(e^{-Mr}) \tag{3.1.16}$$

uniformly with respect to β from any bounded subset of $\mathbb{R}^{d-1}$. Next, in (3.1.3) we now make the change of variable $\mu \leftrightarrow s^2 \mu$ and

$$x(t + \alpha) = \gamma(t) + s^2 z(t) + \sum_{i=1}^{d-1} s\beta_i u_{i+d}(t) + \frac{1}{2(m+S)} b(s\beta, m + S) t,$$

where $1 > s > 0$, $\alpha, \beta_i \in \mathbb{R}$, $m \in \mathbb{Z}_+$, $S = [1/s]$ and $[\tilde{s}]$ is the integer part of $\tilde{s}$. The function b is constructed so that if $z \in Z^p_{m+S}$ then $x \in Z^p_{m+S}$. The differential inclusion for z is

$$\dot{z}(t) - Df(\gamma(t))z(t) \in g_{m,s}(z(t), \alpha, \beta, \mu, t) \quad \text{a.e. on} \quad [-m - S, m + S],$$

where

$$g_{m,s}(x, \alpha, \beta, \mu, t) =$$
$$\left\{ v \in \mathbb{R}^n : v \in \frac{1}{s^2} \left\{ f\left(s^2 x + \gamma(t) + s \sum_{i=1}^{d-1} \beta_i u_{i+d}(t) + \frac{1}{2(m+S)} b(s\beta, m + S) t \right) \right. \right.$$
$$- f(\gamma(t)) - s \sum_{i=1}^{d-1} \beta_i \dot{u}_{i+d}(t) - \frac{1}{2(m+S)} b(s\beta, m + S) - Df(\gamma(t))s^2 x \right\}$$
$$\left. + \sum_{j=1}^{k} \mu_j f_j \left(s^2 x + \gamma(t) + s \sum_{i=1}^{d-1} \beta_i u_{i+d}(t) + \frac{1}{2(m+S)} b(s\beta, m + S) t, s^2 \mu, t + \alpha \right) \right\}.$$

Using $g_{m,s}$ we define a multivalued mapping

$$G_{m,s} : Z_{m+S} \times \mathbb{R} \times \mathbb{R}^{d-1} \times \mathbb{R}^k \to 2^{Y_{m+S}}$$

by the formula

$$G_{m,s}(z, \alpha, \beta, \mu) =$$
$$\left\{ h \in Y_{m+S} : h(t) \in g_{m,s}(z(t), \alpha, \beta, \mu, t) \quad \text{a.e. on} \quad [-m - S, m + S] \right\},$$

so the multivalued equation for z can be written

$$\dot{z} - Df(\gamma)z \in G_{m,s}(z, \alpha, \beta, \mu). \tag{3.1.17}$$

Since $g_{m,s} : \mathbb{R}^n \times \mathbb{R} \times \mathbb{R}^{d-1} \times \mathbb{R}^k \times \mathbb{R} \to 2^{\mathbb{R}^n} \setminus \{\emptyset\}$ is upper–semicontinuous with compact and convex values, according to Theorem 2.4.3, each of these sets $G_{m,s}(z, \alpha, \beta, \mu)$ is non–empty. Moreover, these sets $G_{m,s}(z, \alpha, \beta, \mu)$ are all closed convex and bounded in $Y_{m+S} \subset L^2([-m-S, m+S], \mathbb{R}^n)$. So by Theorem 2.1.1, all $G_{m,s}(z, \alpha, \beta, \mu)$ are weakly compact in the Hilbert space L^2 $([-m-S, m+S], \mathbb{R}^n)$.

We can not solve directly (3.1.17), so first of all, we insert it to the homotopy

$$G_{m,s,\lambda}(z, \alpha, \beta, \mu) = \left\{ h \in Y_{m+S} : h(t) \in g_{m,s,\lambda}(z(t), \alpha, \beta, \mu, t) \text{ a.e. on } [-m-S, m+S] \right\},$$

for $\lambda \in [0, 1]$, where

$$g_{m,s,\lambda}(x, \alpha, \beta, \mu, t) = \left\{ v \in \mathbb{R}^n : \right.$$

$$v \in \frac{\lambda}{s^2} \left\{ f\left(s^2 x + \gamma(t) + s \sum_{i=1}^{d-1} \beta_i u_{i+d}(t) + \frac{1}{2(m+S)} b(s\beta, m+S)t \right) - f(\gamma(t)) \right.$$

$$\left. - s \sum_{i=1}^{d-1} \beta_i \dot{u}_{i+d}(t) - \frac{1}{2(m+S)} b(s\beta, m+S) - Df(\gamma(t))s^2 x \right\}$$

$$+ \lambda \sum_{j=1}^{k} \mu_j f_j \left(s^2 x + \gamma(t) + s \sum_{i=1}^{d-1} \beta_i u_{i+d}(t) + \frac{1}{2(m+S)} b(s\beta, m+S)t, s^2\mu, t + \alpha \right)$$

$$\left. + \frac{1-\lambda}{2} \sum_{i,j=1}^{d-1} \beta_i \beta_j D^2 f(\gamma(t))(u_{d+i}(t), u_{d+j}(t)) + (1 - \lambda) \sum_{j=1}^{k} \mu_j f_j(\gamma(t), 0, t + \alpha) \right\}.$$

Now based on Theorem 3.1.4 and Lemma 3.1.5, we apply the Lyapunov-Schmidt decomposition to (3.1.17) and put it also in the addition homotopy in the following way

$$0 \in \left(z - F_{m,s,\lambda}(z, \alpha, \beta, \mu), B_{m,s,\lambda}(z, \alpha, \beta, \mu) \right), \quad \lambda \in [0, 1], \tag{3.1.18}$$

where

$$\left(F_{m,s,\lambda}(z, \alpha, \beta, \mu), B_{m,s,\lambda}(z, \alpha, \beta, \mu) \right)$$
$$= \left\{ \left(K_{m+S}(\mathbb{I} - \Pi_{m+S})h, L_{m+S}h \right) : h \in G_{m,s,\lambda}(z, \alpha, \beta, \mu) \right\},$$

and

$$L_{m+S}v = \int_{-m-S}^{m+S} P_{uu} U(t)^{-1} v(t)\, dt + P_{uu} \mathbf{L}_{m+S}v.$$

To solve (3.1.18), we introduce the new homotopy

$$0 \in \left(z - \lambda F_{m,s,\lambda}(z, \alpha, \beta, \mu), B_{m,s,\lambda}(z, \alpha, \beta, \mu) \right), \quad \lambda \in [0, 1]. \tag{3.1.19}$$

Since $\|P_{uu}\mathbf{L}_{m+S}\| = O(e^{-2M(m+S)})$, we consider the decomposition and homotopy

$$(B_{m1,s,\lambda} + \lambda B_{m2,s,\lambda})(z, \alpha, \beta, \mu), \quad \lambda \in [0, 1],$$

where

$$(B_{m1,s,\lambda} + \lambda B_{m2,s,\lambda})(z,\alpha,\beta,\mu) = \{L_{m1,s}h + \lambda L_{m2,s}h \; : \; h \in G_{m,s,\lambda}(z,\alpha,\beta,\mu)\},$$

and

$$L_{m1,s}v = \int_{-m-S}^{m+S} P_{uu}U(t)^{-1}v(t)\,dt\,, \quad L_{m2,s}v = P_{uu}\mathbf{L}_{m+S}v\,.$$

Summarizing we obtain that the solvability of (3.1.18–3.1.19) can be replaced by the solvability of the following multivalued equation

$$
\begin{aligned}
0 \in \quad & H_{m,s}(z,\alpha,\beta,\mu,\lambda) := \\
& \Big(z - \lambda F_{m,s,\lambda}(z,\alpha,\beta,\mu), \big(B_{m1,s,\lambda} + \lambda B_{m2,s,\lambda}\big)(z,\alpha,\beta,\mu)\Big),
\end{aligned}
\tag{3.1.20}
$$

when $H_{m,s} : Z_{m+S} \times \mathbb{R} \times \mathbb{R}^{d-1} \times \mathbb{R}^k \times [0,1] \to 2^{Z_{m+S} \times \mathbb{R}^d} \setminus \{\emptyset\}$, while $1 > s > 0$ is sufficiently small and fixed, and $\lambda \in [0,1]$ is a homotopy parameter. Since clearly the multivalued mapping

$$g_{m,s,\cdot} : [0,1] \times \mathbb{R}^n \times \mathbb{R} \times \mathbb{R}^{d-1} \times \mathbb{R}^k \times \mathbb{R} \to 2^{\mathbb{R}^n} \setminus \{\emptyset\}$$

is upper–semicontinuous with compact and convex values, using standard arguments based on Mazur's Theorem 2.1.2 (see arguments below (3.1.21)), the mapping $H_{m,s}$ is also upper–semicontinuous. Moreover, according to the compactness of K_m, $\mathbb{I}_{Z_{m+S} \times \mathbb{R}^d} - H_{m,s}$ has compact convex values and maps bounded sets into relatively compact ones. Hence topological degree methods of Section 2.4.3 can be applied to (3.1.20). Furthermore, ranges of $H_{m,s}$ are bounded provided that z, β, μ are bounded, $s > 0$ is small fixed and $m \in \mathbb{Z}_+$, $\alpha \in \mathbb{R}$, $\lambda \in [0,1]$ are arbitrary.

To state the main result, we introduce a multivalued Melnikov mapping for our problem

$$M_\mu : \mathbb{R}^d \to 2^{\mathbb{R}^d} \setminus \{\emptyset\}, \quad M_\mu = (M_{\mu 1}, \cdots, M_{\mu d})$$

$$M_{\mu l}(\alpha,\beta) = \Big\{ \int_{-\infty}^{\infty} \langle h(s), u_l^\perp(s) \rangle \, ds \; : \; h \in L_{\mathrm{loc}}^2(\mathbb{R},\mathbb{R}^n) \text{ satisfying a.e. on } \mathbb{R}$$

$$h(t) \in \big(\tfrac{1}{2} \sum_{i,j=1}^{d-1} \beta_i \beta_j D^2 f(\gamma(t))(u_{d+i}(t), u_{d+j}(t)) + \sum_{j=1}^{k} \mu_j f_j(\gamma(t), 0, t+\alpha)\big) \Big\}.$$

$$\tag{3.1.21}$$

Here $L_{\mathrm{loc}}^2(\mathbb{R},\mathbb{R}^n) := \cap_{m \in \mathbb{N}} L^2([-m,m],\mathbb{R}^n)$. The mapping M_μ is again upper–semicontinuous with compact convex values and maps bounded sets into bounded ones. The boundedness is clear, so we show the upper–semicontinuity:

Let

$$\mu_p \to \mu_0, \quad (\alpha_p,\beta_p) \to (\alpha_0,\beta_0), \quad p \in \mathbb{N}, \quad h_p \in L_{\mathrm{loc}}^2(\mathbb{R},\mathbb{R}^n)$$

$$h_p(t) \in \big(\tfrac{1}{2} \sum_{i,j=1}^{d-1} \beta_{pi} \beta_{pj} D^2 f(\gamma(t))(u_{d+i}(t), u_{d+j}(t)) + \sum_{j=1}^{k} \mu_{pj} f_j(\gamma(t), 0, t+\alpha_p)\big)$$

a.e. on $\mathbb{R}$, and

$$\int_{-\infty}^{\infty} \langle h_p(s), u_i^{\perp}(s) \rangle \, ds \to \bar{M}_{0i} \in \mathbb{R}, \quad \forall i = 1, 2, \cdots, d \, .$$

Since $\sup\limits_{p \in \mathbb{N}} |h_p|_{\infty} < \infty$, where $| \cdot |_{\infty}$ is the norm on the Banach space $L^{\infty}(\mathbb{R}, \mathbb{R}^n)$, the sequence $\{h_p\}_1^{\infty}$ is bounded in the Hilbert space $L^2([-l, l], \mathbb{R}^n)$ for any $l \in \mathbb{N}$. From Theorem 2.1.1 we can assume, by using the Cantor diagonal procedure and by passing to a subsequence of the original one, that there is an $h_0 \in L_{\text{loc}}^2(\mathbb{R}, \mathbb{R}^n)$ such that $\{h_p\}_1^{\infty}$ tends weakly to h_0 in $L^2([-l, l], \mathbb{R}^n)$ for any $l \in \mathbb{N}$. Next fix l. By Mazur's Theorem 2.1.2 we take $\tilde{h}_p \in \text{con} \, [\{h_p, h_{p+1}, \cdots \}]$ such that $\tilde{h}_p \to h_0$ in $L^2([-l, l], \mathbb{R}^n)$ and so by passing to a subsequence that $\tilde{h}_{p_k} \to h_0$ a.e. on $[-l, l]$. Let us fix for a while $t \in [-l, l]$ such that $\tilde{h}_{p_k}(t) \to h_0(t)$ and the above assumptions hold. Let $O_t \subset \mathbb{R}^n$ be an open convex neighborhood of

$$\Gamma_t := \frac{1}{2} \sum_{i,j=1}^{d-1} \beta_{0i} \beta_{0j} D^2 f(\gamma(t))(u_{d+i}(t), u_{d+j}(t)) + \sum_{j=1}^{k} \mu_{0j} f_j(\gamma(t), 0, t + \alpha_0) \, .$$

By assumption (i) of Subsection 3.1.1, there is an $p_0 \in \mathbb{N}$ such that

$$\frac{1}{2} \sum_{i,j=1}^{d-1} \beta_{pi} \beta_{pj} D^2 f(\gamma(t))(u_{d+i}(t), u_{d+j}(t)) + \sum_{j=1}^{k} \mu_{pj} f_j(\gamma(t), 0, t + \alpha_p) \subset O_t$$

for any $p \geq p_0$. Hence $h_p(t) \in O_t$, $\forall p \geq p_0$, and so $\tilde{h}_{p_k}(t) \in O_t$ for any $p_k \geq p_0$. Consequently, we obtain $h_0(t) \in \bar{O}_t$. By assumption (i) of Subsection 3.1.1 and taking O_t arbitrarily close to Γ_t, we obtain $h_0(t) \in \Gamma_t$. Thus we have

$$h_0(t) \in \left(\frac{1}{2} \sum_{i,j=1}^{d-1} \beta_{0i} \beta_{0j} D^2 f(\gamma(t))(u_{d+i}(t), u_{d+j}(t)) + \sum_{j=1}^{k} \mu_{0j} f_j(\gamma(t), 0, t + \alpha_0) \right)$$

a.e on $\mathbb{R}$. Finally, clearly

$$\int_{-\infty}^{\infty} \langle h_0(s), u_i^{\perp}(s) \rangle \, ds = \bar{M}_{0i}, \quad \forall i = 1, 2, \cdots, d \, .$$

The upper–semicontinuity of M_{μ} follows from Theorem 2.4.2.

Let $S^{\tilde{k}-1} = \{ b \in \mathbb{R}^{\tilde{k}} : |b| = 1 \}$ be the $(\tilde{k} - 1)$–dimensional sphere. We are ready to prove the main theorems of this section [73].

Theorem 3.1.6. Let $d > 1$. If there is a non–empty open bounded set $\mathbf{B} \subset \mathbb{R}^d$ and $\mu_0 \in S^{k-1}$ such that

(i) $0 \notin M_{\mu_0}(\partial \mathbf{B})$

(ii) $\deg(M_{\mu_0}, \mathbf{B}, 0) \neq 0$

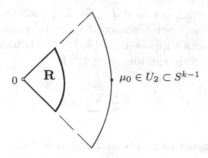

Figure 3.2: The wedge-shaped region $\mathbf{R}$

Then there is a constant $K > 0$ and a wedge–shaped region in $\mathbb{R}^k$ for μ of the form (see Fig. 3.2)

$$\mathbf{R} = \left\{ s^2 \tilde{\mu} : s > 0, \quad \text{respectively } \tilde{\mu}, \text{ is from an open small connected} \right.$$
$$\left. \text{neighborhood } U_1, \text{ respectively } U_2 \subset S^{k-1}, \text{ of } 0 \in \mathbb{R}, \text{ respectively of } \mu_0 \right\}$$

such that for any $\mu \in \mathbf{R}$ of the form $\mu = s^2 \tilde{\mu}$, $0 < s \in U_1$, $\tilde{\mu} \in U_2$, the differential inclusion (3.1.3) possesses a subharmonic solution x_m of period $2m$ for any $m \in \mathbb{N}$, $m \geq [1/s]$ satisfying, according to the change of variable below (3.1.16),

$$\sup_{-m \leq t \leq m} \left| x_m(t) - \gamma(t - \alpha_m) \right| \leq Ks,$$

where $\alpha_m \in \mathbb{R}$ and $|\alpha_m| \leq K$.

Proof. We need to solve (3.1.17), which is inserted into the homotopy (3.1.20). To handle (3.1.20), we use the following Lemmas.

Lemma 3.1.7. *The above condition* (i) *implies for any $A > 0$ the existence of constants $\delta > 0$, $1 > s_0 > 0$ such that*

$$|B_{m1,s,\lambda}(z, \alpha, \beta, \mu)| \geq \delta$$

for any $0 < s < s_0$, $m \in \mathbb{Z}_+$, $\lambda \in [0, 1]$, $\|z\|_{m+s} \leq A$, $(\alpha, \beta) \in \partial \mathbf{B}$, $|\mu - \mu_0| \leq \delta$.

Proof. Assume the contrary. So there is an $A > 0$ and

$$s_p \to 0, \quad m_p \in \mathbb{Z}_+, \quad \lambda_p \to \lambda_0, \quad \|z_p\|_{m_p+s_p} \leq A, \quad p \in \mathbb{N}$$
$$\mu_p \to \mu_0, \quad \partial \mathbf{B} \ni (\alpha_p, \beta_p) \to (\alpha_0, \beta_0) \in \partial \mathbf{B}, \quad h_p \in G_{m_p,s_p,\lambda_p}(z_p, \alpha_p, \beta_p, \mu_p)$$

such that

$$L_{m_p1,s_p} h_p \to 0 \quad \text{as} \quad p \to \infty.$$

Since $\sup_{p \in \mathbb{N}} |h_p|_{\infty, m_p+S_p} < \infty$, where $|\cdot|_{\infty,m}$, $m \in \mathbb{N}$ is the norm on the Banach space $L^\infty([-m, m], \mathbb{R}^n)$, we can assume like above, by using the Cantor diagonal procedure, that there is an $h_0 \in L^2_{\text{loc}}(\mathbb{R}, \mathbb{R}^n)$ such that $\{h_p\}_1^\infty$ tends weakly to h_0 in any $[-l, l]$, $l \in \mathbb{N}$ with respect to the L^2 norm. Next, again by Mazur's Theorem 2.1.2 we can suppose that $\{h_p\}_1^\infty$ tends to h_0 almost everywhere.

Furthermore, clearly the following holds

$$\lim_{s \to 0_+} G_{m,s,\lambda}(z, \alpha, \beta, 0) = \frac{1}{2} \sum_{i,j=1}^{d-1} \beta_i \beta_j D^2 f(\gamma) (u_{i+d}, u_{j+d}) \tag{3.1.22}$$

uniformly with respect to z, β bounded and $\alpha \in \mathbb{R}, m \in \mathbb{Z}_+, \lambda \in [0, 1]$ arbitrary. Moreover, by Theorem 3.1.4 and the properties (ii) and (iv) of Theorem 3.1.2, we have

$$\lim_{s \to 0_+} L_{m1,s} v = \int_{-\infty}^{\infty} P_{uu} U(t)^{-1} v(t) \, dt, \quad \lim_{s \to 0_+} L_{m2,s} v = 0 \tag{3.1.23}$$

uniformly with respect to v bounded and $m \in \mathbb{Z}_+$ arbitrary.

Finally using (3.1.22–3.1.23), we obtain

$$h_0(t) \in \left(\frac{1}{2} \sum_{i,j=1}^{d-1} \beta_{0i} \beta_{0j} D^2 f(\gamma(t))(u_{d+i}(t), u_{d+j}(t)) \right.$$
$$\left. + \sum_{j=1}^{k} \mu_{0j} f_j(\gamma(t), 0, t + \alpha_0) \right) \quad \text{a.e. on} \quad \mathbb{R},$$

and $\int_{-\infty}^{\infty} P_{uu} U(t)^{-1} h_0(t) \, dt = 0$. This contradicts to (i) of this theorem. $\qquad \square$

Using Lemma 3.1.7, the next result follows directly from the construction of $H_{m,s}$.

Lemma 3.1.8. *There are open small connected neighborhoods $\tilde{U}_1 \subset \mathbb{R}$, $U_2 \subset S^{k-1}$ of 0, respectively of μ_0, and a constant $K_1 > 0$ such that*

$$0 \notin H_{m,s}(\partial \Omega, \mu, \lambda)$$

for any $0 < s \in \tilde{U}_1$, $\mu \in U_2$, $\lambda \in [0, 1]$, $m \in \mathbb{Z}_+$, where

$$\Omega = \left\{ (z, \alpha, \beta) \in Z_{m+S} \times \mathbb{R}^d : \|z\|_{m+S} < K_1, (\alpha, \beta) \in \mathbf{B} \right\}.$$

From the homotopy invariance property of the Leray-Schauder topological degree, by Lemma 3.1.8 for any $0 < s \in \tilde{U}_1$, $\mu \in U_2$, $m \in \mathbb{Z}_+$, we obtain

$$\deg\left(H_{m,s}(\cdot, \mu, 1), \Omega, 0 \right) = \deg\left(H_{m,s}(\cdot, \mu_0, 0), \Omega, 0 \right).$$

Since

$$H_{m,s}(z, \alpha, \beta, \mu_0, 0) = \left(z, \{ L_{m1,s} h : h \in Y_{m+S} \text{ satisfying a.e. on } [-m-S, m+S] \right.$$
$$h(t) \in \left(\frac{1}{2} \sum_{i,j=1}^{d-1} \beta_i \beta_j D^2 f(\gamma(t))(u_{d+i}(t), u_{d+j}(t)) + \sum_{j=1}^{k} \mu_{0j} f_j(\gamma(t), 0, t+\alpha)) \} \right),$$

in order to compute deg $\left(H_{m,s}(\cdot,\mu_0,0),\Omega,0\right)$, we take the upper–semicontinuous homotopy

$$\left\{\lambda L_{m1,s}h + (1-\lambda)\int_{-\infty}^{\infty} P_{uu}U(t)^{-1}h(t)\,dt \; : \; h \in L^2_{\text{loc}}(\mathbb{R},\mathbb{R}^n)\text{ satisfying a.e. on }\mathbb{R}\right.$$
$$\left. h(t) \in \left(\tfrac{1}{2}\sum_{i,j=1}^{d-1}\beta_i\beta_j D^2 f(\gamma(t))(u_{d+i}(t),u_{d+j}(t)) + \sum_{j=1}^{k}\mu_{0j}f_j(\gamma(t),0,t+\alpha)\right)\right\}.$$

Then using (3.1.23), the assumption (ii) of Theorem 3.1.6 as well as formula (2.2.5) and Lemma 3.1.7, when s_0 is shrunk if necessary, we arrive at

$$\deg\left(H_{m,s}(\cdot,\mu_0,0),\Omega,0\right) = \deg(M_{\mu_0},\mathbf{B},0) \neq 0.$$

Hence (3.1.20) has a solution in Ω for any $0 < s \in U_1$, $\mu \in U_2$, $m \in \mathbb{Z}_+$ and $\lambda = 1$, where $U_1 = \tilde{U}_1 \cap \{s \in \mathbb{R} : |s| < s_0\}$. This solution gives a solution of (3.1.17) according to the definition of (3.1.20). The proof of Theorem 3.1.6 is completed. □

Now we concentrate on the case $d = 1$ since it has a specific feature. Then $M_\mu : \mathbb{R} \to 2^{\mathbb{R}} \setminus \{\emptyset\}$ has the form

$$M_\mu(\alpha) = \left\{ \begin{array}{l} \int_{-\infty}^{\infty}\langle h(s), u_1^{\perp}(s)\rangle\,ds \; : \; h \in L^2_{\text{loc}}(\mathbb{R},\mathbb{R}); \\[2mm] h(t) \in \sum_{j=1}^{k}\mu_j f_j(\gamma(t),0,t+\alpha) \quad \text{a.e. on}\quad \mathbb{R} \end{array} \right\}. \tag{3.1.24}$$

For any $A, B \subset \mathbb{R}$ we set $AB := \{ab \mid a \in A, b \in B\}$.

Theorem 3.1.9. *Let $d = 1$. If there are constants $a < b$ and $\mu_0 \in S^{k-1}$ such that*

$$M_{\mu_0}(a)M_{\mu_0}(b) \subset (-\infty, 0),$$

then there is a constant $K > 0$ and a wedge–shaped region in $\mathbb{R}^k$ for μ of the form

$$\mathbf{R} = \left\{ \pm s^2\tilde{\mu} : s > 0 \text{ respectively } \tilde{\mu},\text{ is from an open small connected} \right.$$
$$\left. \text{neighborhood } U_1,\text{ respectively } U_2 \subset S^{k-1},\text{ of } 0 \in \mathbb{R},\text{ respectively of } \mu_0 \right\}$$

such that for any $\mu \in \mathbf{R}$ of the form $\mu = \pm s^2\tilde{\mu}$, $0 < s \in U_1$, $\tilde{\mu} \in U_2$, the differential inclusion (3.1.3) possesses a subharmonic solution x_m of period $2m$ for any $m \in \mathbb{N}$, $m \geq [1/s]$ satisfying, according to the change of variable below (3.1.16),

$$\sup_{-m \leq t \leq m} |x_m(t) - \gamma(t - \alpha_m)| \leq Ks^2,$$

where $\alpha_m \in (a, b)$.

Proof. We apply Theorem 3.1.6 with $\mathbf{B} = (a, b)$ by considering both M_{μ_0} and $M_{-\mu_0}$. The assumption (i) of Theorem 3.1.6 is clearly satisfied. To prove (ii), it is enough to consider the case (the remaining one is similar) that $M_{\mu_0}(a)$ contains positive and $M_{\mu_0}(b)$ negative numbers, and then to take the homotopy

$$M^\lambda(\alpha) := \lambda M_{\mu_0}(\alpha) + (1 - \lambda)\left(\frac{a + b}{2} - \alpha\right).$$

From $0 \notin M^\lambda(b)$, $0 \notin M^\lambda(a)$ for any $\lambda \in [0, 1]$ it follows $\deg(M_{\mu_0}, (a, b), 0) = \deg(M^1, (a, b), 0) = \deg(M^0, (a, b), 0) \neq 0$. Similarly for $M_{-\mu_0}$.   □

Remark 3.1.10. The restriction $|\mu_0| = 1$ is not essential, because of M_μ in both (3.1.21)–(3.1.24) is homogeneous with respect to the variables β and μ.

Remark 3.1.11. If $d = 1$, $k = 1$ and f_1 is C^2-smooth in (3.1.3) then the existence of a simple root of $M_1(\alpha) = 0$ implies chaos for such systems by the Smale-Birkhoff Homoclinic Theorem 2.5.4. This is mentioned above in Chapter 1 (Introduction) and Sections 2.5.3, 3.1.1. In particular, they have subharmonic solutions of all large periods. But the existence of a simple root α_0 of $M_1(\alpha) = 0$ gives a small $\widetilde{\delta} > 0$ such that $a = \alpha_0 - \widetilde{\delta}$, $b = \alpha_0 + \widetilde{\delta}$ satisfy the assumption of Theorem 3.1.9. Consequently, Theorem 3.1.9 is an extension of the Smale-Birkhoff Homoclinic Theorem 2.5.4 concerning subharmonics to the multivalued case (3.1.3) (see also Subsection 4.2.1).

3.1.4   Subharmonics for Singular Periodic Perturbations

Now we directly extend a method of Subsection 3.1.3 to singularly perturbed differential inclusions of the form

$$\varepsilon \dot{x}(t) \in f(x(t)) + \varepsilon h(x(t), \varepsilon, t) \quad \text{a.e. on} \quad \mathbb{R}, \tag{3.1.25}$$

where $\varepsilon \neq 0$ is small, as well as to the modification of (3.1.25) given by

$$\varepsilon \dot{x}(t) \in f(x(t)) + \varepsilon^2 h(x(t), \varepsilon, t) \quad \text{a.e. on} \quad \mathbb{R}, \tag{3.1.26}$$

where $\varepsilon > 0$ is small, f has the above properties (i–iii) of Subsection 3.1.1 and $h : \mathbb{R}^n \times \mathbb{R} \times \mathbb{R} \to 2^{\mathbb{R}^n} \setminus \{\emptyset\}$ is 2–periodic in t as well as upper–semicontinuous with compact convex values. We show the existence of $2m$–periodic solutions of (3.1.25)–(3.1.26) for any $m \in \mathbb{N}$, not just for large m like for (3.1.3), provided f, h satisfy additional conditions. We are motivated to study (3.1.25–3.1.26) by similar results for differential equations with slowly varying coefficients [29, 90].

Searching for periodic solutions of (3.1.25) with periods $2m$, $m = 1, 2, \cdots$ is equivalent to finding $\frac{2m}{\varepsilon}$–periodic solutions of the differential inclusion

$$\dot{x}(t) \in f(x(t)) \pm \varepsilon h(x(t), \pm\varepsilon, \pm\varepsilon t) \quad \text{a.e. on} \quad \mathbb{R}, \quad \varepsilon > 0. \tag{3.1.27}$$

Inclusion (3.1.27) has a slowly varying coefficient represented by the term $\pm\varepsilon t$. Now we can apply the procedure of Subsection 3.1.3 to (3.1.27) with the following exchanges

$$\varepsilon \leftrightarrow \varepsilon^2, \quad m \leftrightarrow \frac{m}{\varepsilon^2}, \quad S \leftrightarrow 0, \quad m \in \mathbb{N} \text{ is arbitrary}.$$

This induces in (3.1.27) the change of variables

$$x\left(t \pm \frac{\alpha}{\varepsilon^2}\right) = \gamma(t) + \varepsilon^2 z(t) + \sum_{i=1}^{d-1} \varepsilon \beta_i u_{i+d}(t) + \frac{\varepsilon^2}{2m} b(\varepsilon\beta, \frac{m}{\varepsilon^2})t\,,$$

where $-\frac{m}{\varepsilon^2} \le t \le \frac{m}{\varepsilon^2}$ and $1 > \varepsilon > 0$, α, $\beta_i \in \mathbb{R}$. Then after insertion it into (3.1.27) the differential inclusion for z is

$$\dot{z}(t) - Df(\gamma(t))z(t) \in g_{m,\varepsilon}(z(t), \alpha, \beta, \pm, t) \quad \text{a.e. on} \quad [-\frac{m}{\varepsilon^2}, \frac{m}{\varepsilon^2}]\,,$$

where

$$g_{m,\varepsilon}(x, \alpha, \beta, \pm, t) =$$
$$\left\{ v \in \mathbb{R}^n : v \in \frac{1}{\varepsilon^2}\left\{ f\left(\varepsilon^2 x + \gamma(t) + \varepsilon \sum_{i=1}^{d-1} \beta_i u_{i+d}(t) + \frac{\varepsilon^2}{2m} b(\varepsilon\beta, \frac{m}{\varepsilon^2})t\right) \right.\right.$$
$$\left. -f(\gamma(t)) - \varepsilon \sum_{i=1}^{d-1} \beta_i \dot{u}_{i+d}(t) - \frac{\varepsilon^2}{2m} b(\varepsilon\beta, \frac{m}{\varepsilon^2}) - Df(\gamma(t))\varepsilon^2 x \right\}$$
$$\left. \pm h\left(\varepsilon^2 x + \gamma(t) + \varepsilon \sum_{i=1}^{d-1} \beta_i u_{i+d}(t) + \frac{\varepsilon^2}{2m} b(\varepsilon\beta, \frac{m}{\varepsilon^2})t, \pm\varepsilon^2, \pm\varepsilon^2 t + \alpha\right) \right\}.$$

Consequently, the formulas (3.1.17)–(3.1.23) can be straightforwardly modified to (3.1.27). The multivalued mappings of (3.1.21)–(3.1.24) possess the forms

$$M_\pm : \mathbb{R}^d \to 2^{\mathbb{R}^d} \setminus \{\emptyset\}, \quad M_\pm = (M_{\pm 1}, \cdots, M_{\pm d})$$

$$M_{\pm l}(\alpha, \beta) = \left\{ \int_{-\infty}^{\infty} \langle p(s), u_l^\perp(s)\rangle \, ds \ : \ p \in L^2_{loc}(\mathbb{R}, \mathbb{R}^n); \right.$$

$$p(t) \in (\frac{1}{2} \sum_{i,j=1}^{d-1} \beta_i\beta_j D^2 f(\gamma(t))(u_{d+i}(t), u_{d+j}(t)) \pm h(\gamma(t), 0, \alpha)) \quad \text{a.e. on} \quad \mathbb{R} \left. \right\},$$

$$\tag{3.1.28}$$

and

$$M : \quad \mathbb{R} \to 2^{\mathbb{R}} \setminus \{\emptyset\}$$

$$M(\alpha) = \left\{ \int_{-\infty}^{\infty} \langle p(s), u_1^\perp(s)\rangle \, ds \ : \ p \in L^2_{loc}(\mathbb{R}, \mathbb{R}); \right.$$

$$\tag{3.1.29}$$

$$p(t) \in h(\gamma(t), 0, \alpha) \quad \text{a.e. on} \quad \mathbb{R} \left. \right\}.$$

Then Theorems 3.1.6–3.1.9 have the following analogies.

Theorem 3.1.12. *Let $d > 1$. If there is a non–empty open bounded set $\mathbf{B} \subset \mathbb{R}^d$ and $* \in \{-, +\}$ such that $0 \notin M_*(\partial\mathbf{B})$ and $\deg(M_*, \mathbf{B}, 0) \ne 0$. Then there is a constant $K > 0$ such that for any sufficiently small $\varepsilon \ne 0$ with $\operatorname{sgn}\varepsilon = *1$, the differential inclusion (3.1.25) possesses a subharmonic solution $x_{m,\varepsilon}$ of period $2m$ for any $m \in \mathbb{N}$ satisfying*

$$\sup_{-m \le t \le m} \left| x_{m,\varepsilon}(t) - \gamma\left(\frac{t - \alpha_{m,\varepsilon}}{\varepsilon}\right) \right| \le K\sqrt{|\varepsilon|}\,,$$

where $\alpha_{m,\varepsilon} \in \mathbb{R}$ and $|\alpha_{m,\varepsilon}| \le K$.

Theorem 3.1.13. *Let $d=1$. If there are constants $a < b$ such that $M(a)M(b) \subset (-\infty, 0)$. Then there is a constant $K > 0$ such that for any sufficiently small $\varepsilon \neq 0$, the differential inclusion (3.1.25) possesses a subharmonic solution $x_{m,\varepsilon}$ of period $2m$ for any $m \in \mathbb{N}$ satisfying*

$$\sup_{m \leq t \leq m} \left| x_{m,\varepsilon}(t) - \gamma\left(\frac{t - \alpha_{m,\varepsilon}}{\varepsilon}\right) \right| \leq K|\varepsilon|,$$

where $\alpha_{m,\varepsilon} \in (a, b)$.

Finally we consider the following differential inclusion equivalent to (3.1.26) of the form

$$\dot{x}(t) \in f(x(t)) + \varepsilon^2 h(x(t), \varepsilon, \varepsilon t) \quad \text{a.e. on} \quad \mathbb{R}, \quad \varepsilon > 0. \tag{3.1.30}$$

Now we make in (3.1.30) the change of variables

$$x\left(t + \frac{\alpha}{\varepsilon}\right) = \gamma(t) + \varepsilon^2 z(t) + \sum_{i=1}^{d-1} \varepsilon \beta_i u_{i+d}(t) + \frac{\varepsilon}{2m} b(\varepsilon\beta, \frac{m}{\varepsilon})t,$$

where $-\frac{m}{\varepsilon} \leq t \leq \frac{m}{\varepsilon}$ and $1 > \varepsilon > 0$, $\alpha, \beta_i \in \mathbb{R}$. Then the differential inclusion for z is

$$\dot{z}(t) - Df(\gamma(t))z(t) \in q_{m,\varepsilon}(z(t), \alpha, \beta, t) \quad \text{a.e. on} \quad [-\frac{m}{\varepsilon}, \frac{m}{\varepsilon}],$$

where

$$q_{m,\varepsilon}(x, \alpha, \beta, t) =$$
$$\left\{ v \in \mathbb{R}^n : v \in \frac{1}{\varepsilon^2}\left\{ f\left(\varepsilon^2 x + \gamma(t) + \varepsilon \sum_{i=1}^{d-1} \beta_i u_{i+d}(t) + \frac{\varepsilon}{2m} b(\varepsilon\beta, \frac{m}{\varepsilon})t\right) \right.\right.$$
$$\left. -f(\gamma(t)) - \varepsilon \sum_{i=1}^{d-1} \beta_i \dot{u}_{i+d}(t) - \frac{\varepsilon}{2m} b(\varepsilon\beta, \frac{m}{\varepsilon}) - Df(\gamma(t))\varepsilon^2 x \right\}$$
$$\left. +h\left(\varepsilon^2 x + \gamma(t) + \varepsilon \sum_{i=1}^{d-1} \beta_i u_{i+d}(t) + \frac{\varepsilon}{2m} b(\varepsilon\beta, \frac{m}{\varepsilon})t, \varepsilon, \varepsilon t + \alpha\right) \right\}.$$

We see that the procedure for (3.1.27) can be applied to (3.1.30). Consequently, we obtain the following result.

Theorem 3.1.14. *If either for $d > 1$, there is a non–empty open bounded set $\mathbf{B} \subset \mathbb{R}^d$ such that $0 \notin M_+(\partial\mathbf{B})$ and $\deg(M_+, \mathbf{B}, 0) \neq 0$, or for $d = 1$, there are constants $a < b$ such that $M(a)M(b) \subset (-\infty, 0)$. Then there is a constant $K > 0$ such that for any sufficiently small $\varepsilon > 0$, the differential inclusion (3.1.26) possesses a subharmonic solution $x_{m,\varepsilon}$ of period $2m$ for any $m \in \mathbb{N}$ satisfying*

$$\sup_{-m \leq t \leq m} \left| x_{m,\varepsilon}(t) - \gamma\left(\frac{t - \alpha_{m,\varepsilon}}{\varepsilon}\right) \right| \leq K\varepsilon \quad \text{for} \quad d > 1 \quad (\leq K\varepsilon^2 \quad \text{for} \quad d = 1),$$

where $\alpha_{m,\varepsilon} \in \mathbb{R}$ and $|\alpha_{m,\varepsilon}| \leq K$. Moreover, $\alpha_{m,\varepsilon} \in (a, b)$ for $d = 1$.

3.1.5 Subharmonics for Regular Autonomous Perturbations

In this subsection, we study (3.1.3) when the nonlinearities are independent of t, i.e.,

$$\dot{x}(t) \in f(x(t)) + \sum_{i=1}^{k} \mu_i f_i(x(t), \mu) \quad \text{a.e. on} \quad \mathbb{R}, \qquad (3.1.31)$$

and $k \geq 2$. So (3.1.31) is autonomous and then a time shifted solution is again its solution. For this reason, we take $\alpha = 0$ in Subsection 3.1.3. We know that to solve the problem of existence of periodic solutions for (3.1.31) near γ is reduced to the solvability of multivalued equation (3.1.20) with $m \in [0, \infty)$. Since we lose α in (3.1.21–3.1.24) by setting $\alpha = 0$, we need to replace it. For this reason, first of all, we divide the sets $\{1, \cdots, d-1\}$ and $\{1, \cdots, k\}$ into two complementary subsets $\{i_1, \cdots, i_{d_1}\}$, $\{i_{d_1+1}, \cdots, i_{d-1}\}$ and $\{j_1, \cdots, j_{k_1}\}$, $\{j_{k_1+1}, \cdots, j_k\}$ such that $d_1 + k_1 = d$. These subsets may be empty. We put

$$\xi = (\beta_{i_1}, \cdots, \beta_{i_{d_1}}, \mu_{j_1}, \cdots, \mu_{j_{k_1}}) \in \mathbb{R}^d$$
$$\tau = (\mu_{j_1}, \cdots, \mu_{j_{k_1}}) \in \mathbb{R}^{k_1}$$
$$\rho = (\mu_{j_{k_1}+1}, \cdots, \mu_{j_k}) \in \mathbb{R}^{k-k_1}$$
$$\theta = (\beta_{i_{d_1}+1}, \cdots, \beta_{i_{d-1}}, \mu_{j_{k_1}+1}, \cdots, \mu_{j_k}) \in \mathbb{R}^{k-1}.$$

The projections $\theta \to \rho$, $\xi \to \tau$ are denoted by P_1, P_2, respectively, so $\rho = P_1\theta$, $\tau = P_2\xi$. Instead of (3.1.21–3.1.24), we consider

$$M_\theta : \mathbb{R}^d \to 2^{\mathbb{R}^d} \setminus \{\emptyset\}, \quad M_\theta = (M_{\theta 1}, \cdots, M_{\theta d})$$

$$M_{\theta l}(\xi) = \left\{ \int_{-\infty}^{\infty} \langle h(s), u_l^{\perp}(s) \rangle \, ds \; : \; h \in L^2_{\text{loc}}(\mathbb{R}, \mathbb{R}^n); \right.$$

$$\left. h(t) \in \left(\frac{1}{2} \sum_{i,j=1}^{d-1} \beta_i \beta_j D^2 f(\gamma(t))(u_{d+i}(t), u_{d+j}(t)) + \sum_{j=1}^{k} \mu_j f_j(\gamma(t), 0) \right) \text{ a.e. on } \mathbb{R} \right\}.$$

$$(3.1.32)$$

Theorem 3.1.15. *Let $k > k_1$. Assume the existence of a non–empty open bounded set $\mathbf{B} \subset \mathbb{R}^d$ and $\theta_0 \in \mathbb{R}^{k-1}$ such that $0 \notin M_{\theta_0}(\partial \mathbf{B})$ and $\deg(M_{\theta_0}, \mathbf{B}, 0) \neq 0$. Then there is a constant $K > 0$, an open bounded neighborhood U_3 of $(\mathbb{I} - P_1)\theta_0 \in \mathbb{R}^{d-1-d_1}$ and an open bounded region in $\mathbb{R}^{k-k_1}$ for ρ of the form*

$$\mathbf{R} = \left\{ s^2 \tilde{\rho} \colon s > 0, \text{ respectively } \tilde{\rho}, \text{ is from an open small connected neighborhoods} \right.$$

$$\left. U_1, \text{ respectively } U_2, \text{ of } 0 \in \mathbb{R}, \text{ respectively of } \rho_0 = P_1\theta_0 \in \mathbb{R}^{k-k_1} \right\}$$

such that for any $\rho \in \mathbf{R}$ of the form $\rho = s^2\tilde{\rho}$, $0 < s \in U_1$, $\tilde{\rho} \in U_2$, and any $T \geq [1/s]$, there is a $(d-1-d_1)$–parametric family $\tau_{(T,\rho,p)} \in P_2(\mathbf{B})$, $p \in U_3$ such that the differential inclusion (3.1.31) with $\mu = \mu_p = (s^2\tau_{(T,\rho,p)}, \rho)$, $p \in U_3$ possesses a T–periodic solution $x_{T,p}$ satisfying, according to the change of variable below (3.1.16),

$$\sup_{-T \leq t \leq T} \left| x_{T,p}(t) - \gamma(t) \right| \leq Ks \quad \text{for} \quad d > 1, \quad (\leq Ks^2 \quad \text{for} \quad d = 1).$$

If $d = d_1 + 1$ then U_3 is omitted.

Proof. The proof is the same as for Theorem 3.1.6. The only difference is now that (3.1.20) has to be solved in z and $\xi \in \mathbb{R}^d$ while $\theta \in \mathbb{R}^{k-1}$ is a parameter. We note that now $\alpha = 0$ and $m \in [0, \infty)$. Hence we have

$$0 \in H_{m,s}(z, 0, \beta, \mu, \lambda) = H_{m,s}(z, \xi, \theta, \lambda) \qquad (3.1.33)$$

on the set

$$\Omega = \left\{ (z, \xi) \in Z_{m+S} \times \mathbb{R}^d : \|z\|_{m+S} < K_1, \, \xi \in \mathbf{B} \right\}$$

for a constant $K_1 > 0$. Consequently from the proof of Theorem 3.1.6, (3.1.33), with $\lambda = 1$, has a solution in Ω for any $0 < s \in U_1$, $\theta \in \tilde{U}_2 = U_3 \times U_2 \subset \mathbb{R}^{d-1-d_1} \times \mathbb{R}^{k-k_1}$ and $m \in [0, \infty)$, where U_1, respectively $\tilde{U}_2$ is an open small connected neighborhood of $0 \in \mathbb{R}$, respectively of $\theta_0 \in \mathbb{R}^{k-1}$. On the other hand, according to the definition of $H_{m,s}$, Theorem 3.1.4 and the change of variable below (3.1.16), any solution of (3.1.33) satisfies the relation $P_{ss}U(0)^{-1}z(0) = 0$. Consequently, solutions are different for different $p \in U_3$.   □

Remark 3.1.16. For $\rho_0 = 0$ in Theorem 3.1.15 we can take

$$\mathbf{R} = \left\{ \rho \in \mathbb{R}^{k-k_1} : |\rho| < r_1 \right\}$$

for $r_1 > 0$ sufficiently small. Indeed, since now $0 \in U_2$, by expressing any $\rho \in \mathbf{R}$ in the form $\rho = s^2\tilde{\rho}$, $0 < s \in U_1$, $\tilde{\rho} \in U_2$, Theorem 3.1.15 can be applied. Moreover, there is a $s_0 > 0$ such that any $\mu \in \mathbf{R}$ does possess the previous form with $s = s_0$ and $\tilde{\rho} \in U_2$. Consequently, (3.1.31) has in this case a T–periodic solution for any $T \geq [1/s_0]$, i.e., the lower bound of the period T is independent of $\mu \in \mathbf{R}$.

For $d = 1$, like in Theorem 3.1.9 we can take a symmetric $\mathbf{R}$ with respect to $0 \in \mathbb{R}^{k-k_1}$ in Theorem 3.1.15.

3.1.6   Applications to Discontinuous O.D.Eqns

In this subsection we use abstract bifurcation results of the previous Subsections 3.1.3–3.1.5 to study concrete discontinuous differential equations. First of all, we apply Theorem 3.1.9 to (3.1.2) of the form

$$\ddot{x} + g(x) + \mu_1 \operatorname{sgn} \dot{x} = \mu_2 \psi(t), \qquad (3.1.34)$$

where $\mu_{1,2} \in \mathbb{R}$ are small parameters, $g \in C^2(\mathbb{R}, \mathbb{R})$, $g(0) = 0$, $g'(0) < 0$, $\psi \in C^1(\mathbb{R}, \mathbb{R})$ and ψ is periodic. We assume the existence of a homoclinic solution ω of $\ddot{x} + g(x) = 0$ such that $\lim_{t \to \pm\infty} \omega(t) = 0$ and $\omega(0) > 0$ (see Fig. 3.3).

In further calculations we need the following well-known property of $\omega(t)$.

Lemma 3.1.17. *There is a unique $t_0 \in \mathbb{R}$ satisfying $\dot{\omega}(t_0) = 0$. Consequently, $\dot{\omega}(t) > 0$, $\forall t < t_0$ and $\dot{\omega}(t) < 0$, $\forall t > t_0$.*

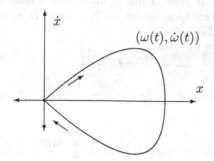

Figure 3.3: The homoclinic structure of $\ddot{x} + g(x) = 0$

Proof. If $t_0 \neq t_1$ are such that $\dot{w}(t_0) = \dot{w}(t_1) = 0$, then $x_0(t) = w(2t_0 - t)$ and $x_1(t) = w(2t_1 - t)$ are both solutions of $\ddot{x} + g(x) = 0$ satisfying

$$x_0(t_0) = w(t_0), \quad \dot{x}_0(t_0) = \dot{w}(t_0)$$
$$x_1(t_1) = w(t_1), \quad \dot{x}_1(t_1) = \dot{w}(t_1).$$

Hence $x_0(t) = w(t) = x_1(t)$ and then

$$w(2(t_0 - t_1) + t) = w(t).$$

But w can not be periodic. On the other hand, since $\lim_{t \to \pm\infty} w(t) = 0$, there is a t_0 such that $\dot{w}(t_0) = 0$. So there is a unique $t_0 \in \mathbb{R}$ such that $\dot{w}(t_0) = 0$. Since $w(0) > 0$, we obtain that $\dot{w}(t) > 0$, $\forall t < t_0$ and $\dot{w}(t) < 0$, $\forall t > t_0$.   □

Rewriting (3.1.34) in the form

$$\dot{x} = y, \quad \dot{y} = -g(x) - \mu_1 \operatorname{sgn} y + \mu_2 \psi(t), \qquad (3.1.35)$$

in the notations of Subsection 3.1.1, we put

$$f(x,y) = (y, -g(x)), \quad f_1(x,y,\mu,t) = (0, -\operatorname{Sgn} y)$$
$$f_2(x,y,\mu,t) = (0, \psi(t)), \quad \gamma = (w, \dot{w}),$$

where $\operatorname{Sgn} r$ is defined by (2.4.1). Since $n = 2$, we have $d = 1$ and

$$u_2 = (\dot{w}, \ddot{w}), \quad u_1^{\perp} = (-\ddot{w}, \dot{w}).$$

Since $w(0) > 0$, Lemma 3.1.17 implies $w(t_0) > 0$. Then easy calculation in (3.1.24) leads to

$$M_\mu(\alpha) = -\mu_1 \left(\int_{-\infty}^{t_0} \dot{w}(s)\,ds - \int_{t_0}^{\infty} \dot{w}(s)\,ds \right) + \mu_2 \int_{-\infty}^{\infty} \dot{w}(s)\psi(s+\alpha)\,ds$$

$$= A(\alpha)\mu_2 - 2w(t_0)\mu_1$$

with $A(\alpha) = \int_{-\infty}^{\infty} \dot{w}(s)\psi(s+\alpha)\,ds$. Since A is periodic and C^1–smooth, there are constants $\bar{m} := \min A$ and $\bar{M} := \max A$. By applying Theorem 3.1.9, we obtain the following theorem.

Theorem 3.1.18. *Assume that A has critical points only at maximums and minimums. Then there is an open, wedge–shaped subset* **R** *of* $\{(\mu_2, \mu_1) : \mu_{1,2} \in \mathbb{R},$ $\mu_2 \neq 0\}$ *with the limits slopes*

$$\frac{\bar{m}}{2\omega(t_0)} \quad and \quad \frac{\bar{M}}{2\omega(t_0)},$$

on which the (3.1.34) *has subharmonic solutions with all sufficiently large periods.*

Proof. For any $(\mu_2, \mu_1) \in \mathbb{R}^2$ such that

$$\mu_1^2 + \mu_2^2 = 1, \quad \mu_2 \neq 0, \quad \bar{m} < \frac{2\omega(t_0)\mu_1}{\mu_2} < \bar{M},$$

M_μ does have a simple root. Then the assumptions of Theorem 3.1.9 are satisfied for a small open interval (a, b) containing this simple root. $\square$

Remark 3.1.19. (a) Since any element (μ_2, μ_1) of **R** satisfies

$$|\mu_2| > \frac{2\omega(t_0)}{\max\{|\bar{M}|, |\bar{m}|\}}|\mu_1|,$$

the driving force term in (3.1.34) has to be sufficiently large with respect to the dry friction force for the applicability of Theorem 3.1.9 to (3.1.34).

(b) If $\omega(0) < 0$ then $M_\mu(\alpha) = 2\omega(t_0)\mu_1 + A(\alpha)\mu_2$.

For more concreteness, we consider the Duffing–type equation

$$\ddot{x} - x + 2x^3 + \mu_1 \operatorname{sgn} \dot{x} = \mu_2 \cos t. \tag{3.1.36}$$

Then $\omega(t) = \operatorname{sech} t$ and $A(\alpha) = \int_{-\infty}^{\infty} \operatorname{sech} s \cos(s + \alpha)\, ds = \pi \operatorname{sech} \frac{\pi}{2} \sin \alpha$. So $\bar{M} = -\bar{m} = \pi \operatorname{sech} \frac{\pi}{2}$, $t_0 = 0$, $\omega(t_0) = 1$. Theorem 3.1.18 gives the following

Corollary 3.1.20. *Equation* (3.1.36) *has subharmonic solutions with all sufficiently large periods provided that the parameters μ_1, μ_2 are sufficiently small satisfying*

$$\frac{\pi}{2} \operatorname{sech} \frac{\pi}{2} \cdot |\mu_2| > |\mu_1|.$$

Now we apply Theorem 3.1.14 to a modification of (3.1.34) of the form

$$\ddot{x} + \delta g(x) + \frac{\psi(t)}{\sqrt{\delta}} \dot{x} + \eta \operatorname{sgn} \dot{x} = 0, \tag{3.1.37}$$

where $g \in C^2(\mathbb{R}, \mathbb{R})$, $g(0) = 0$, $g'(0) < 0$, $\delta > 0$ is a large parameter, $\psi \in C^1(\mathbb{R}, (0, \infty))$ is periodic and η is a constant. Now we assume the existence of a homoclinic solution ω of $\ddot{x} + g(x) = 0$ such that $\lim_{t \to \pm\infty} \omega(t) = 0$ and $\omega(0) < 0$.

Then again there is a unique $t_0 \in \mathbb{R}$ satisfying $\dot{\omega}(t_0) = 0$, and $\dot{\omega}(t) < 0$, $\forall t < t_0$, $\dot{\omega}(t) > 0$, $\forall t > t_0$ and $\omega(t_0) < 0$. Since δ is large, we set $\varepsilon = \sqrt{1/\delta}$, and rewrite (3.1.37) in the form

$$\varepsilon \dot{x} = y, \quad \varepsilon \dot{y} = -g(x) - \varepsilon^2 (\psi(t)y + \eta \operatorname{sgn} y), \tag{3.1.38}$$

so Theorem 3.1.14 can be applied for (3.1.38) to obtain the following result.

Theorem 3.1.21. *If the function $M(\alpha) = 2\eta\omega(t_0) - \psi(\alpha)\int_{-\infty}^{\infty}\dot{\omega}(s)^2\,ds$ has a simple root, then for any $\delta > 0$ sufficiently large, (3.1.37) has subharmonic solutions of all periods.*

Remark 3.1.22. (a) It is clear that for the existence of a simple root of $M(\alpha)$ in Theorem 3.1.21, it is necessary to assume

$$\eta \in \left[\tilde{M}\int_{-\infty}^{\infty}\dot{\omega}(s)^2\,ds/2\omega(t_0), \tilde{m}\int_{-\infty}^{\infty}\dot{\omega}(s)^2\,ds/2\omega(t_0)\right],$$

where $\tilde{m} = \min\psi$, $\tilde{M} = \max\psi$. On the other hand, if ψ has critical points only at minimums and maximums, then the condition

$$\eta \in \left(\tilde{M}\int_{-\infty}^{\infty}\dot{\omega}(s)^2\,ds/2\omega(t_0), \tilde{m}\int_{-\infty}^{\infty}\dot{\omega}(s)^2\,ds/2\omega(t_0)\right)$$

is sufficient for the existence of a simple root of $M(\alpha)$.

(b) If $\eta \geq 0$ and x is a T–periodic solution of (3.1.37) then

$$0 = \int_0^T\left(\ddot{x}(s)\dot{x}(s) + \delta g(x(s))\dot{x}(s) + \frac{\psi(s)}{\sqrt{\delta}}\dot{x}(s)^2 + \eta|\dot{x}(s)|\right)ds$$

$$\geq \int_0^T\frac{\psi(s)}{\sqrt{\delta}}\dot{x}(s)^2\,ds \geq \frac{\tilde{m}}{\sqrt{\delta}}\int_0^T\dot{x}(s)^2\,ds.$$

So x is constant. Hence (3.1.37) with $\eta \geq 0$ has at most constant subharmonic solutions.

For more concreteness, we consider again the Duffing–type equation

$$\ddot{x} + \delta(-x + 2x^3) + \frac{(2 + \cos t)}{\sqrt{\delta}}\dot{x} + \eta\operatorname{sgn}\dot{x} = 0 \qquad (3.1.39)$$

with $g(x) = -x + 2x^3$, $\psi(t) = 2 + \cos t$, $\omega(t) = -\operatorname{secht}$. Then $\tilde{m} - 1$, $\tilde{M}{=}3$, $t_0{=}0$, $\omega(t_0) = -1$ and $\int_{-\infty}^{\infty}\dot{\omega}(s)^2\,ds = \int_{-\infty}^{\infty}(\operatorname{sech}s)^2\,ds = 2/3$. Consequently, Remark 3.1.22 a) gives the next result.

Corollary 3.1.23. *Equation (3.1.39) has subharmonic solutions with all periods provided that $\delta > 0$ is sufficiently large and $\eta \in (-1, -1/3)$.*

Now we deal with more difficult problem represented by the following coupled discontinuous differential equations

$$\begin{aligned}\ddot{x}_1 + \delta g(x_1) + \eta_1\operatorname{sgn}\dot{x}_1 &= \eta_2 x_2\\\ddot{x}_2 + \delta g(x_2) &= \psi(t)x_1,\end{aligned} \qquad (3.1.40)$$

where $\eta_1 > 0$, $\eta_2 \neq 0$ are constants, $\delta > 0$ is a large parameter, $\psi \in C^1(\mathbb{R}, \mathbb{R})$ is periodic and g satisfies the properties of (3.1.34). Setting $\varepsilon = \sqrt{1/\delta}$ and rewriting system (3.1.40) in the form

$$\begin{aligned}\varepsilon\dot{x}_1 = y_1, \quad \varepsilon\dot{y}_1 &= -g(x_1) + \varepsilon^2\big(-\eta_1\operatorname{sgn}y_1 + \eta_2 x_2\big)\\\varepsilon\dot{x}_2 = y_2, \quad \varepsilon\dot{y}_2 &= -g(x_2) + \varepsilon^2\psi(t)x_1,\end{aligned} \qquad (3.1.41)$$

we have in the notation of (3.1.30)

$$f(x_1, y_1, x_2, y_2) = (y_1, -g(x_1), y_2, -g(x_2))$$
$$h(x_1, y_1, x_2, y_2, \varepsilon, t) = (0, -\eta_1 \operatorname{Sgn} y_1 + \eta_2 x_2, 0, \psi(t)x_1).$$

Since $\dot{x} = f(x)$ with $x = (x_1, y_1, x_2, y_2)$ is $\ddot{x}_1 + g(x_1) = 0$ and $\ddot{x}_2 + g(x_2) = 0$, so it is decoupled, it is better to take

$$\gamma_\sigma = (\omega, \dot\omega, \omega_\sigma, \dot\omega_\sigma), \quad \omega_\sigma(t) = \omega(t - \sigma), \quad \sigma \in \mathbb{R}$$
$$u_{\sigma 1}^\perp = (-\ddot\omega, \dot\omega, 0, 0), \quad u_{\sigma 2}^\perp = (0, 0, -\ddot\omega_\sigma, \dot\omega_\sigma)$$
$$u_{\sigma 3} = (\dot\omega, \ddot\omega, 0, 0), \quad u_{\sigma 4} = (0, 0, \dot\omega_\sigma, \ddot\omega_\sigma).$$

Note $\{\gamma_\sigma(t) \mid \sigma, t \in \mathbb{R}\}$ represents a non-degenerate homoclinic manifold of $\dot{x} = f(x)$ to $x = 0$ (see Subsection 4.1.4 for similar results). So β is replaced with σ. Then we use formula (3.1.29) twice for $u_{\sigma 1}^\perp$ and $u_{\sigma 2}^\perp$ instead of (3.1.28) to get

$$M(\alpha, \sigma) = \begin{pmatrix} -2\eta_1 \omega(t_0) + \eta_2 B(\sigma) \\ -\psi(\alpha) B(\sigma) \end{pmatrix}$$

with $B(\sigma) = \int_{-\infty}^{\infty} \omega(t - \sigma)\dot\omega(t)\, dt$. Next we make in (3.1.30) the change of variables

$$x\left(t + \frac{\alpha}{\varepsilon}\right) = \gamma_\sigma(t) + \varepsilon^2 z(t) + \frac{\varepsilon}{2m}\left(\gamma_\sigma\left(-\frac{m}{\varepsilon}\right) - \gamma_\sigma\left(\frac{m}{\varepsilon}\right)\right) t.$$

In this way, we can incorporate parameter σ in the differential inclusion above Theorem 3.1.14 concerning (3.1.30). Consequently, it is enough to find a simple zero point of the map

$$M : \begin{pmatrix} \alpha \\ \sigma \end{pmatrix} \to \begin{pmatrix} -2\eta_1 \omega(t_0) + \eta_2 B(\sigma) \\ -\psi(\alpha) B(\sigma) \end{pmatrix}.$$

Summarizing, we obtain the next result.

Theorem 3.1.24. *If there is a simple root σ_0 of $B(\sigma_0) - 2\eta_1\omega(t_0)/\eta_2 = 0$ and a simple root α_0 of $\psi(\alpha) = 0$ as well, then (3.1.40) has subharmonic solutions with all periods provided that $\delta > 0$ is sufficiently large.*

Proof. According to the above arguments, it is enough to observe that (α_0, σ_0) is a simple zero point of M, i.e. $M(\alpha_0, \sigma_0) = 0$ and $\det DM(\alpha_0, \sigma_0) \neq 0$, which is obvious. $\square$

Corollary 3.1.25. *There is a constant $K > 0$ such that if $0 < \eta_1/|\eta_2| < K$ and $\psi(\alpha) = 0$ has a simple root, then (3.1.40) has subharmonic solutions with all periods provided that $\delta > 0$ is sufficiently large.*

Proof. Since $B(0) = 0$ and $B'(0) \neq 0$, there is a constant $K > 0$ such that for any $0 < \eta_1/|\eta_2| < K$, there is a small simple root of $B(\sigma_0) = 2\eta_1\omega(t_0)/\eta_2$. The proof is finished by Theorem 3.1.24. $\square$

For more concreteness, we again take $g(x) = -x + 2x^3$, $\psi(t) = \cos t$, $\omega(t) =$ sech t, $t_0 = 0$, $\omega(t_0) = 1$. Then

$$B(\sigma) = 2\,\frac{\sinh\sigma - \sigma\cosh\sigma}{\sinh^2\sigma}\,.$$

Function $B(\sigma)$ has the only critical points: one maximum about 0.6196336 at $\sigma \doteq -1.6061152$ and one minimum about -0.6196336 at $\sigma \doteq 1.6061152$. Moreover, $B(\sigma) = 0$ if and only if $\sigma = 0$. Hence we obtain the following result.

Corollary 3.1.26. *The equation*

$$\ddot{x}_1 + \delta(-x_1 + 2x_1^3) + \eta_1\operatorname{sgn}\dot{x}_1 = \eta_2 x_2$$
$$\ddot{x}_2 + \delta(-x_2 + 2x_2^3) = x_1\cos t$$

has subharmonic solutions with all periods provided that $\delta > 0$ is sufficiently large and $0 < \eta_1/|\eta_2| < 0.3098167$.

Finally, we consider the following autonomous version of (3.1.34)

$$\ddot{x} + g(x) + \mu_1\operatorname{sgn}\dot{x} + \mu_2\dot{x} = 0\,, \tag{3.1.42}$$

where $\mu_{1,2} \in \mathbb{R}$ are small parameters and g satisfies the properties either of (3.1.34) or (3.1.37).

Theorem 3.1.27. *There is a $r_1 > 0$ and mappings*

$$\Pi_1, \Pi_2 : (-r_1, r_1) \times (1/r_1, \infty) \to \mathbb{R}$$

such that for any $T \in (1/r_1, \infty)$, (3.1.42) has a T-periodic solution near ω with either for any $\mu_2 \in (-r_1, r_1)$ and $\mu_1 = \Pi_1(\mu_2, T)$ or for any $\mu_1 \in (-r_1, r_1)$ and $\mu_2 = \Pi_2(\mu_1, T)$. Moreover, $\lim\limits_{T\to\infty,\mu_2\to 0} \Pi_1(\mu_2, T) = 0$ and $\lim\limits_{T\to\infty,\mu_1\to 0} \Pi_2(\mu_1, T) = 0$.

Proof. We apply Theorem 3.1.15 with either $\xi = \mu_1$ and $\theta = \mu_2$ or $\xi = \mu_2$ and $\theta = \mu_1$. Then according to the above computations

$$M_\theta(\xi) = -2|\omega(t_0)|\mu_1 - \mu_2\int_{-\infty}^{\infty}\dot{\omega}(s)^2\,ds\,.$$

So we take in Theorem 3.1.15: $\mathbf{B} = (-1, 1)$, $\theta_0 = 0$, $d = 1$, $d_1 = 0$ and by Remark 3.1.16, we obtain the desired mappings $\Pi_{1,2}$. Finally, the limits $\lim\limits_{T\to\infty,\mu_2\to 0} \Pi_1(\mu_2, T) = 0$ and $\lim\limits_{T\to\infty,\mu_1\to 0} \Pi_2(\mu_1, T) = 0$ follow from the fact that the only zero point of $M_{\theta_0}(\xi) = 0$ is $\xi = 0$. $\qquad\square$

Remark 3.1.28. Like in Remark 3.1.22b), it is clear that (3.1.42) with $\mu_1\mu_2 \geq 0$, $\mu_1^2 + \mu_2^2 \neq 0$ has at most constant periodic solutions.

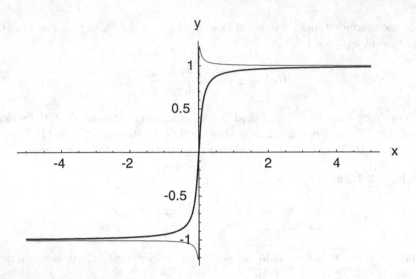

Figure 3.4: The graphs of functions $y = \frac{2}{\pi} \arctan kx$ (thick line) and $y = \Phi(x)$ with $k = 10$

Remark 3.1.29. The discontinuous function $\operatorname{sgn} r$ modeling dry friction in the Coulomb's friction law is often approximated [9,12] by the mathematically convenient approximation of the form

$$r \to \frac{2}{\pi} \arctan kr, \quad k \gg 1.$$

On the other hand, its physically more relevant approximation is given by (see Fig. 3.4)

$$\Phi(r) = \frac{1}{\pi}(7 \arctan 8kr - 5 \arctan 4kr), \quad k \gg 1.$$

The function Φ has two symmetric spikes at $r = \pm \frac{\sqrt{6}}{8k}$ of the values

$$\pm \frac{1}{\pi}\left(7 \arctan \sqrt{6} - 5 \arctan \frac{\sqrt{6}}{2}\right) \doteq \pm 1,2261344.$$

Moreover, $\Phi(r)$ is quickly near 1, respectively -1, when $r > 0$, respectively $r < 0$, tends off 0. Then a multivalued version of Φ can be taken as

$$\operatorname{Sgn}_{\eta,\zeta,\kappa} r = \begin{cases} -1 & \text{for } r < -\eta, \\ [-\zeta, -\kappa] & \text{for } -\eta \leq r < 0, \\ [-\zeta, \zeta] & \text{for } r = 0, \\ [\kappa, \zeta] & \text{for } 0 < r \leq \eta, \\ 1 & \text{for } r > \eta \end{cases}$$

for some constants $\eta \geq 0, \zeta \geq 1, 0 < \kappa \leq 1$. The term $\operatorname{Sgn}_{\eta,\zeta,\kappa} \dot{x}$ can be considered as an extension for modeling dry friction forces.

To apply Remark 3.1.29, let us consider the problem

$$\ddot{x} + g(x) - \mu_2 \psi(t) \in -\mu_1 \mathrm{Sgn}_{\eta,\zeta,\kappa} \dot{x}, \qquad (3.1.43)$$

where $\mu_1 > 0$, $\mu_2 > 0$ are small parameters, g has the properties of either (3.1.34) or (3.1.37) and $\psi \in C(\mathbb{R}, \mathbb{R})$ is periodic. Then, for this case with $\eta < \max |\dot{\omega}|$, the multivalued mapping (3.1.24) has the form

$$M_\mu(\alpha) = \left[-\mu_1 \left(\zeta \int_{I_{\eta-}} |\dot{\omega}(t)| \, dt + \int_{I_{\eta+}} |\dot{\omega}(t)| \, dt \right) + \mu_2 A(\alpha), \right.$$

$$\left. -\mu_1 \left(\kappa \int_{I_{\eta-}} |\dot{\omega}(t)| \, dt + \int_{I_{\eta+}} |\dot{\omega}(t)| \, dt \right) + \mu_2 A(\alpha) \right],$$

where A is the function defined above Theorem 3.1.18 and

$$I_{\eta-} = \{t \in \mathbb{R} : |\dot{\omega}(t)| \le \eta\}, \quad I_{\eta+} = \{t \in \mathbb{R} : |\dot{\omega}(t)| > \eta\}.$$

Then Theorem 3.1.9 gives the following result.

Theorem 3.1.30. *If $\eta < \max |\dot{\omega}|$ and*

$$0 < \bar{M}, \quad (\bar{m}\zeta - \bar{M}\kappa) \int_{I_{\eta-}} |\dot{\omega}(t)| \, dt < (\bar{M} - \bar{m}) \int_{I_{\eta+}} |\dot{\omega}(t)| \, dt, \qquad (3.1.44)$$

where $\bar{m} = \min A$ and $\bar{M} = \max A$, then (3.1.43) has subharmonic solutions with all large periods provided that $\mu_1 > 0$, $\mu_2 > 0$ are sufficiently small satisfying

$$\mu_1/\mu_2 \in \left(\max\left\{ 0, \bar{m} \middle/ \left(\kappa \int_{I_{\eta-}} |\dot{\omega}(t)| \, dt + \int_{I_{\eta+}} |\dot{\omega}(t)| \, dt \right) \right\}, \right.$$

$$\left. \bar{M} \middle/ \left(\zeta \int_{I_{\eta-}} |\dot{\omega}(t)| \, dt + \int_{I_{\eta+}} |\dot{\omega}(t)| \, dt \right) \right). \qquad (3.1.45)$$

Proof. We take $a < b$ such that $A(a) = \bar{m}$, $A(b) = \bar{M}$. Then (3.1.45) implies that $r_1 \in M_\mu(a)$, $r_2 \in M_\mu(b)$ give $r_1 < 0$, $r_2 > 0$. Hence the condition of Theorem 3.1.9 is satisfied. $\qquad \square$

To be more concrete, we consider the problem

$$\ddot{x} - x + 2x^3 - \mu_2 \cos t \in -\mu_1 \mathrm{Sgn}_{\eta,\zeta,\kappa}\, \dot{x}\,. \tag{3.1.46}$$

Corollary 3.1.31. *Inclusion* (3.1.46), *with* $1/2 > \eta \geq 0$, *has subharmonic solutions with all large periods provided that* $\mu_1 > 0$, $\mu_2 \neq 0$ *are sufficiently small satisfying*

$$\frac{\mu_1}{|\mu_2|} < \frac{\pi}{2} \operatorname{sech} \frac{\pi}{2} \Big/ \Big(\zeta + (1-\zeta) \operatorname{sech} t_{1,\eta} + (\zeta - 1) \operatorname{sech} t_{2,\eta} \Big),$$

where $0 \leq t_{1,\eta} < t_{2,\eta} \leq \infty$ *are the solutions of* $\operatorname{sech} t = -\eta$.

Proof. Now $g(x) = -x + 2x^3$, $\omega(t) = \operatorname{sech} t$, $\psi(t) = \pm \cos t$. We note $\max_{t \in \mathbb{R}} |\dot{\operatorname{sech}} t| = 1/2$. The equation $|\dot{\operatorname{sech}} t| = \eta$ for $1/2 > \eta \geq 0$ has the solutions

$$-\infty \leq -t_{2,\eta} < -t_{1,\eta} \leq 0 \leq t_{1,\eta} < t_{2,\eta} \leq \infty,$$

when clearly there are strict inequalities for $\eta > 0$. Consequently, we have

$$I_{\eta-} = (-\infty, -t_{2,\eta}] \cup [-t_{1,\eta}, t_{1,\eta}] \cup [t_{2,\eta}, \infty), \quad I_{\eta+} = (-t_{2,\eta}, -t_{1,\eta}) \cup (t_{1,\eta}, t_{2,\eta}),$$

and

$$\int_{I_{\eta-}} |\dot{\omega}(t)|\, dt = 2(1 - \operatorname{sech} t_{1,\eta} + \operatorname{sech} t_{2,\eta}), \quad \int_{I_{\eta+}} |\dot{\omega}(t)|\, dt = 2(\operatorname{sech} t_{1,\eta} - \operatorname{sech} t_{2,\eta}).$$

We also know by the results for (3.1.36) that $A(\alpha) = \pm \pi \operatorname{sech} \frac{\pi}{2} \sin \alpha$; hence $\bar{M} = -\bar{m} = \pi \operatorname{sech} \frac{\pi}{2}$. By using these computations and Theorem 3.1.30, the corollary is proved. $\qquad \square$

Remark 3.1.32. In spite of the fact that the aim of this section is to deal with multivalued perturbation problems, we note that our method is clearly applied to piecewise smoothly perturbed problems. For simple illustration of this, let us consider the following concrete piecewise linearly perturbed problem

$$\ddot{x} - x + 2x^3 + \mu_1 \dot{x}^+ + \mu_2 \dot{x}^- = \mu_3 \cos t\,, \tag{3.1.47}$$

where $\mu_{1,2,3} \in \mathbb{R}$ are small parameters and $z^+ = \max\{0, z\}$, $z^- = \min\{0, z\}$. Then we have similarly like for (3.1.36) that (3.1.24) possesses for this case the form

$$M_\mu(\alpha) = -\frac{\mu_1 + \mu_2}{3} + \mu_3 \pi \operatorname{sech} \frac{\pi}{2} \sin \alpha\,.$$

If $|\mu_1 + \mu_2|/|\mu_3| < 3\pi \operatorname{sech} \frac{\pi}{2}$ then $M_\mu(\alpha)$ has a simple root, and consequently, (3.1.47) possesses subharmonic solutions with all large periods provided that $\mu_{1,2,3}$ are sufficiently small.

3.1.7 Bounded Solutions Close to Homoclinics

We show in this subsection the existence of bounded solutions on $\mathbb{R}$ of (3.1.3), (3.1.25) and (3.1.26) which are near to γ and on which subharmonics, found in the previous subsections, accumulate.

Theorem 3.1.33. *The assumptions of Theorem 3.1.6 imply for any $\mu \in \mathbf{R}$ of the form $\mu = s^2 \tilde{\mu}$, $0 < s \in U_1$, $\tilde{\mu} \in U_2$, where $\mathbf{R}$, U_1, U_2 are from Theorem 3.1.6, the existence of a solution x_μ of (3.1.3) on $\mathbb{R}$ satisfying*

$$\sup_{t \in \mathbb{R}} |x_\mu(t) - \gamma(t - \alpha_\mu)| \le Ks, \tag{3.1.48}$$

where $\alpha_\mu \in \mathbb{R}$, $|\alpha_\mu| \le K$ and K is a constant from Theorem 3.1.6. Moreover, a subsequence of the subharmonics $\{x_m\}_{m \ge [1/s]}$ from Theorem 3.1.6 accumulates on x_μ.

Proof. Let us fix $\mu \in \mathbf{R}$ and let $\{x_m\}_{m \ge [1/s]}$ be the sequence of subharmonics from Theorem 3.1.6. Since $\{x_m\}_{m \ge [1/s]}$ is a bounded sequence in the Sobolev space $W^{1,\infty}(\mathbb{R}, \mathbb{R}^n)$, then by the Arzela-Ascolli Theorem [199], there is a subsequence $\{x_{m_j}\}_{j=1}^\infty$ of $\{x_m\}_{m \ge [1/s]}$ and a function

$$x_\mu \in W^{1,2}_{\text{loc}}(\mathbb{R}, \mathbb{R}^n) := \cap_{m \in \mathbb{N}} W^{1,2}([-m, m], \mathbb{R}^n)$$

with $|x_\mu|_\infty < \infty$ and such that for any $p \in \mathbb{N}$ the sequence $\{x_{m_j}\}_{j=1}^\infty$ converges to x_μ in Z_p. Moreover, we can assume $\alpha_{m_j} \to \alpha_\mu \in \mathbb{R}$ as $j \to \infty$.

We note that $\dot{x}_\mu$ exist a.e. on $\mathbb{R}$. Now let $t_0 \in \mathbb{R}$ be such that $\dot{x}_\mu(t_0)$ exists. Since f is continuous as well as f_i, $i = 1, \cdots, k$ are upper–semicontinuous with compact values, for any $\varepsilon > 0$ there are $j_0 \in \mathbb{N}$ and $\delta > 0$ with the following property for any t, $|t - t_0| < \delta$, $j \ge j_0$:

$$f(x_{m_j}(t)) + \sum_{i=1}^k \mu_i f_i(x_{m_j}(t), \mu, t) \subset \mathbf{K}_\varepsilon,$$

where

$$\mathbf{K}_\varepsilon = \left\{ z \in \mathbb{R}^n : \text{dist}\left(z, f(x_\mu(t_0)) + \sum_{i=1}^k \mu_i f_i(x_\mu(t_0), \mu, t_0) \right) \le \varepsilon \right\}.$$

The set $\mathbf{K}_\varepsilon$ is compact and convex. Let us choose a $y \in \mathbb{R} \setminus \mathbf{K}_\varepsilon$. Then there is at least one closest point k_y of $\mathbf{K}_\varepsilon$ to y since $\mathbf{K}_\varepsilon$ is compact. The uniqueness of k_y follows from the convexity of $\mathbf{K}_\varepsilon$, since if there could be two such k'_y, k''_y then $(k'_y + k''_y)/2 \in \mathbf{K}_\varepsilon$ would be closer to y as either k'_y or k''_y. Moreover from the triangular with vertexes $k \in \mathbf{K}_\varepsilon$, k_y and y it follows that the angle between $k - k_y$ and $y - k_y$ is obtuse (see also [47, 199]). So there is a unique $k_y \in \mathbf{K}_\varepsilon$ such that

$$\langle k - k_y, y - k_y \rangle \le 0, \quad \forall k \in \mathbf{K}_\varepsilon. \tag{3.1.49}$$

Next from

$$\dot{x}_{m_j}(t) \in f(x_{m_j}(t)) + \sum_{i=1}^{k} \mu_i f_i(x_{m_j}(t), \mu, t) \quad \text{a.e. on} \quad \{t : |t - t_0| \le \delta\}$$

for any $j \ge j_0$, and (3.1.49) we derive

$$\frac{2}{p} \int_{t_0-(1/p)}^{t_0+(1/p)} \langle \dot{x}_{m_j}(s) - k_y, y - k_y \rangle \, ds$$

$$= \left\langle \frac{x_{m_j}(t_0 + (1/p)) - x_{m_j}(t_0 - (1/p))}{2/p} - k_y, y - k_y \right\rangle \le 0$$

for any $p \in \mathbb{N}$ satisfying $p > 1/\delta$. By passing to the limit first with $j \to \infty$ and then with $p \to \infty$, we obtain

$$\langle \dot{x}_\mu(t_0) - k_y, y - k_y \rangle \le 0, \quad \forall y \in \mathbb{R}^n \setminus \mathbf{K}_\varepsilon.$$

It is clear that

$$\mathbf{K}_\varepsilon = \bigcap_{y \in \mathbb{R}^n \setminus \mathbf{K}_\varepsilon} \left\{ z \in \mathbb{R}^n : \langle z - k_y, y - k_y \rangle \le 0 \right\}.$$

Consequently, we have

$$\dot{x}_\mu(t_0) \in \mathbf{K}_\varepsilon. \tag{3.1.50}$$

Since (3.1.50) holds for any $\varepsilon > 0$ and the set $f(x_\mu(t_0)) + \sum_{i=1}^{k} \mu_i f_i(x_\mu(t_0), \mu, t_0)$
is compact, we obtain

$$\dot{x}_\mu(t_0) \in f(x_\mu(t_0)) + \sum_{i=1}^{k} \mu_i f_i(x_\mu(t_0), \mu, t_0).$$

The estimate (3.1.48) follows from the similar one of Theorem 3.1.6. $\qquad \square$

We can analogously find accumulation of subharmonics on bounded solutions in all above results. Finally the method of this Section is extended in Chapter 4 to show chaotic behavior of (3.1.3), (3.1.25) and (3.1.26).

3.2   Bifurcation of Periodics from Homoclinics II

3.2.1   Singular Discontinuous O.D.Eqns

In this section we proceed with the study of bifurcations of subharmonics from homoclinics in discontinuously perturbed ordinary differential equations. But we study more complicated equations. To motivate our interest in this problem, let us consider a mass m located in a forcing field $-q(x)$, putting horizontally on a moving two–dimensional ribbon with a speed v_0 and periodically excited by a force $mp(t)$. The viscous damping is $-\widetilde{k}\dot{x}$, and the dry friction between the ribbon and the mass is $-\mu m\,\mathrm{sgn}\,(\dot{x}+v_0)$. Here $\widetilde{k}$ and μ are positive constants, and $\mathrm{sgn}\,r = \frac{r}{|r|_2}$, $r \in \mathbb{R}^2 \setminus \{0\}$ (see Fig. 3.5). By taking the balance of these forces, we arrive at the equation

$$m\ddot{x} + \widetilde{k}\dot{x} + q(x) + \mu m\,\mathrm{sgn}\,(\dot{x}+v_0) = mp(t), \quad x \in \mathbb{R}^2. \tag{3.2.1}$$

Rewriting (3.2.1) as a first-order system, we get

$$\dot{x} = y, \quad m\dot{y} = -\widetilde{k}y - q(x) + m\big(p(t) - \mu\,\mathrm{sgn}\,(y+v_0)\big). \tag{3.2.2}$$

Like in the previous section, we consider (3.2.2) with m small as a singularly perturbed differential inclusion of the form

$$\dot{x} = y, \quad m\dot{y} \in -\widetilde{k}y - q(x) + m\big(p(t) - \mu\,\mathrm{Sgn}\,(y+v_0)\big), \tag{3.2.3}$$

where

$$\mathrm{Sgn}\,r = \begin{cases} \mathrm{sgn}\,r & \text{for } r \neq 0, \\ \{x \in \mathbb{R}^2 : |x|_2 \leq 1\} & \text{for } r = 0. \end{cases}$$

For more generality, we consider singularly perturbed differential inclusions of the following form

$$\begin{aligned} \dot{x}(t) &\in f(x(t), y(t)) + \varepsilon h_1(x(t), y(t), t) \quad \text{a.e. on } \mathbb{R} \\ \varepsilon\dot{y}(t) &\in g(x(t), y(t)) + \varepsilon h_2(x(t), y(t), t) \quad \text{a.e. on } \mathbb{R} \end{aligned} \tag{3.2.4}$$

with $x \in \mathbb{R}^n$, $y \in \mathbb{R}^k$ and $\varepsilon > 0$ is small. Let $\langle \cdot, \cdot \rangle_i$ be the inner product on $\mathbb{R}^i$, $i \in \mathbb{N}$ with the corresponding norm $|\cdot|_i$. Now we set the main assumptions about (3.2.4):

(i) $f \in C^2(\mathbb{R}^n \times \mathbb{R}^k, \mathbb{R}^n)$, $g \in C^2(\mathbb{R}^n \times \mathbb{R}^k \times \mathbb{R}, \mathbb{R}^k)$, and $h_1 : \mathbb{R}^n \times \mathbb{R}^k \times \mathbb{R} \to 2^{\mathbb{R}^n} \setminus \{\emptyset\}$, $h_2 : \mathbb{R}^n \times \mathbb{R}^k \times \mathbb{R} \to 2^{\mathbb{R}^k} \setminus \{\emptyset\}$ are upper–semicontinuous with compact and convex values.

(ii) $f(0,0) = 0$ and $\Re\tau \neq 0$ for any $\tau \in \sigma\big(D_x f(0,0)\big)$.

(iii) $g(\cdot,0) = 0$, $g(x,y) = A(x)y + o(|y|_k)$ for $A(x) \in L(\mathbb{R}^k)$ satisfying

$$B(x)A(x)B^{-1}(x) = (D_1(x), D_2(x)) \quad \forall x \in \mathbb{R}^n,$$

where $B : \mathbb{R}^n \to L(\mathbb{R}^k)$, $D_1 : \mathbb{R}^n \to L(\mathbb{R}^{k_1})$, $D_2 : \mathbb{R}^n \to L(\mathbb{R}^{k_2})$ are C^1–smooth mappings, $k = k_1 + k_2$ and

$$\langle D_1(x)v, v\rangle_{k_1} \geq a|v|_{k_1}^2, \quad \langle D_2(x)w, w\rangle_{k_2} \leq -a|w|_{k_2}^2$$
$$\forall x \in \mathbb{R}^n, \forall v \in \mathbb{R}^{k_1}, \forall w \in \mathbb{R}^{k_2},$$

where $a > 0$ is a constant.

(iv) The reduced equation of (3.2.4) of the form $\dot{x} = f(x,0)$ has a homoclinic solution: There is a nonzero differentiable function $t \to \gamma(t)$ such that $\lim_{t\to+\infty} \gamma(t) = \lim_{t\to-\infty} \gamma(t) = 0$ and $\dot{\gamma}(t) = f(\gamma(t), 0)$.

(v) $h_i(x, y, t+2) = h_i(x, y, t)$ for $t \in \mathbb{R}$ and $i = 1, 2$.

When the perturbations $h_{1,2}(x, y, t)$ in (3.2.4) are smooth functions, then we have singularly perturbed ordinary differential equations

$$\begin{aligned} \dot{x}(t) &= f(x(t), y(t)) + \varepsilon h_1(x(t), y(t), t) \\ \varepsilon \dot{y}(t) &= g(x(t), y(t)) + \varepsilon h_2(x(t), y(t), t) \,. \end{aligned} \tag{3.2.5}$$

Setting $\varepsilon = 0$ in the (3.2.5) we obtain the *reduced equation*

$$\dot{x} = f(x, 0) \,. \tag{3.2.6}$$

The study of dynamics between (3.2.5) and (3.2.6) is started from 1952 [187] by showing that under certain conditions, for given $T > 0$ the solutions of (3.2.5) are at a $O(\varepsilon)$-distance from the corresponding solutions of (3.2.6), for t in any compact subset of $(0, T]$. This result was improved in [116, 117]. Later, in [97], a geometric theory of singular systems was developed, where this theory is applied to the autonomous case and states, under certain hypotheses, the existence of a center manifold for (3.2.5) defined on compact subsets on which system (3.2.5) is a regular perturbation of the reduced system (3.2.6). This was used to improve previous results in [100] concerning the existence of periodic solutions of (3.2.5). Geometric theory has been used in [184, 185] to study the problem of bifurcation from a heteroclinic orbit of the reduced system towards a heteroclinic orbit of the overall system (3.2.5). Later, in [28, 72], the non-autonomous case have been handled, together with the homoclinic case. These results were improved in [21] by showing that the system (3.2.5) has a global center manifold on which the system (3.2.5) is a regular perturbation of the reduced system (3.2.6). Then applying regular perturbation theory a smooth Melnikov function was derived to obtain homoclinic/heteroclinic solutions. Also there was studied a case when the reduced system (3.2.6) has a heteroclinic orbit joining semi-hyperbolic fixed points. Finally let us mention some other related results in this direction. Attractive invariant manifolds of (3.2.5) are studied in [155] when h_1, h_2 are independent of t. In [183] the same problem is investigated as in [155] when h_1, h_2 do depend on t. Asymptotic expansions of solutions for (3.2.5) are derived in [193]. Summarizing, a geometric theory for (3.2.5) is rather well-developed, not like for (3.2.4). On the other hand, there are several papers dealing with singularly perturbed differential inclusions (see [69, 104] and references therein). Singularly perturbed boundary value problems are studied in [77, 118, 119, 136, 194].

Next from the above-mentioned results on (3.2.5) it follows that (3.2.5) under certain conditions is chaotic (see [28, 72]) with an infinite number of subharmonics. The purpose of this section is to find conditions ensuring that even in multivalued case (3.2.4) there are still infinitely many subharmonics accumulating on a bounded solution of (3.2.4) on $\mathbb{R}$ which is near to the reduced homoclinic solution $(\gamma, 0)$. But approaches already used to (3.2.5) are not possible to apply for differential inclusions such as (3.2.4). To prove our results, we extend the method of Section 3.1 to system (3.2.4).

3.2.2   Linearized Equations

Let W_0^s, W_0^u be the stable and unstable manifolds, respectively, of the hyperbolic origin for (3.2.6), and $d_s = \dim W_0^s$, $d_u = \dim W_0^u$. Next, for the variational equation of (3.2.6) along γ:

$$\dot{u}(t) = D_x f(\gamma(t), 0)u(t), \qquad\qquad (3.2.7)$$

we have Theorem 3.1.2 of Subsection 3.1.2 when (3.1.5) is replaced with (3.2.7).

We use functional-analytic approach on certain Banach spaces like in the previous section. So we fix m, $j \in \mathbb{N}$ and define the following Banach spaces:

$$Z_{m,j} = C\left([-m, m], \mathbb{R}^j\right),$$
$$Z_{m,j}^p = \{z \in Z_{m,j} : z(-m) = z(m)\},$$
$$Y_{m,j} = L^\infty\left([-m, m], \mathbb{R}^j\right)$$

with the supremum norms $\|\cdot\|_{m,j}$ for $Z_{m,j}$, respectively $|\cdot|_{m,j}$ for $Y_{m,j}$. Next, for the non–homogeneous linear equation

$$\dot{z} = D_x f(\gamma, 0)z + h \qquad\qquad (3.2.8)$$

we have the following analogy of Theorem 3.1.4.

Theorem 3.2.1. *There exist $m_0 > 0$, $A > 0$, $B > 0$ such that for every $m > m_0$, $m \in \mathbb{N}$ there exists a linear function $\mathbf{L}_m : Y_{m,n} \to \mathbb{R}^n$ with $\|P_{uu}\mathbf{L}_m\| \le A\,\mathrm{e}^{-2Mm}$ and if $h \in Y_{m,n}$ satisfies*

$$\int_{-m}^m P_{uu} U(t)^{-1} h(t)\, dt + P_{uu} \mathbf{L}_m h = 0$$

then (3.2.8) has a solution in $z \in Z_{m,n}^p$ with $P_{ss} U(0)^{-1} z(0) = 0$ and $\|z\|_{m,n} \le B|h|_{m,n}$. Moreover, this solution z depends linearly on h.

Setting

$$\tilde{Y}_{m,n} = \left\{z \in Y_{m,n} : \int_{-m}^m P_{uu} U(t)^{-1} z(t)\, dt + P_{uu} \mathbf{L}_m z = 0\right\},$$

we define a variation of constants map $K_m : \tilde{Y}_{m,n} \to Z_{m,n}$ by taking $K_m h$ to be the solution in $Z_{m,n}^p$ to (3.2.8) from Theorem 3.2.1. Then the norm $\|K_m\|$ is uniformly bounded with respect to m and K_m is a compact linear operator. We have the following analogy of Lemma 3.1.5.

Theorem 3.2.2. *There exist $A > 0$, $m_0 > 0$ and for each $m > m_0$, $m \in \mathbb{N}$ a projection $\Pi_m : Y_{m,n} \to Y_{m,n}$ such that*

(i) $\|\Pi_m\| < A$ for all $m > m_0$

(ii) $\mathcal{R}(\mathbb{I} - \Pi_m) = \tilde{Y}_{m,n}$

Now we study the non–homogeneous variational equation

$$\varepsilon \dot{y} = A(\gamma(t))y + h(t). \tag{3.2.9}$$

Theorem 3.2.3. *There is a constant $c > 0$ such that for any $m \in \mathbb{N}$ and $0 < \varepsilon \leq 1$, (3.2.9) has a unique solution $y \in Z^p_{m,k}$ provided that $h \in Y_{m,k}$ and moreover, it is satisfying $\|y\|_{m,k} \leq c|h|_{m,k}$.*

Proof. By taking the transformation $z = B(\gamma)y$, (3.2.9) becomes to the form

$$\varepsilon \dot{z} = (D_1(\gamma), D_2(\gamma))z + \varepsilon(D_x B(\gamma)\dot{\gamma})B^{-1}(\gamma)z + B(\gamma)h. \tag{3.2.10}$$

Let us first consider the equation

$$\varepsilon \dot{z} = D_1(\gamma)z + h, \quad h \in Y_{m,k_1}. \tag{3.2.11}$$

Let Z_ε be the fundamental solution of $\varepsilon \dot{z} = D_1(\gamma)z$. Then $z(t) = Z_\varepsilon(t)Z_\varepsilon^{-1}(s)z_0$, $z_0 \in \mathbb{R}^{k_1}$ solves $\varepsilon \dot{z} = D_1(\gamma)z$, $z(s) = z_0$. Using assumption (iii) of Subsection 3.2.1, we derive

$$\frac{d}{dt}\left(|z(t)|^2_{k_1}\right) = 2\langle \dot{z}(t), z(t)\rangle_{k_1} = \frac{2}{\varepsilon}\langle D_1(\gamma(t))z(t), z(t)\rangle_{k_1} \geq \frac{2a}{\varepsilon}|z(t)|^2_{k_1}.$$

So we get $|z(t)|_{k_1} \leq e^{-a(s-t)/\varepsilon}|z_0|_{k_1}$ for $t \leq s$, that is Z_ε satisfies the following property

$$|Z_\varepsilon(t)Z_\varepsilon^{-1}(s)| \leq e^{-a(s-t)/\varepsilon}, \quad t \leq s. \tag{3.2.12}$$

The general solution of (3.2.11) has the form

$$z(t) = Z_\varepsilon(t)Z_\varepsilon^{-1}(m)z(m) - \frac{1}{\varepsilon}\int_t^m Z_\varepsilon(t)Z_\varepsilon^{-1}(s)h(s)\,ds. \tag{3.2.13}$$

Consequently, this solution is from Z^p_{m,k_1} if and only if

$$z(m) = Z_\varepsilon(-m)Z_\varepsilon^{-1}(m)z(m) - \frac{1}{\varepsilon}\int_{-m}^m Z_\varepsilon(-m)Z_\varepsilon^{-1}(s)h(s)\,ds.$$

Since by (3.2.12) for $0 < \varepsilon \leq 1$ and $m \in \mathbb{N}$:

$$|Z_\varepsilon(-m)Z_\varepsilon^{-1}(m)| \leq e^{-2am/\varepsilon} < 1$$
$$|Z_\varepsilon(-m)Z_\varepsilon^{-1}(s)| \leq e^{-a(s+m)/\varepsilon}, \quad s \geq -m,$$

we obtain

$$z(m) = -\left(\mathbb{I} - Z_\varepsilon(-m)Z_\varepsilon^{-1}(m)\right)^{-1}\frac{1}{\varepsilon}\int_{-m}^m Z_\varepsilon(-m)Z_\varepsilon^{-1}(s)h(s)\,ds$$

$$|z(m)|_{k_1} \leq (1 - e^{-2a})^{-1}\frac{1}{a}|h|_{m,k_1}.$$

Moreover, for $-m \leq t \leq m$ we have

$$|z(t)|_{k_1} \leq |z(m)|_{k_1} + \frac{1}{a}|h|_{m,k_1} \leq \frac{1}{a}\left(1 + (1 - e^{-2a})^{-1}\right)|h|_{m,k_1}.$$

Hence

$$\|z\|_{m,k_1} \leq \frac{1}{a}\left(1 + (1 - e^{-2a})^{-1}\right)|h|_{m,k_1}.$$

We have a similar result for the equation

$$\varepsilon \dot{z} = D_2(\gamma)z + h, \quad h \in Y_{m,k_2}. \tag{3.2.14}$$

Since $\sup |(D_x B(\gamma)\dot{\gamma})B^{-1}(\gamma)| < \infty$ and $\sup |B^{-1}(\gamma)| < \infty$, $\sup |B(\gamma)| < \infty$, the statement of this theorem follows from (3.2.10) by using the above results for (3.2.11) and (3.2.14). $\qquad\square$

Let $K_{m,\varepsilon}h$ be the unique solution of Theorem 3.2.3. Because $K_{m,\varepsilon}: Y_{m,k} \to W^{1,2}([-m,m], \mathbb{R}^k)$ is bounded linear and $W^{1,2}([-m,m], \mathbb{R}^k)$ is compactly embedded into $Z_{m,k}$, we see that $K_{m,\varepsilon}: Y_{m,k} \to Z_{m,k}$ is a linear compact operator. Moreover, the norm $\|K_{m,\varepsilon}\|$ is uniformly bounded for $m \in \mathbb{N}$, $0 < \varepsilon \leq 1$. We do not know the limit of $K_{m,\varepsilon}$ as $\varepsilon \to 0_+$ and $m \to \infty$. On the other hand, we have the following result.

Theorem 3.2.4. *Let $X_{m,k} = W^{1,\infty}([-m,m], \mathbb{R}^k)$ with the usual norm denoted by $\||\cdot\||_{m,k}$. Then for a fixed $b \in \mathbb{N}$ and any $h \in X_{m,k}$ satisfying $\||h\||_{m,k} \leq 1$, the function $K_{m,\varepsilon}h$ tends on $Z_{b,k}$ to $-A^{-1}(\gamma)h$ uniformly by h as $\varepsilon \to 0_+$ and $m \to \infty$.*

Proof. For any $h \in X_{m,k}$ satisfying $\||h\||_{m,k} \leq 1$, we take $z_{h,m} \in Z_{m,k}^p$ given by

$$z_{h,m}(t) = -A^{-1}(\gamma(t))h(t) - t\left(A^{-1}(\gamma(-m))h(-m) - A^{-1}(\gamma(m))h(m)\right)/2m.$$

By taking $y = z + z_{h,m}$ in (3.2.9), we arrive at the equation

$$\begin{aligned} \varepsilon \dot{z}(t) = A(\gamma(t))z(t) - \varepsilon \dot{z}_{h,m} \\ -tA(\gamma(t))\left(A^{-1}(\gamma(-m))h(-m) - A^{-1}(\gamma(m))h(m)\right)/2m. \end{aligned} \tag{3.2.15}$$

According to the construction of $z_{h,m}$, there is a constant $c > 0$ such that

$$\||z_{h,m}\||_{m,k} \leq c, \quad \|A(\gamma(t))\left(A^{-1}(\gamma(-m))h(-m) - A^{-1}(\gamma(m))h(m)\right)\|_{m,k} \leq c.$$

It is enough to study (3.2.11). In view of (3.2.13), (3.2.15) has the form

$$z(t) = Z_\varepsilon(t)Z_\varepsilon^{-1}(m)z(m) - \frac{1}{\varepsilon}\int\limits_t^m Z_\varepsilon(t)Z_\varepsilon^{-1}(s)\bar{g}_{h,m}(s)\,ds$$

$$z(m) = Z_\varepsilon(-m)Z_\varepsilon^{-1}(m)z(m) - \frac{1}{\varepsilon}\int\limits_{-m}^m Z_\varepsilon(-m)Z_\varepsilon^{-1}(s)\bar{g}_{h,m}(s)\,ds,$$

where

$$\bar{g}_{h,m}(t) = -\varepsilon \dot{z}_{h,m} - tA(\gamma(t))\big(A^{-1}(\gamma(-m))h(-m) - A^{-1}(\gamma(m))h(m)\big)/2m \,.$$

According to (3.2.12), there is a constant $c_1 > 0$ such that for any $t \in [-b, b]$, $0 < b < m$, we have

$$|Z_\varepsilon(t)Z_\varepsilon^{-1}(m)z(m)|_{k_1} \le e^{-a(m-t)/\varepsilon}c_1 \le e^{-a(m-b)/\varepsilon}c_1$$

$$\left| \frac{1}{\varepsilon} \int_t^m Z_\varepsilon(t)Z_\varepsilon^{-1}(s)\bar{g}_{h,m}(s)\,ds \right|_{k_1} \le \frac{1}{\varepsilon}\int_t^m e^{-a(s-t)/\varepsilon}c_1\Big(\varepsilon + \frac{b}{2m}\Big)\,ds \le \frac{c_1}{a}\Big(\varepsilon + \frac{b}{2m}\Big).$$

By letting $\varepsilon \to 0_+$ and $m \to \infty$ in the above inequalities for $b \in \mathbb{N}$ fixed, the proof is finished. $\square$

Corollary 3.2.5. *Let $h \in L^\infty((-\infty, \infty), \mathbb{R}^k)$ and let there be a finite sequence of numbers $-\infty < t_1 < t_2 < \cdots < t_i < \infty$ such that*

$$h/[t_j, t_{j+1}] \in W^{1,\infty}([t_j, t_{j+1}], \mathbb{R}^k) \quad \forall j = 1, \cdots, i$$
$$h/(-\infty, t_1] \in W^{1,\infty}((-\infty, t_1], \mathbb{R}^k)$$
$$h/[t_i, \infty) \in W^{1,\infty}([t_i, \infty), \mathbb{R}^k) \,.$$

For any sufficiently small $\varepsilon > 0$, let us consider $h_\varepsilon \in W^{1,\infty}((-\infty, \infty), \mathbb{R}^k)$ defined as follows

$$h_\varepsilon(t) = h(t) \quad for \quad t \in \bigcup_{j=1}^{i-1} [t_j, t_{j+1} - \sqrt{\varepsilon}] \bigcup (-\infty, t_1 - \sqrt{\varepsilon}] \bigcup [t_i, \infty)$$
$$h_\varepsilon(t) \quad is\ linear\ on \quad \bigcup_{j=1}^{i} [t_j - \sqrt{\varepsilon}, t_j] \,.$$

Then for any fixed $b \in \mathbb{N}$ and any $\delta > 0$, there are $\varepsilon_\delta > 0$ and $b < m_\delta \in \mathbb{N}$ such that

$$\|K_{m,\varepsilon}h_\varepsilon + A^{-1}(\gamma)h_\varepsilon\|_{b,k} < \delta$$

for any $m \ge m_\delta$ and $0 < \varepsilon < \varepsilon_\delta$.

Proof. Clearly $|\dot{h}_\varepsilon|_{m,k} = O(1/\sqrt{\varepsilon})$ and $|h_\varepsilon|_{m,k} = O(|h|_{m,k})$ for any $m \in \mathbb{N}$ and $0 < \varepsilon \le 1$. By following the proof of Theorem 3.2.4 (see (3.2.15) when h is replaced by h_ε), the statement of this corollary is proved. $\square$

3.2.3   Bifurcation of Subharmonics

To find periodic solutions to (3.2.4) of very large periods which are near $(\gamma, 0)$, we follow the method of Subsection 3.1.3. By hypothesis, the multivalued vector field in (3.2.4) has period $2m$. We look for solutions to (3.2.4) in $Z_{m,n}^p \times Z_{m,k}^p$.

In (3.2.4) we change the variable

$$\varepsilon \leftrightarrow \varepsilon^2$$
$$y(t+\alpha) = \varepsilon^2 w(t)$$
$$x(t+\alpha) = \gamma(t) + \varepsilon^2 z(t) + \sum_{i=1}^{d-1} \varepsilon \beta_i u_{i+d}(t) + \frac{1}{2(m+E)} b(\varepsilon\beta, m+E)t,$$

where $1 > \varepsilon > 0$, $\alpha, \beta_i \in \mathbb{R}$, $m \in \mathbb{Z}_+$ and $E = [1/\varepsilon]$. Then the differential inclusions for (z,w) are

$$\dot{z}(t) - D_x f(\gamma(t),0)z(t) \in g_{m,\varepsilon}(z(t),w(t),\alpha,\beta,t) \quad \text{a.e. on} \quad [-m-E, m+E]$$
$$\varepsilon^2 \dot{w}(t) - A(\gamma(t))w \in h_{m,\varepsilon}(z(t),w(t),\alpha,\beta,t) \quad \text{a.e. on} \quad [-m-E, m+E],$$

where

$$g_{m,\varepsilon}(x,y,\alpha,\beta,t)=$$
$$\left\{ v \in \mathbb{R}^n : v \in \frac{1}{\varepsilon^2}\left\{ f\left(\varepsilon^2 x + \gamma(t) + \varepsilon \sum_{i=1}^{d-1} \beta_i u_{i+d}(t) + \frac{1}{2(m+E)} b(\varepsilon\beta, m+E)t, \varepsilon^2 y\right) \right. \right.$$
$$\left. -f(\gamma(t),0) - \varepsilon \sum_{i=1}^{d-1} \beta_i \dot{u}_{i+d}(t) - \frac{1}{2(m+E)} b(\varepsilon\beta, m+E) - D_x f(\gamma(t),0)\varepsilon^2 x \right\}$$
$$\left. + h_1\left(\varepsilon^2 x + \gamma(t) + \varepsilon \sum_{i=1}^{d-1} \beta_i u_{i+d}(t) + \frac{1}{2(m+E)} b(\varepsilon\beta, m+E)t, \varepsilon^2 y, t+\alpha\right) \right\},$$

$$h_{m,\varepsilon}(x,y,\alpha,\beta,t)=$$
$$\left\{ u \in \mathbb{R}^k : u \in \frac{1}{\varepsilon^2}\left\{ g\left(\varepsilon^2 x + \gamma(t) + \varepsilon \sum_{i=1}^{d-1} \beta_i u_{i+d}(t) + \frac{1}{2(m+E)} b(\varepsilon\beta, m+E)t, \varepsilon^2 y\right) \right. \right.$$
$$\left. -A(\gamma(t))\varepsilon^2 y \right\}$$
$$\left. + h_2\left(\varepsilon^2 x + \gamma(t) + \varepsilon \sum_{i=1}^{d-1} \beta_i u_{i+d}(t) + \frac{1}{2(m+E)} b(\varepsilon\beta, m+E)t, \varepsilon^2 y, t+\alpha\right) \right\}.$$

Introducing multivalued mappings

$$G^1_{m,\varepsilon} : Z_{m+E,n} \times Z_{m+E,k} \times \mathbb{R} \times \mathbb{R}^{d-1} \to 2^{Y_{m+E,n}}$$
$$G^2_{m,\varepsilon} : Z_{m+E,n} \times Z_{m+E,k} \times \mathbb{R} \times \mathbb{R}^{d-1} \to 2^{Y_{m+E,k}}$$

by the formulas

$$G^1_{m,\varepsilon}(z,w,\alpha,\beta) = \left\{ h \in Y_{m+E,n} : \ h(t) \in g_{m,\varepsilon}(z(t),w(t),\alpha,\beta,t) \right.$$
$$\left. \text{a.e. on} \quad [-m-E, m+E] \right\},$$
$$G^2_{m,\varepsilon}(z,w,\alpha,\beta) = \left\{ h \in Y_{m+E,k} : \ h(t) \in h_{m,\varepsilon}(z(t),w(t),\alpha,\beta,t) \right.$$
$$\left. \text{a.e. on} \quad [-m-E, m+E] \right\},$$

the above differential inclusions for z, w can be written to

$$\dot{z} - D_x f(\gamma,0)z \in G^1_{m,\varepsilon}(z,w,\alpha,\beta)$$
$$\varepsilon^2 \dot{w} - A(\gamma)w \in G^2_{m,\varepsilon}(z,w,\alpha,\beta). \tag{3.2.16}$$

Since $g_{m,\varepsilon} : \mathbb{R}^n \times \mathbb{R}^k \times \mathbb{R} \times \mathbb{R}^{d-1} \times \mathbb{R} \to 2^{\mathbb{R}^n} \setminus \{\emptyset\}$ and $h_{m,\varepsilon} : \mathbb{R}^n \times \mathbb{R}^k \times \mathbb{R} \times$ $\mathbb{R}^{d-1} \times \mathbb{R} \to 2^{\mathbb{R}^k} \setminus \{\emptyset\}$ are upper–semicontinuous with compact and convex values, like for (3.1.17), $G^j_{m,\varepsilon}(z, w, \alpha, \beta)$, $j = 1, 2$ are non–empty, closed, convex and bounded in $Y_{m+E,n}$, $Y_{m+E,k}$, and they are also weakly compact in $L^2([-m - E, m + E], \mathbb{R}^n)$, $L^2([-m - E, m + E], \mathbb{R}^k)$, respectively.

To proceed, we assume that the following condition holds.

(H) There is an upper–semicontinuous mapping $C : \mathbb{R} \times \mathbb{R} \to 2^{\mathbb{R}^k} \setminus \{\emptyset\}$ with compact convex values such that $C(\mathbb{R} \times \mathbb{R})$ is bounded and $C(t, \alpha + 2) = C(t, \alpha)$. Moreover, for any $\delta > 0$, $l \in \mathbb{N}$, $l > l_0$, where $l_0 \in \mathbb{N}$ is fixed, and $\alpha \in \mathbb{R}$, there are $\varepsilon_{\delta,l} > 0$, $l < m_{\delta,l} \in \mathbb{N}$, $\zeta_{\delta,l} > 0$ such that for any $\mathbb{N} \ni m \geq m_{\delta,l}$ and $h \in Y_{m,k}$ satisfying

$$|h|_{m,k} \leq 1 + \sup_{s,t \in \mathbb{R}} \left\{ \max \left\{ |v|_k : v \in h_2(\gamma(t), 0, s) \cup h_2(0, 0, s) \right\} \right\}$$

$$(t, h(t)) \in \left\{ u \in \mathbb{R}^{k+1} : \exists s \in \mathbb{R}; \; \text{dist} \left\{ u, \left(s, h_2(\gamma(s), 0, s + \alpha) \right) \right\} < \zeta_{\delta,l} \right\}$$

$$\text{a.e. on} \quad [-l - 1, l + 1],$$

the solution y of (3.2.9) with $0 < \varepsilon < \varepsilon_{\delta,l}$ satisfies

$$(t, y(t)) \in \left\{ u \in \mathbb{R}^{n+1} : \exists s \in \mathbb{R}; \; \text{dist} \left\{ u, \left(s, C(s, \alpha) \right) \right\} < \delta \right\} \quad \text{on} \quad [-l, l].$$

System (3.2.16) can not be solved directly. We need it to modify. For this reason, for $(z, w, \tilde{w}, \alpha, \beta) \in Z_{m+E,n} \times Z_{m+E,k} \times Z_{m+E,k} \times \mathbb{R} \times \mathbb{R}^{d-1}$, we take the homotopy

$$G^1_{m,\varepsilon,\lambda}(z, w, \tilde{w}, \alpha, \beta) = \left\{ h \in Y_{m+E,n} : h(t) \in g_{m,\varepsilon,\lambda}(z(t), w(t), \tilde{w}(t), \alpha, \beta, t) \right.$$
$$\left. \text{a.e. on} \; [-m - E, m + E] \right\}, \; \lambda \in [0, 1],$$

$$G^2_{m,\varepsilon,\lambda}(z, w, \alpha, \beta) = \left\{ h \in Y_{m+E,k} : h(t) \in h_{m,\varepsilon,\lambda}(z(t), w(t), \alpha, \beta, t) \right.$$
$$\left. \text{a.e. on} \; [-m - E, m + E] \right\}, \; \lambda \in [0, 1],$$

where

$$g_{m,\varepsilon,\lambda}(x, y, \tilde{y}, \alpha, \beta, t) =$$
$$\left\{ v \in \mathbb{R}^n : v \in \tfrac{\lambda}{\varepsilon^2} \left\{ f \left(\varepsilon^2 x + \gamma(t) + \varepsilon \sum_{i=1}^{d-1} \beta_i u_{i+d}(t) + \tfrac{1}{2(m+E)} b(\varepsilon\beta, m + E) t, 0 \right) \right. \right.$$
$$\left. - f(\gamma(t), 0) - \varepsilon \sum_{i=1}^{d-1} \beta_i \dot{u}_{i+d}(t) - \tfrac{1}{2(m+E)} b(\varepsilon\beta, m + E) - D_x f(\gamma(t), 0) \varepsilon^2 x \right\}$$
$$+ \tfrac{\lambda}{\varepsilon^2} \left\{ f \left(\varepsilon^2 x + \gamma(t) + \varepsilon \sum_{i=1}^{d-1} \beta_i u_{i+d}(t) + \tfrac{1}{2(m+E)} b(\varepsilon\beta, m + E) t, \varepsilon^2 y \right) \right.$$

$$-f\Big(\varepsilon^2 x+\gamma(t)+\varepsilon\sum_{i=1}^{d-1}\beta_i u_{i+d}(t)+\tfrac{1}{2(m+E)}b(\varepsilon\beta,m+E)t,0\Big)-D_y(f(\gamma(t),0)\varepsilon^2 y\Big\}$$

$$+\lambda h_1\Big(\varepsilon^2 x+\gamma(t)+\varepsilon\sum_{i=1}^{d-1}\beta_i u_{i+d}(t)+\tfrac{1}{2(m+E)}b(\varepsilon\beta,m+E)t,\varepsilon^2 y,t+\alpha\Big)$$

$$+\tfrac{1-\lambda}{2}\sum_{i,j=1}^{d-1}\beta_i\beta_j D_x^2 f(\gamma(t),0)(u_{d+i}(t),u_{d+j}(t))$$

$$+D_y f(\gamma(t),0)(\lambda\tilde{y}+(1-\lambda)C(t,\alpha))+(1-\lambda)h_1(\gamma(t),0,t+\alpha)\Big\},$$

and

$$h_{m,\varepsilon,\lambda}(x,y,\alpha,\beta,t)=$$
$$\Big\{u\in\mathbb{R}^k:u=\tfrac{\lambda}{\varepsilon^2}\Big\{g\Big(\varepsilon^2 x+\gamma(t)+\varepsilon\sum_{i=1}^{d-1}\beta_i u_{i+d}(t)+\tfrac{1}{2(m+E)}b(\varepsilon\beta,m+E)t,\varepsilon^2 y\Big)$$

$$-A(\gamma(t))\varepsilon^2 y\Big\}+(1-\lambda)h_2(\gamma(t),0,t+\alpha)$$

$$+\lambda h_2\Big(\varepsilon^2 x+\gamma(t)+\varepsilon\sum_{i=1}^{d-1}\beta_i u_{i+d}(t)+\tfrac{1}{2(m+E)}b(\varepsilon\beta,m+E)t,\varepsilon^2 y,t+\alpha\Big)\Big\}.$$

We note that if $a+b=1$, $a\geq 0$, $b\geq 0$ then

$$aG^1_{m,\varepsilon,\lambda}(z,w,\tilde{w}_1,\alpha,\beta)+bG^1_{m,\varepsilon,\lambda}(z,w,\tilde{w}_2,\alpha,\beta)\subset G^1_{m,\varepsilon,\lambda}(z,w,a\tilde{w}_1+b\tilde{w}_2,\alpha,\beta).$$

Next based on Theorems 3.2.1, 3.2.2 and 3.2.3, we use the Lyapunov-Schmidt approach to decompose and put (3.2.16) in the homotopy as follows

$$0\in\Big(z-F^1_{m,\varepsilon,\lambda}(z,w,\alpha,\beta),w-F^2_{m,\varepsilon,\lambda}(z,w,\alpha,\beta),B_{m,\varepsilon,\lambda}(z,w,\alpha,\beta)\Big)\quad(3.2.17)$$

for $\lambda\in[0,1]$, where

$$\Big(F^1_{m,\varepsilon,\lambda}(z,w,\alpha,\beta),F^2_{m,\varepsilon,\lambda}(z,w,\alpha,\beta),B_{m,\varepsilon,\lambda}(z,w,\alpha,\beta)\Big)$$
$$=\Big\{\Big(K_{m+E}(\mathbb{I}-\Pi_{m+E})h,K_{m+E,\varepsilon^2}v,L_{m+E}h\Big):$$
$$h\in G^1_{m,\varepsilon,\lambda}(z,w,K_{m+E,\varepsilon^2}v,\alpha,\beta),v\in G^2_{m,\varepsilon,\lambda}(z,w,\alpha,\beta)\Big\},$$

and

$$L_{m+E}h=\int_{-m-E}^{m+E}P_{uu}U(t)^{-1}h(t)\,dt+P_{uu}\mathbf{L}_{m+E}h.$$

To handle (3.2.17), we consider the new homotopy

$$0\in\Big(z-\lambda F^1_{m,\varepsilon,\lambda}(z,w,\alpha,\beta),w-\lambda F^2_{m,\varepsilon,\lambda}(z,w,\alpha,\beta),B_{m,\varepsilon,\lambda}(z,w,\alpha,\beta)\Big)\quad(3.2.18)$$

for $\lambda\in[0,1]$. Using $\|P_{uu}\mathbf{L}_{m+E}\|=O(e^{-2M(m+E)})$, we consider the decomposition and homotopy

$$(B_{m1,\varepsilon,\lambda}+\lambda B_{m2,\varepsilon,\lambda})(z,w,\alpha,\beta),\quad\lambda\in[0,1],$$

where

$$\left(B_{m1,\varepsilon,\lambda} + \lambda B_{m2,\varepsilon,\lambda}\right)(z, w, \alpha, \beta) = \left\{ L_{m1,\varepsilon}h + \lambda L_{m2,\varepsilon}h : \right.$$
$$\left. h \in G^1_{m,\varepsilon,\lambda}(z, w, K_{m+E,\varepsilon^2}v, \alpha, \beta),\ v \in G^2_{m,\varepsilon,\lambda}(z, w, \alpha, \beta) \right\},$$

and

$$L_{m1,\varepsilon}h = \int_{-m-E}^{m+E} P_{uu}U(t)^{-1}h(t)\, dt\,, \quad L_{m2,\varepsilon}v = P_{uu}\mathbf{L}_{m+E}h\,.$$

Summarizing, the solvability of (3.2.17)–(3.2.18) can be replaced by the solvability of the following multivalued equation

$$0 \in H_{m,\varepsilon}(z, w, \alpha, \beta, \lambda) \tag{3.2.19}$$

where $H_{m,\varepsilon} : Z_{m+E,n} \times Z_{m+E,k} \times \mathbb{R} \times \mathbb{R}^{d-1} \times [0,1] \to 2^{Z_{m+E,n} \times Z_{m+E,k} \times \mathbb{R}^d} \setminus \{\emptyset\}$
is given by

$$H_{m,\varepsilon}(z, w, \alpha, \beta, \lambda) = \Big(z - \lambda F^1_{m,\varepsilon,\lambda}(z, w, \alpha, \beta),\ w - \lambda F^2_{m,\varepsilon,\lambda}(z, w, \alpha, \beta),$$
$$\left(B_{m1,\varepsilon,\lambda} + \lambda B_{m2,\varepsilon,\lambda}\right)(z, w, \alpha, \beta) \Big)$$

for $1 > \varepsilon > 0$ sufficiently small and fixed, and $\lambda \in [0,1]$ is a homotopy parameter. Like in Subsection 3.1.3, it is not difficult to observe that the mapping $\mathbb{I}_{Z_{m+E,n} \times Z_{m+E,k} \times \mathbb{R}^d} - H_{m,\varepsilon}$ is upper–semicontinuous with compact convex values and maps bounded sets into relatively compact ones. So topological degree methods of Section 2.4.3 can be applied to (3.2.19).

Finally, we introduce a multivalued mapping

$$M : \mathbb{R}^d \to 2^{\mathbb{R}^d} \setminus \{\emptyset\}\,, \quad M = (M_1, \cdots, M_d) \tag{3.2.20}$$

given by

$$M_l(\alpha, \beta) = \left\{ \int_{-\infty}^{\infty} \langle h(s), u_l^{\perp}(s) \rangle_n\, ds : h \in L^2_{\text{loc}}(\mathbb{R}, \mathbb{R}^n) \text{ satisfying a.e. on } \mathbb{R} \right.$$
$$\text{the relation}\quad h(t) \in \Big(\tfrac{1}{2} \sum_{i,j=1}^{d-1} \beta_i \beta_j D^2_x f(\gamma(t), 0)(u_{d+i}(t), u_{d+j}(t))$$
$$\left. + D_y f(\gamma(t), 0)(C(t, \alpha)) + h_1(\gamma(t), 0, t+\alpha) \Big) \right\}.$$

Following the arguments below (3.1.21), we see that mapping M is upper–semicontinuous with compact convex values and maps bounded sets into bounded ones. Now we are ready to prove the main theorems of this section [74].

Theorem 3.2.6. *Let $d > 1$. If there is a non–empty open bounded set $\mathbf{B} \subset \mathbb{R}^d$ such that* *(i)* $0 \notin M(\partial \mathbf{B})$

(ii) $\deg(M, \mathbf{B}, 0) \neq 0$

Then there are constants $K > 0$ and $\varepsilon_0 > 0$ such that for any $0 < \varepsilon < \varepsilon_0$, the differential inclusion (3.2.4) possesses a subharmonic solution (x_m, y_m) of period $2m$ for any $m \in \mathbb{N}$, $m \geq [1/\sqrt{\varepsilon}]$ satisfying

$$\sup_{-m \leq t \leq m} |y_m(t)|_k \leq K\varepsilon, \qquad \sup_{-m \leq t \leq m} |x_m(t) - \gamma(t - \alpha_m)|_n \leq K\sqrt{\varepsilon},$$

where $\alpha_m \in \mathbb{R}$ and $|\alpha_m| \leq K$.

Proof. We need to solve (3.2.16) which is plugged into the homotopy (3.2.19). In order to handle (3.2.19), we need the following results.

Lemma 3.2.7. *Let $D^i : [a, b] \to 2^{\mathbb{R}^m} \setminus \{\emptyset\}$, $i = 1, 2$ be upper-semicontinuous mappings with convex and compact values. Here $a, b \in \mathbb{R}$, $a < b$. Let Γ_i be the graph of D^i. Take $c, d \in \mathbb{R}$, $a \leq c < d \leq b$. Then $\forall \zeta > 0$, $\exists \delta > 0$ such that*

$$\lambda \in [0, 1], \quad y_i, z_i \in \mathbb{R}^m, \quad i = 1, 2, \quad t \in [c, d]$$
$$\mathrm{dist}\,\{(t, y_i), \Gamma_i\} < \delta, \quad \mathrm{dist}\,\{(t, z_i), \Gamma_i\} < \delta$$

imply

$$\mathrm{dist}\,\{(t, \lambda(y_1 + y_2) + (1 - \lambda)(z_1 + z_2)), \Gamma_1 + \Gamma_2\} < \zeta,$$

where $\Gamma_1 + \Gamma_2 = \{(s, d_1 + d_2) : d_i \in D^i(s), i = 1, 2, s \in [a, b]\}$.

Proof. Assume the contrary. So there is $\zeta_0 > 0$ and for all $p \in \mathbb{N}$, $i = 1, 2$:

$$\lambda_p \in [0, 1], \quad y_{i,p}, z_{i,p} \in \mathbb{R}^m, \quad t_p \in [c, d]$$
$$s_{i,p}, \tilde{s}_{i,p} \in [a, b], \quad d_{i,p} \in D^i(s_{i,p}), \quad \tilde{d}_{i,p} \in D^i(\tilde{s}_{i,p})$$
$$|t_p - s_{i,p}| + |y_{i,p} - d_{i,p}|_m < 1/p, \quad |t_p - \tilde{s}_{i,p}| + |z_{i,p} - \tilde{d}_{i,p}|_m < 1/p$$
$$\mathrm{dist}\,\{(t_p, \lambda_p(y_{1,p} + y_{2,p}) + (1 - \lambda_p)(z_{1,p} + z_{2,p})), \Gamma_1 + \Gamma_2\} \geq \zeta_0.$$

We can assume

$$\lambda_p \to \lambda_0, \quad y_{i,p} \to y_{i,0}, \quad z_{i,p} \to z_{i,0}, \quad t_p \to t_0, \quad s_{i,p} \to t_0$$
$$\tilde{s}_{i,p} \to t_0, \quad d_{i,p} \to y_{i,0}, \quad \tilde{d}_{i,p} \to z_{i,0}, \quad i = 1, 2.$$

Hence $y_{i,0} \in D^i(t_0) \ni z_{i,0}$, $i = 1, 2$ and

$$\mathrm{dist}\,\{(t_0, \lambda_0(y_{1,0} + y_{2,0}) + (1 - \lambda_0)(z_{1,0} + z_{2,0})), \Gamma_1 + \Gamma_2\} \geq \zeta_0.$$

This contradicts to

$$\lambda_0(y_{1,0} + y_{2,0}) + (1 - \lambda_0)(z_{1,0} + z_{2,0}) \in D^1(t_0) + D^2(t_0),$$

since $D^1(t_0)$, $D^2(t_0)$ are convex and

$$\lambda_0(y_{1,0} + y_{2,0}) + (1 - \lambda_0)(z_{1,0} + z_{2,0}) = \lambda_0 y_{1,0} + (1 - \lambda_0)z_{1,0} + \lambda_0 y_{2,0} + (1 - \lambda_0)z_{2,0}.$$

The proof is finished. $\qquad \square$

Lemma 3.2.8. *Let $D : [a, b] \to 2^{\mathbb{R}^m} \setminus \{\emptyset\}$ be an upper–semicontinuous mapping with compact values. Here $a, b \in \mathbb{R}$, $a < b$. Let Γ be the graph of D. Then $\forall t \in [a, b]$, $\forall \zeta > 0$, $\exists \delta > 0$ such that*

$$y \in \mathbb{R}^m, \quad \text{dist}\,\{(t, y), \Gamma\} < \delta \quad \Longrightarrow \quad \text{dist}\,\{y, D(t)\} < \zeta\,.$$

Proof. Assume the contrary. So there are $\zeta_0 > 0$, $t_0 \in [a, b]$ and for all $p \in \mathbb{N}$:

$$y_p \in \mathbb{R}^m, \quad s_p \in [a, b], \quad d_p \in D(s_p)$$
$$|t_0 - s_p| + |y_p - d_p|_m < 1/p, \quad \text{dist}\,\{y_p, D(t_0)\} \geq \zeta_0\,.$$

We can assume $y_p \to y_0$, $s_p \to t_0$, $d_p \to y_0$. Then $y_0 \in D(t_0)$ and dist $\{y_0, D(t_0)\} \geq \zeta_0$. We arrive at the contradiction. $\qquad\square$

Lemma 3.2.9. *Condition (i) of Theorem 3.2.6 implies for any $A > 0$ the existence of constants $\delta > 0$, $1 > \varepsilon_0 > 0$ such that*

$$|B_{m1,\varepsilon,\lambda}(z, w, \alpha, \beta)|_d \geq \delta$$

for any $0 < \varepsilon < \varepsilon_0$, $m \in \mathbb{Z}_+$, $\lambda \in [0, 1]$, $\|z\|_{m+E,n} \leq A$, $\|w\|_{m+E,k} \leq A$, $(\alpha, \beta) \in \partial\mathbf{B}$.

Proof. First we note

$$\lim_{\varepsilon \to 0_+} \frac{1}{\varepsilon^2} \Big\{ f\Big(\varepsilon^2 x + \gamma(t) + \varepsilon \sum_{i=1}^{d-1} \beta_i u_{i+d}(t) + \tfrac{1}{2(m+E)} b(\varepsilon\beta, m + E)t, 0\Big)$$
$$-f(\gamma(t), 0) - \varepsilon \sum_{i=1}^{d-1} \beta_i \dot{u}_{i+d}(t) - \tfrac{1}{2(m+E)} b(\varepsilon\beta, m + E) - D_x f(\gamma(t), 0)\varepsilon^2 x \Big\}$$
$$= \frac{1}{2} \sum_{i,j=1}^{d-1} \beta_i \beta_j D_x^2 f(\gamma(t), 0)\big(u_{i+d}(t), u_{j+d}(t)\big)\,,$$
$$\lim_{\varepsilon \to 0_+} \frac{1}{\varepsilon^2} \Big\{ f\Big(\varepsilon^2 x + \gamma(t) + \varepsilon \sum_{i=1}^{d-1} \beta_i u_{i+d}(t) + \tfrac{1}{2(m+E)} b(\varepsilon\beta, m + E)t, \varepsilon^2 y\Big)$$
$$-f\Big(\varepsilon^2 x + \gamma(t) + \varepsilon \sum_{i=1}^{d-1} \beta_i u_{i+d}(t) + \tfrac{1}{2(m+E)} b(\varepsilon\beta, m + E)t, 0\Big) \Big\} = D_y f(\gamma, 0)y$$

$$(3.2.21)$$

and

$$\lim_{\varepsilon \to 0_+} \frac{\lambda}{\varepsilon^2} \Big\{ f\Big(\varepsilon^2 x + \gamma(t) + \varepsilon \sum_{i=1}^{d-1} \beta_i u_{i+d}(t) + \tfrac{1}{2(m+E)} b(\varepsilon\beta, m + E)t, 0\Big)$$
$$-f(\gamma(t), 0) - \varepsilon \sum_{i=1}^{d-1} \beta_i \dot{u}_{i+d}(t) - \tfrac{1}{2(m+E)} b(\varepsilon\beta, m + E) - D_x f(\gamma(t), 0)\varepsilon^2 x \Big\}$$
$$+\frac{\lambda}{\varepsilon^2} \Big\{ f\Big(\varepsilon^2 x + \gamma(t) + \varepsilon \sum_{i=1}^{d-1} \beta_i u_{i+d}(t) + \tfrac{1}{2(m+E)} b(\varepsilon\beta, m + E)t, \varepsilon^2 y\Big)$$

$$-f\Big(\varepsilon^2 x + \gamma(t) + \varepsilon \sum_{i=1}^{d-1} \beta_i u_{i+d}(t) + \tfrac{1}{2(m+E)} b(\varepsilon\beta, m+E)t, 0\Big) - D_y f(\gamma(t),0)\varepsilon^2 y\Big\}$$

$$= \tfrac{\lambda}{2} \sum_{i,j=1}^{d-1} \beta_i\beta_j D_x^2 f(\gamma(t),0)(u_{d+i}(t), u_{d+j}(t)),$$

$$\lim_{\varepsilon\to 0_+} \tfrac{\lambda}{\varepsilon^2}\Big\{ g\Big(\varepsilon^2 x + \gamma(t) + \varepsilon \sum_{i=1}^{d-1} \beta_i u_{i+d}(t) + \tfrac{1}{2(m+E)} b(\varepsilon\beta, m+E)t, \varepsilon^2 y\Big)$$

$$- A(\gamma(t))\varepsilon^2 y\Big\} = 0$$

$$(3.2.22)$$

uniformly with respect to x, y, β bounded and $t \in [-m-E, m+E]$, $m \in \mathbb{Z}_+, \lambda \in [0,1]$.

Now assume the contrary in Lemma 3.2.9. So there is an $A > 0$ and

$$\varepsilon_p \to 0, \quad m_p \in \mathbb{Z}_+, \quad \lambda_p \to \lambda_0, \quad \|z_p\|_{m_p+E_p, n} \le A$$

$$\|w_p\|_{m_p+E_p, k} \le A, \quad p \in \mathbb{N}\setminus\{1,2\}, \quad \partial\mathbf{B} \ni (\alpha_p, \beta_p) \to (\alpha_0, \beta_0) \in \partial\mathbf{B}$$

$$h_p \in G^1_{m_p,\varepsilon_p,\lambda_p}(z_p, w_p, K_{m_p+E_p,\varepsilon_p^2} v_p, \alpha_p, \beta_p), \quad v_p \in G^2_{m_p,\varepsilon_p,\lambda_p}(z_p, w_p, \alpha_p, \beta_p)$$

such that

$$L_{m_p 1, \varepsilon_p} h_p \to 0 \quad \text{as} \quad p \to \infty.$$

$$(3.2.23)$$

Since $\displaystyle\sup_{p\in\mathbb{N}\setminus\{1,2\}} |h_p|_{m_p+E_p, n} < \infty$, we can assume, by using the Cantor diagonal procedure together with Theorem 2.1.1, that there is an $h_0 \in L^2_{\mathrm{loc}}(\mathbb{R}, \mathbb{R}^n)$ such that $\{h_p\}_3^\infty$ tends weakly to h_0 in any $L^2([-l, l], \mathbb{R}^n)$, $l \in \mathbb{N}$. Now let us fix a sufficiently large $l \in \mathbb{N}$ and let $\delta > 0$. Then for any $\delta_1 > 0$, by using the upper–semicontinuity of h_2, (3.2.3) and Lemma 3.2.7, there is a $p_0 > 2$ such that for any $\mathbb{N} \ni p > p_0$, the graph of $h_{m_p,\varepsilon_p,\lambda_p}(z_p, w_p, \alpha_p, \beta_p, \cdot)$ in the set $[-l-1, l+1] \times \mathbb{R}^k$ is in the $\zeta_{\delta_1, l}$–neighborhood of the graph of $h_2(\gamma(\cdot), 0, \cdot + \alpha_0)$. By **(H)**, the graph of $K_{m_p+E_p,\varepsilon_p^2} v_p$ in the set $[-l, l] \times \mathbb{R}^k$ is in the δ_1–neighborhood of the graph of $C(\cdot, \alpha_0)$ provided that $p > p_0$ is sufficiently large. Then according to Lemma 3.2.7, the upper–semicontinuity of h_1 and C, (3.2.21) and (3.2.3) we have that the graph of $g_{m_p,\varepsilon_p,\lambda_p}(z_p, w_p, K_{m_p+E_p,\varepsilon_p^2} v_p, \alpha_p, \beta_p, \cdot)$ in the set $[-l+1, l-1] \times \mathbb{R}^n$ is located in the δ–neighborhood $O_{\delta, l}$ of the graph $\tilde{\Gamma}$ of

$$\frac{1}{2} \sum_{i,j=1}^{d-1} \beta_{0i}\beta_{0j} D_x^2 f(\gamma(\cdot), 0)(u_{d+i}(\cdot), u_{d+j}(\cdot))$$

$$+ D_y f(\gamma(\cdot), 0)(C(\cdot, \alpha_0)) + h_1(\gamma(\cdot), 0, \cdot + \alpha_0)$$

for a fixed sufficiently small $\delta_1 > 0$ and for any large $p \in \mathbb{N}$. Hence $(t, h_p(t)) \in O_{\delta, l}$ a.e. on $[-l+1, l-1]$ for large $p \in \mathbb{N}$. By Mazur's Theorem 2.1.2 there are $\tilde{h}_p \in \mathrm{con}\,[\{h_i : i \ge p\}], \forall p \in \mathbb{N}\setminus\{1,2\}$ such that $\tilde{h}_p \to h_0$ on $L^2([-l, l], \mathbb{R}^n)$. Consequently, we may assume that $\tilde{h}_p(t) \to h_0(t)$ a.e. on $[-l, l]$. Since we can take arbitrarily small convex neighborhood of a compact convex subset of $\mathbb{R}^n$, by Lemma 3.2.8 and letting $\delta \to 0_+$, we have $(t, h_0(t)) \in \tilde{\Gamma}$ a.e. on $[-l+1, l-1]$.

By letting $l \to \infty$, we obtain

$$h_0(t) \in \Big(\frac{1}{2} \sum_{i,j=1}^{d-1} \beta_{0i}\beta_{0j} D_x^2 f(\gamma(t),0)(u_{d+i}(t), u_{d+j}(t))$$

$$+ D_y f(\gamma(t),0)(C(t,\alpha_0)) + h_1(\gamma(t),0,t+\alpha_0) \Big) \quad \text{a.e. on } \ \mathbb{R}.$$

Finally, by Theorem 3.2.1 and the properties (ii) and (iv) of Theorem 3.1.2, we have

$$\lim_{\varepsilon \to 0_+} L_{m1,\varepsilon} h = \int_{-\infty}^{\infty} P_{uu} U(t)^{-1} h(t)\, dt, \quad \lim_{\varepsilon \to 0_+} L_{m2,\varepsilon} h = 0 \qquad (3.2.24)$$

uniformly with respect to h bounded and $m \in \mathbb{Z}_+$ arbitrary. Then (3.2.23) and (3.2.24) give $\int_{-\infty}^{\infty} P_{uu} U(t)^{-1} h_0(t)\, dt = 0$. This contradicts to (i) of this theorem. $\qquad \square$

The next result is a simple consequence of Lemma 3.2.9 and the construction of $H_{m,\varepsilon}$.

Lemma 3.2.10. *There is an open small connected neighborhood $\tilde{U} \subset \mathbb{R}$ of 0 and a constant $K_1 > 0$ such that*

$$0 \notin H_{m,\varepsilon}(\partial\Omega, \lambda)$$

for any $0 < \varepsilon \in \tilde{U}$, $\lambda \in [0,1]$, $m \in \mathbb{Z}_+$, where

$$\Omega = \Big\{ (z,w,\alpha,\beta) \in Z_{m+E,n} \times Z_{m+E,k} \times \mathbb{R}^d :$$

$$\|z\|_{m+E,n} < K_1, \|w\|_{m+E,k} < K_1, (\alpha,\beta) \in \mathbf{B} \Big\}.$$

From Lemma 3.2.10 for any $0 < \varepsilon \in \tilde{U}$, $m \in \mathbb{Z}_+$, we get

$$\deg\left(H_{m,\varepsilon}(\cdot,1), \Omega, 0 \right) = \deg\left(H_{m,\varepsilon}(\cdot,0), \Omega, 0 \right),$$

where according to (3.2.19):

$$H_{m,\varepsilon}(z,w,\alpha,\beta,0) = \Big(z,w,$$
$$\Big\{ L_{m1,\varepsilon} h : h \in L^2([-m-E, m+E], \mathbb{R}^n) \text{ satisfying a.e. on } [-m-E, m+E]$$
$$\text{the relation} \quad h(t) \in \Big(\frac{1}{2} \sum_{i,j=1}^{d-1} \beta_i\beta_j D_x^2 f(\gamma(t),0)(u_{d+i}(t), u_{d+j}(t))$$
$$+ D_y f(\gamma(t),0)(C(t,\alpha)) + h_1(\gamma(t),0,t+\alpha)) \Big\} \Big).$$

In order to compute $\deg\left(H_{m,\varepsilon}(\cdot,0), \Omega, 0 \right)$, we consider the homotopy

$$\Big\{ \lambda L_{m1,\varepsilon} h + (1-\lambda) \int_{-\infty}^{\infty} P_{uu} U(t)^{-1} h(t)\, dt : h \in L^2_{loc}(\mathbb{R}, \mathbb{R}^n) \text{ satisfying a.e. on } \mathbb{R}$$
$$\text{the relation} \quad h(t) \in \Big(\frac{1}{2} \sum_{i,j=1}^{d-1} \beta_i\beta_j D_x^2 f(\gamma(t),0)(u_{d+i}(t), u_{d+j}(t))$$
$$+ D_y f(\gamma(t),0)(C(t,\alpha)) + h_1(\gamma(t),0,t+\alpha)) \Big\}.$$

Using (3.2.24) with Lemma 3.2.9, when ε_0 is shrunk if necessary, we derive

$$\deg\left(H_{m,\varepsilon}(\cdot,0),\Omega,0\right) = \deg(M,\mathbf{B},0) \neq 0.$$

Consequently (3.2.19) has a solution in Ω for any $0 < \varepsilon \in U_1$, $m \in \mathbb{Z}_+$ and $\lambda = 1$, where $U_1 = \tilde{U}_1 \cap \{\varepsilon \in \mathbb{R} : 0 < \varepsilon < \varepsilon_0\}$. This solution is a solution of (3.2.16) according to the definition of (3.2.19). □

For $d = 1$ we have $M : \mathbb{R} \to 2^{\mathbb{R}} \setminus \{\emptyset\}$ with

$$M(\alpha) = \left\{ \int_{-\infty}^{\infty} \langle h(s), u_1^{\perp}(s)\rangle_n \, ds : h \in L^2_{\text{loc}}(\mathbb{R},\mathbb{R}); \right.$$

$$\left. h(t) \in D_y f(\gamma(t),0)(C(t,\alpha)) + h_1(\gamma(t),0,t+\alpha) \quad \text{a.e. on} \quad \mathbb{R} \right\}.$$

$$(3.2.25)$$

Theorem 3.2.11. *Let $d = 1$. If there are constants $a < b$ such that $M(a)M(b) \subset (-\infty,0)$, then there are constants $K > 0$ and $\varepsilon_0 > 0$ such that for any $0 < \varepsilon < \varepsilon_0$, the differential inclusion (3.2.4) possesses a subharmonic solution (x_m,y_m) of period $2m$ for any $m \in \mathbb{N}$, $m \geq [1/\sqrt{\varepsilon}]$ satisfying*

$$\sup_{-m \leq t \leq m} |y_m(t)|_k \leq K\varepsilon, \qquad \sup_{-m \leq t \leq m} \left|x_m(t) - \gamma(t - \alpha_m)\right|_n \leq K\varepsilon,$$

where $\alpha_m \in (a,b)$.

Proof. We use Theorem 3.2.6 with $\mathbf{B} = (a,b)$: assumption (i) is clearly satisfied. For showing (ii), we can directly repeat arguments in the proof of Theorem 3.1.9. The proof is finished. □

Remark 3.2.12. When $d = 1$ and h_1, h_2 are C^2–smooth in all variables, then according to Theorem 3.2.4, we can take

$$C(t,\alpha) = \left\{ - A^{-1}(\gamma(t))h_2(\gamma(t),0,t+\alpha) \right\}$$

and then $M \in C^1(\mathbb{R},\mathbb{R})$. A simple root α_0 of $M(\alpha) = 0$ implies chaos for such systems with subharmonic solutions of all large periods. This is mentioned in Subsection 3.2.1. But the existence of a simple root α_0 of $M(\alpha) = 0$ implies also the validity of the assumption of Theorem 3.2.11 with $a = \alpha_0 - \tilde{\delta}$ and $b = \alpha_0 + \tilde{\delta}$ for $\tilde{\delta} > 0$ small. Consequently, Theorem 3.2.11 is a generalization of results in [28,72] concerning subharmonics to the multivalued case (3.2.4) (see also Subsection 4.2.1).

Finally we note that like in Subsection 3.1.7 periodic solutions of (3.2.4) found above, accumulate on bounded solutions on $\mathbb{R}$ of (3.2.4), i.e. repeating the proof of Theorem 3.1.33 we can derive the following

Theorem 3.2.13. *The assumptions of Theorems 3.2.6 and 3.2.11 imply for any $\varepsilon > 0$ sufficiently small, the existence of a solution $(x_\varepsilon, y_\varepsilon)$ of (3.2.4) on $\mathbb{R}$ satisfying*

either $\quad \sup_{t \in \mathbb{R}} \left| x_\varepsilon(t) - \gamma(t - \alpha_\varepsilon) \right|_n \leq K\sqrt{\varepsilon} \quad$ *for Theorem 3.2.6*

or $\quad \sup_{t \in \mathbb{R}} \left| x_\varepsilon(t) - \gamma(t - \alpha_\varepsilon) \right|_n \leq K\varepsilon \quad$ *for Theorem 3.2.11 and* $\quad$ (3.2.26)

$\quad \sup_{t \in \mathbb{R}} \left| y_\varepsilon(t) \right|_k \leq K\varepsilon \quad$ *in the both Theorems 3.2.6 and 3.2.11,*

where $\alpha_\varepsilon \in \mathbb{R}$, $|\alpha_\varepsilon| \leq K$ and K is a constant from Theorem 3.2.6, respectively Theorem 3.2.11. Moreover, a subsequence of the subharmonics $\{(x_m, y_m)\}_{m \geq [1/\varepsilon]}$ from Theorem 3.2.6, respectively Theorem 3.2.11, accumulates on $(x_\varepsilon, y_\varepsilon)$.

3.2.4   Applications to Singular Discontinuous O.D.Eqns

As a first application of the above results we consider (3.2.3) with $\widetilde{k} = 1$, for simplicity, and $q \in C^2(\mathbb{R}^2, \mathbb{R}^2)$, $p \in C^1(\mathbb{R}, \mathbb{R}^2)$ and $p(t+2) = p(t)$, $\forall t \in \mathbb{R}$. To get the form of (3.2.4), we exchange the variables

$$y \leftrightarrow y + q(x), \quad x \leftrightarrow x, \quad t \leftrightarrow -t,$$

and then (3.2.3) possesses the form

$$\dot{x} = -y + q(x)$$
$$m\dot{y} \in y + m\big(\mu \operatorname{Sgn}(y - q(x) + v_0) - p(-t) - D_x q(x)(y - q(x))\big) \quad (3.2.27)$$

with

$$f(x,y) = -y + q(x), \quad h_1 = 0, \quad g(x,y) = y, \quad A(x) = \mathbb{I}, \quad \varepsilon = m$$
$$h_2(x,y,t) = \mu \operatorname{Sgn}(y - q(x) + v_0) - p(-t) - D_x q(x)(y - q(x))$$

in the notation of (3.2.4). Moreover we assume $\dot{x} = q(x)$ has a homoclinic solution γ to the hyperbolic fixed point $x = 0$. Then

$$h_2(\gamma(t), 0, t + \alpha) = \mu \operatorname{Sgn}(-\dot{\gamma}(t) + v_0) - p(-t - \alpha) + D_x q(\gamma(t)) q(\gamma(t))$$
$$= \mu \operatorname{Sgn}(-\dot{\gamma}(t) + v_0) - p(-t - \alpha) + \ddot{\gamma}(t).$$

Lemma 3.2.14. *If there is only a finite sequence of numbers $-\infty < t_1 < t_2 < \cdots < t_i < \infty$ such that*

$$-\dot{\gamma}(t) + v_0 = 0 \quad \text{if and only if} \quad t = t_1, \cdots, t_i,$$

and $\quad \displaystyle\sup_{[t_1-1, t_i+1] \ni t \neq t_1, \cdots, t_i} \left| \frac{d}{dt} \big(\operatorname{sgn}(-\dot{\gamma}(t) + v_0) \big) \right|_2 < \infty,$ $\quad$ (3.2.28)

then we can take in the condition **(H)**

$$C(t, \alpha) = \begin{cases} \big\{ -\mu \operatorname{sgn}(-\dot{\gamma}(t) + v_0) + p(-t - \alpha) - \ddot{\gamma}(t) \big\} \\ \qquad \text{for} \quad t \neq t_1, \cdots, t_i, \\ \big\{ v \in \mathbb{R}^2 : |v|_2 \leq \max_{t \in [0,2]} |p(t)|_2 + \mu + 2 + \sup_{t \in \mathbb{R}} |\ddot{\gamma}(t)|_2 \big\} \\ \qquad \text{for} \quad t = t_1, \cdots, t_i. \end{cases}$$

Proof. Let $1 > \delta > 0$ be small and $l \in \mathbb{N}$ satisfying $l > 3 + \max\{|t_j| \mid j = 1, \cdots, i\}$. To verify **(H)**, we take $h \in Y_{m,2}$ such that

$$(t, h(t)) \in \left\{ u \in \mathbb{R}^3 : \exists s \in \mathbb{R}; \operatorname{dist}\left\{ u, \left(s, h_2(\gamma(s), 0, s + \alpha)\right)\right\} < \delta \right\}$$

$$\text{a.e. on} \quad [-l-1, l+1] \tag{3.2.29}$$

$$|h|_{m,2} \leq K_1, \quad K_1 = \max_{t \in [0,2]} |p(t)|_2 + \mu + 2 + \sup_{t \in \mathbb{R}} |\ddot{\gamma}(t)|_2$$

for $\mathbb{N} \ni m > l$. According to (3.2.28), $h_2(\gamma(t), 0, t + \alpha)$, $t \neq t_1, \cdots, t_i$ satisfies the assumptions of Corollary 3.2.5. So for any sufficiently small $1 \geq \varepsilon > 0$, we take $h_{2,\varepsilon}$ from this corollary such that

$$|h(t) - h_{2,\varepsilon}(t)|_2 < \delta \quad \text{a.e. on} \quad [-l-1, l+1] \setminus \cup_{j=1}^{i} [t_j - \delta, t_j + \delta],$$

$$|h - h_{2,\varepsilon}|_{m,2} \leq 2K_1, \quad \text{where} \quad m > \max\{m_\delta, l + 2\}.$$

We note that (3.2.9) has now the form (3.2.11) with $D_1 = \mathbb{I}$, so $Z_\varepsilon(t) = e^{t/\varepsilon}\mathbb{I}$. We have for $t \in [-l, l]$ the following cases:

If $t_i + 2\delta \leq t \leq l$, then

$$\frac{1}{\varepsilon} \int_t^{l+1} e^{-(s-t)/\varepsilon} |h(s) - h_{2,\varepsilon}(s)|_2 \, ds \leq \delta.$$

If $t_j - 2\delta \leq t \leq t_j + 2\delta$ for some $j = 1, \cdots, i$, then

$$\frac{1}{\varepsilon} \int_t^{l+1} e^{-(s-t)/\varepsilon} |h(s)|_2 \, ds \leq K_1.$$

If $t_j + 2\delta \leq t \leq t_{j+1} - 2\delta$ for some $j = 1, \cdots, i-1$, then

$$\frac{1}{\varepsilon} \int_t^{l+1} e^{-(s-t)/\varepsilon} |h(s) - h_{2,\varepsilon}(s)|_2 \, ds = \frac{1}{\varepsilon} \int_t^{t_{j+1}-\delta} e^{-(s-t)/\varepsilon} |h(s) - h_{2,\varepsilon}(s)|_2 \, ds$$

$$+ \frac{1}{\varepsilon} \int_{t_{j+1}-\delta}^{l+1} e^{-(s-t)/\varepsilon} |h(s) - h_{2,\varepsilon}(s)|_2 \, ds \leq \delta + 2K_1 e^{-(t_{j+1}-\delta-t)/\varepsilon}$$

$$\leq \delta + 2K_1 e^{-\delta/\varepsilon}.$$

If $-l \leq t \leq t_1 - 2\delta$, then

$$\frac{1}{\varepsilon} \int_t^{l+1} e^{-(s-t)/\varepsilon} |h(s) - h_{2,\varepsilon}(s)|_2 \, ds = \frac{1}{\varepsilon} \int_t^{t_1-\delta} e^{-(s-t)/\varepsilon} |h(s) - h_{2,\varepsilon}(s)|_2 \, ds$$

$$+ \frac{1}{\varepsilon} \int_{t_1-\delta}^{l+1} e^{-(s-t)/\varepsilon} |h(s) - h_{2,\varepsilon}(s)|_2 \, ds \leq \delta + 2K_1 e^{-(t_1-\delta-t)/\varepsilon}$$

$$\leq \delta + 2K_1 e^{-\delta/\varepsilon}.$$

On the other hand,

$$\frac{1}{\varepsilon} \int_{l+1}^{m} e^{-(s-t)/\varepsilon} |h(s) - h_{2,\varepsilon}(s)|_2 \, ds \leq e^{-(l+1-t)/\varepsilon} 2K_1 \leq 2K_1 e^{-\delta/\varepsilon}$$

for $t \leq l$.

Furthermore, according to (3.2.13), the proof of Theorem 3.2.3 and the above estimates, we know that

$$|K_{m,\varepsilon} h(t) - K_{m,\varepsilon} h_{2,\varepsilon}(t)|_2 \leq 2 e^{-(m-l)/\varepsilon} (1 - e^{-2})^{-1} K_1 + \delta + 4K_1 e^{-\delta/\varepsilon}$$

for $t \in [-l, l] \setminus \cup_{j=1}^{i} [t_j - 2\delta, t_j + 2\delta]$, and

$$|K_{m,\varepsilon} h(t)|_2 \leq 2 e^{-(m-l)/\varepsilon} (1 - e^{-2})^{-1} K_1 + K_1 + 2K_1 e^{-\delta/\varepsilon} \qquad (3.2.30)$$

for $t \in \cup_{j=1}^{i} [t_j - 2\delta, t_j + 2\delta]$. Then Corollary 3.2.5 gives

$$|K_{m,\varepsilon} h(t) + h(t)|_2 \leq |K_{m,\varepsilon} h(t) - K_{m,\varepsilon} h_{2,\varepsilon}(t)|_2 + |K_{m,\varepsilon} h_{2,\varepsilon}(t)(t) + h_{2,\varepsilon}(t)(t)|_2$$

$$+ |h_{2,\varepsilon}(t)(t) - h(t)(t)|_2 \leq 3\delta + 2 e^{-(m-l)/\varepsilon} (1 - e^{-2})^{-1} K_1 + 4K_1 e^{-\delta/\varepsilon}$$
$$(3.2.31)$$

for $t \in [-l, l] \setminus \cup_{j=1}^{i} [t_j - 2\delta, t_j + 2\delta]$ and $m > \max\{m_\delta, l+2\}$. Now it follows from (3.2.30)–(3.2.31) by considering sufficiently small $\varepsilon > 0$ and sufficiently large $m > l$ that

$$|K_{m,\varepsilon} h(t) + h(t)|_2 \leq 4\delta \text{ for } t \in [-l, l] \setminus \cup_{j=1}^{i} [t_j - 2\delta, t_j + 2\delta],$$
$$|K_{m,\varepsilon} h(t)|_2 \leq K_1 + \delta \text{ for } t \in \cup_{j=1}^{i} [t_j - 2\delta, t_j + 2\delta]. \qquad (3.2.32)$$

Recalling (3.2.29) together with (3.2.32) the Lemma 3.2.14 is proved.   □

The variational equation (3.2.7) is now $\dot{u}(t) = D_x q(\gamma(t)) u(t)$. Clearly $d = 1$ and according to [157, p. 253], we can take

$$u_1^{\perp}(t) = e^{-\int_0^t \left(\frac{\partial q_1}{\partial x_1}(\gamma(s)) + \frac{\partial q_2}{\partial x_2}(\gamma(s)) \right) ds} (\dot{\gamma}_2(t), -\dot{\gamma}_1(t)),$$

where $q = (q_1, q_2)$ and $\gamma = (\gamma_1, \gamma_2)$. Then (3.2.25) has now the form

$$M(\alpha) = \int_{-\infty}^{\infty} \left\langle -\mathbb{I}(p(-s - \alpha) - \mu\,\mathrm{sgn}\,(-\dot\gamma(t) + v_0) - \ddot\gamma(s), u_1^{\perp}(s) \right\rangle_2 ds$$

$$= \int_{-\infty}^{\infty} \left\langle \ddot\gamma(s) + \mu\frac{-\dot\gamma(s) + v_0}{|-\dot\gamma(s) + v_0|_2}, u_1^{\perp}(s) \right\rangle_2 ds - \int_{-\infty}^{\infty} \left\langle p(-s - \alpha), u_1^{\perp}(s) \right\rangle_2 ds .$$

$$(3.2.33)$$

The following result follows immediately from Theorems 3.2.11 and 3.2.13.

Theorem 3.2.15. *If the assumptions of Lemma 3.2.14 are satisfied and there are constants $a < b$ such that $M(a)M(b) < 0$, where M is given by (3.2.33), then for any sufficiently small $m > 0$, (3.2.2) possesses a $2i$–periodic solution near $(\gamma, \dot\gamma)$ for all $\mathbb{N} \ni i \geq [\sqrt{1/m}\,]$. Moreover, a subsequence of these solutions accumulates as $i \to \infty$ on a bounded solution on $\mathbb{R}$ of (3.2.2) which is also near $(\gamma, \dot\gamma)$.*

The next result is useful for verifying assumptions of (3.2.28).

Proposition 3.2.16. *If $-\dot\gamma(t_1) + v_0 = 0$, $\ddot\gamma(t_1) \neq 0$ then t_1 is an isolated solution of $-\dot\gamma(t) + v_0 = 0$ and moreover,*

$$\lim_{t \to t_{1\pm}} \frac{-\dot\gamma(t) + v_0}{|-\dot\gamma(t) + v_0|_2} = \mp\frac{\ddot\gamma(t_1)}{|\ddot\gamma(t_1)|_2}, \quad \lim_{t \to t_{1\pm}} \frac{d}{dt}\left(\frac{-\dot\gamma(t) + v_0}{|-\dot\gamma(t) + v_0|_2}\right) =$$

$$\frac{\pm 1}{2|\ddot\gamma(t_1)|_2^3}\left(-\dddot\gamma(t_1)\langle\ddot\gamma(t_1), \ddot\gamma(t_1)\rangle_2 + \ddot\gamma(t_1)\langle\dddot\gamma(t_1), \ddot\gamma(t_1)\rangle_2\right).$$

Proof. Since $\gamma \in C^3(\mathbb{R}, \mathbb{R}^2)$, we get

$$-\dot\gamma(t) + v_0 = -\ddot\gamma(t_1)(t - t_1) - \dddot\gamma(t_1)\frac{(t - t_1)^2}{2} + o(t - t_1)^2 .$$

So we have

$$\frac{-\dot\gamma(t) + v_0}{|-\dot\gamma(t) + v_0|_2} = \frac{-\ddot\gamma(t_1)(t - t_1) - \dddot\gamma(t_1)\frac{(t-t_1)^2}{2} + o(t - t_1)^2}{|-\ddot\gamma(t_1)(t - t_1) - \dddot\gamma(t_1)\frac{(t-t_1)^2}{2} + o(t - t_1)^2|_2}$$

$$= \frac{-\ddot\gamma(t_1) + o(1)}{|-\ddot\gamma(t_1) + o(1)|_2}\frac{t - t_1}{|t - t_1|} \to \mp\frac{\ddot\gamma(t_1)}{|\ddot\gamma(t_1)|_2} \quad \text{as} \quad t \to t_{1\pm} .$$

Similarly we obtain the second limit. $\qquad\square$

Remark 3.2.17. Supposing $\int_{-\infty}^{\infty} \left\langle \frac{-\dot\gamma(s) + v_0}{|-\dot\gamma(s) + v_0|_2}, u_1^{\perp}(s) \right\rangle_2 ds \neq 0$, Theorem 3.2.15 is applicable only for non–large μ, because of then there is a $\mu_0 > 0$ such that for any $\mu > \mu_0$ and $a < b$ it holds $M(a)M(b) > 0$. On the other hand, if $\sup |q(\cdot)|_2 < \infty$ and $v_0 \neq 0$ then for $\mu > \left(\sup |q(\cdot)|_2 + m\,\max |p(\cdot)|_2\right)/m,$

(3.2.1) has no periodic solutions: if x is a $2i$–periodic solution of (3.2.1) for some $i \in \mathbb{R}$ then

$$0 \le \tilde{k} \int\limits_0^{2i} |\dot{x}(s)|_2^2 \, ds = \int\limits_0^{2i} \Big(\langle -q(x(s)) + mp(s), \, \mathrm{sgn}\,(\dot{x}(s) + v_0) \rangle_2$$

$$-\mu m \Big) |\dot{x}(s) + v_0|_2 \, ds \le 0 \,.$$

This implies $\int\limits_0^{2i} |\dot{x}(s)|_2^2 \, ds = \int\limits_0^{2i} |\dot{x}(s) + v_0|_2^2 \, ds = 0$, which is impossible.

Now we consider the following more simple tractable, coupled equations

$$\ddot{z} + 2z^3 - z = \dot{y}$$
$$\varepsilon \ddot{y} + \dot{y} + y + \varepsilon \, \mathrm{sgn}\, \dot{y} = \varepsilon \mu \sin t \,, \tag{3.2.34}$$

where $z, y \in \mathbb{R}$ and $\mu \in \mathbb{R}$ is a parameter. Rewriting again (3.2.34) in the form

$$\dot{x}_1 = x_2, \quad \dot{x}_2 = x_1 - 2x_1^3 + y - x_3$$
$$\dot{x}_3 = y - x_3, \quad \varepsilon \dot{y} = -y + \varepsilon \big(\mu \sin t + y - x_3 - \mathrm{sgn}\,(y - x_3) \big) \,,$$

we put

$$x = (x_1, x_2, x_3), \quad f(x,y) = \big(x_2, x_1 - 2x_1^3 + y - x_3, y - x_3\big), \quad A(x) = -\mathbb{I}$$
$$g(x,y) = -y, \quad h_1(x,y,t) = 0, \quad h_2(x,y,t) = \mu \sin t + y - x_3 - \mathrm{Sgn}\,(y - x_3) \,.$$

Since the reduced equation is

$$\dot{x}_1 = x_2, \quad \dot{x}_2 = x_1 - 2x_1^3 - x_3, \quad \dot{x}_3 = -x_3 \,,$$

we have

$$\gamma(t) = (r(t), \dot{r}(t), 0), \quad r(t) = \mathrm{sech}\, t, \quad u_1^{\perp}(t) = (-\ddot{r}(t), \dot{r}(t), 0), \quad d = 1$$
$$h_2(\gamma(t), 0, t + \alpha) = \big[-1 + \mu \sin(t + \alpha), 1 + \mu \sin(t + \alpha) \big] \,.$$

To find $C(t, \alpha)$, we consider (3.2.9) for this case of the form

$$\varepsilon \dot{y} = -y + h, \quad h \in Y_{m,1}, \quad |h|_{m,1} \le 2 + \mu, \quad \mathbb{N} \ni l < m - 1$$
$$h(t) \in \big[-1 - \zeta + \mu \sin(t + \alpha), 1 + \zeta + \mu \sin(t + \alpha) \big] \text{ a.e. on } [-l - 1, l + 1] \,. \tag{3.2.35}$$

Putting in (3.2.35) $y = z + \mu \sin(t + \alpha)$, $h = w + \mu \sin(t + \alpha)$ and taking $t \leftrightarrow -t$, we have

$$\varepsilon \dot{z} = z + \varepsilon \mu \cos(-t + \alpha) - w(-t), \quad w \in Y_{m,1}, \quad |w|_{m,1} \le 2 + 2\mu$$
$$w(t) \in [-1 - \zeta, 1 + \zeta] \quad \text{a.e. on} \quad [-l - 1, l + 1] \,.$$

Following the proof of Theorem 3.2.3 (see (3.2.13)) and Lemma 3.2.14, we obtain

$$\|z\|_{l,1} \le \Big(\frac{e^{-(m-l)/\varepsilon}}{1 - e^{-2}} + e^{-1/\varepsilon} \Big) (2 + 2\mu + \varepsilon \mu) + 1 + \zeta + \varepsilon \mu \,.$$

Consequently, we can take in **(H)** that

$$C(t, \alpha) = \left[-1 + \mu \sin (t + \alpha), 1 + \mu \sin (t + \alpha) \right],$$

and then (3.2.25) possesses the form

$$
\begin{aligned}
M(\alpha) = \Bigg[& \int_{-\infty}^{0} (\mu \sin (s + \alpha) - 1)\dot{r}(s)\, ds + \int_{0}^{\infty} (\mu \sin (s + \alpha) + 1)\dot{r}(s)\, ds, \\
& \int_{-\infty}^{0} (\mu \sin (s + \alpha) + 1)\dot{r}(s)\, ds + \int_{0}^{\infty} (\mu \sin (s + \alpha) - 1)\dot{r}(s)\, ds \Bigg] \\
= & \left[-\mu\pi \operatorname{sech} \tfrac{\pi}{2} \cos \alpha - 2, -\mu\pi \operatorname{sech} \tfrac{\pi}{2} \cos \alpha + 2 \right].
\end{aligned}
$$

$$(3.2.36)$$

By Theorems 3.2.11, 3.2.13 with $a = 0$, $b = \pi$ and M is given by (3.2.36), we obtain the following result.

Theorem 3.2.18. *If $|\mu| > \frac{2}{\pi} \cosh \frac{\pi}{2} \doteq 1.5973925$ is fixed, then (3.2.34) possesses for any sufficiently small $\varepsilon > 0$ a $2\pi i$–periodic solution for all $\mathbb{N} \ni i \geq [1/\sqrt{\varepsilon}]$. Moreover, a subsequence of these solutions accumulates as $i \to \infty$ on a bounded solution on $\mathbb{R}$ of (3.2.34).*

Finally, our method is clearly applied like in Remark 3.1.32 to piecewise smoothly and singularly perturbed problems. For instance, for

$$
\begin{aligned}
\ddot{z} + 2z^3 - z + \varepsilon \dot{z}^+ &= \dot{y} \\
\varepsilon \ddot{y} + \dot{y} + y &= \varepsilon \mu \sin t,
\end{aligned}
$$

$$(3.2.37)$$

we find that (3.2.25) now has the form

$$M(\alpha) = \mu \int_{-\infty}^{\infty} \sin (s + \alpha)\dot{r}(s)\, ds - \int_{-\infty}^{0} \dot{r}(s)^2 \, ds = -\mu\pi \operatorname{sech} \frac{\pi}{2} \cos \alpha - \frac{1}{3}.$$

Consequently, by Theorems 3.2.11 and 3.2.13, we obtain the following result.

Theorem 3.2.19. *If $|\mu| > \frac{1}{3\pi} \cosh \frac{\pi}{2} \doteq 0.266232$ is fixed, then (3.2.37) possesses for any sufficiently small $\varepsilon > 0$ a $2\pi i$–periodic solution for all $\mathbb{N} \ni i \geq [1/\sqrt{\varepsilon}]$. Moreover, a subsequence of these solutions accumulates as $i \to \infty$ on a bounded solution on $\mathbb{R}$ of (3.2.37).*

The method of this Section is extended in Chapter 4 to show chaotic behavior of (3.2.4).

3.3 Bifurcation of Periodics from Periodics

3.3.1 Discontinuous O.D.Eqns

In this section we continue with the study of bifurcations of periodic solutions for ordinary differential equations with discontinuous periodic perturbations. But

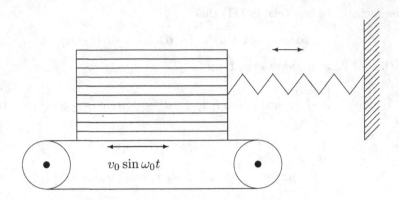

Figure 3.5: A block on a periodically moving ribbon under a force

now we investigate the problem of bifurcations of periodics from periodics, while
in the previous two Sections 3.1 and 3.2 bifurcations of subharmonics are studied
with very large periods from homoclinics. For motivation, we again consider a
mass attached to a spring and putting horizontally on a moving ribbon with a
speed $v_0 \sin \omega t$ (see Fig. 3.5). The resulting differential equation has the form

$$\ddot{x} + q(x) + \mu \operatorname{sgn}\left(\dot{x} + v_0 \sin \omega t\right) = 0 \,, \tag{3.3.1}$$

where $q \in C^2(\mathbb{R}, \mathbb{R})$ and $\mu > 0$, $v_0 > 0$, $\omega > 0$ are constants. Like in the previous
sections, discontinuous equation (3.3.1) is considered as a perturbed differential
inclusion of the form

$$\dot{x} = y, \quad \dot{y} \in -q(x) - \mu \operatorname{Sgn}\left(y + v_0 \sin \omega t\right). \tag{3.3.2}$$

By assuming the existence of a $2\pi/\omega$–periodic solution γ of $\ddot{x} + q(x) = 0$, we study
bifurcation of $2\pi/\omega$–periodic solutions for (3.3.2) from the 1-parametric family
$\{\gamma(t + \theta) \mid \theta \in \mathbb{R}\}$. Of course, we consider more general systems of perturbed
differential inclusions than (3.3.2) which take the form

$$\dot{x}(t) \in f(x(t)) + \sum_{j=1}^{k} \mu_j f_j(x(t), \mu, t) \quad \text{a.e. on} \quad \mathbb{R} \tag{3.3.3}$$

with $x \in \mathbb{R}^n$, $\mu \in \mathbb{R}^k$, $\mu = (\mu_1, \cdots, \mu_k)$. Motivated by (3.3.2), we set the main
assumptions about (3.3.3):

(i) $f \in C^2(\mathbb{R}^n, \mathbb{R}^n)$ and $f_j : \mathbb{R}^n \times \mathbb{R}^k \times \mathbb{R} \to 2^{\mathbb{R}^n} \setminus \{\emptyset\}$, $j = 1, \cdots, k$ are all
upper–semicontinuous with compact and convex values.

(ii) The unperturbed equation $\dot{x} = f(x)$ has a *manifold of 1–periodic solutions*,
i.e. there is an open subset $\mathcal{O} \subset \mathbb{R}^{d-1}$, $d \geq 1$ and a C^2–mapping $\gamma :$
$\mathcal{O} \times \mathbb{R} \to \mathbb{R}^n$ such that $\gamma(\theta, t + 1) = \gamma(\theta, t)$ and $\gamma(\theta, \cdot)$ is a solution of
$\dot{x} = f(x)$.

(iii) $f_j(x, \mu, t+1) = f_j(x, \mu, t)$ for $j = 1, \cdots, k$.

We investigate, if some of these periodic solutions $\gamma(\theta, t)$ persists after perturbation (3.3.3). The case when f_j are all singlevalued and smooth is a classical problem of bifurcation of periodic solutions for o.d.eqns and we refer the reader to [55, 151] for more details. The purpose of this section is to extend some of those results to the multivalued case (3.3.3).

3.3.2 Linearized Problem

In order to study bifurcations of 1-periodic solutions for (3.3.3) from the family $\gamma(\theta, t)$, $\theta \in \mathcal{O}$, first we consider the non–homogeneous variational equation

$$\dot{x} = A(\theta, t)x + h(t) \quad \text{a.e. on} \quad [0, 1],$$
$$A(\theta, t) = D_x f(\gamma(\theta, t)), \quad h \in L^2 := L^2([0, 1], \mathbb{R}^n), \qquad (3.3.4)$$
$$x \in C_p = \{y \in C([0, 1], \mathbb{R}^n) : y(0) = y(1)\}$$

along with the homogeneous one

$$\dot{x} = A(\theta, t)x, \quad x \in C_p. \qquad (3.3.5)$$

Note (3.3.5) is just the linearization of the unperturbed equation $\dot{x} = f(x)$ of (3.3.3) along $\gamma(\theta, t)$, $\theta \in \mathcal{O}$. The aim of this subsection is to derive a Fredholm-like alternative result for (3.3.4) in order to apply the Lyapunov-Schmidt decomposition in the next subsections to (3.3.3). The first step towards this is the differentiation with respect to θ and t of $\dot{\gamma}(\theta, t) = f(\gamma(\theta, t))$ which yields respectively

$$\frac{\partial}{\partial \theta_i} \dot{\gamma}(\theta, t) = A(\theta, t) \frac{\partial}{\partial \theta_i} \gamma(\theta, t), \quad \ddot{\gamma}(\theta, t) = A(\theta, t)\dot{\gamma}(\theta, t),$$

where $\theta = (\theta_1, \cdots, \theta_{d-1})$. We see that (3.3.5) has "trivial" 1-periodic solutions $\frac{\partial}{\partial \theta_i} \gamma(\theta, t)$, $i = 1, \cdots, d-1$ and $\dot{\gamma}(\theta, t)$. We assume that the following condition is satisfied:

(iv) The family $\{\gamma(\theta, t) \mid \theta \in \mathcal{O}\}$ is *non-degenerate*, i.e. the only 1-periodic solutions of $\dot{x} = A(\theta, \cdot)x$ are linear combinations of $\frac{\partial}{\partial \theta_i} \gamma(\theta, t)$, $i = 1, \cdots, d-1$ and $\dot{\gamma}(\theta, t)$. Moreover, $\frac{\partial}{\partial \theta_i} \gamma(\theta, t)$, $i = 1, \cdots, d-1$ and $\dot{\gamma}(\theta, t)$ are linearly independent.

Let $U(\theta, t)$ be the fundamental solution of $\dot{x} = A(\theta, t)x$. Then

$$\mathcal{N}(\mathbb{I} - U(\theta, 1)) = \text{span} \left\{ \frac{\partial}{\partial \theta_i} \gamma(\theta, 0), \ \dot{\gamma}(\theta, 0), \ i = 1, \cdots, d-1 \right\} \qquad (3.3.6)$$

and $\mathcal{R}(\mathbb{I} - U(\theta, 1))$ is a continuous trivial vector bundle over $\mathcal{O}$. The adjoint equation to (3.3.5) has the form

$$\dot{x} = -A(\theta, t)^* x, \quad x \in C_p \qquad (3.3.7)$$

with the fundamental solution $(U^{-1}(\theta, t))^*$. Then any 1-periodic solution of (3.3.7) has the form $(U^{-1}(\theta, t))^* y$, $y \in \mathbb{R}^n$ with $(U^{-1}(\theta, 1))^* y = y$. Since

$$(U^{-1}(\theta, 1))^* y = y \iff U(\theta, 1)^* y = y, \tag{3.3.8}$$

we get

$$\mathcal{N}(\mathbb{I} - (U^{-1}(\theta, 1))^*) = \mathcal{N}(\mathbb{I} - U(\theta, 1)^*) = \left(\mathcal{R}(\mathbb{I} - U(\theta, 1))\right)^{\perp}.$$

According to (3.3.6), $\left(\mathcal{R}(\mathbb{I} - U(\theta, 1))\right)^{\perp}$ is a continuous trivial vector bundle over $\mathcal{O}$. So taking its basis, there are linearly independent continuous mappings $v_i(\theta, t)$, $i = 1, \cdots, d$, $v_i : \mathcal{O} \times \mathbb{R} \to \mathbb{R}^n$, $v_i(\theta, t + 1) = v_i(\theta, t)$ and all v_i are solutions of (3.3.7). Let $\langle \cdot, \cdot \rangle$ be the inner product on $\mathbb{R}^n$ and let $\Pi(\theta) : L^2 \to L^2$ be a projection onto the subspace

$$\left\{ h \in L^2 \mid \int\limits_0^1 \langle h(s), v_i(\theta, s) \rangle \, ds = 0, \ \forall i = 1, \cdots, d \right\}$$

depending continuously on θ in $L(L^2)$ and having uniformly bounded $\|\Pi(\theta)\|$ on any bounded subset of $\mathcal{O}$. Such a projection exists, for instance the orthogonal one. Now we can state the following well–known Fredholm alternative result [110, p. 411].

Lemma 3.3.1. *System (3.3.4) has a solution if and only if* $(\mathbb{I} - \Pi(\theta))h = 0$. *This solution x is unique provided that it satisfies* $\int_0^1 \langle x(s), \frac{\partial}{\partial \theta_i} \gamma(\theta, s) \rangle \, ds = 0$ *for all $i = 1, \cdots, d-1$ along with* $\int_0^1 \langle x(s), \dot{\gamma}(\theta, s) \rangle \, ds = 0$. *Moreover, let $K(\theta) : L^2 \to C_p$ be the linear operator defined so that $x = K(\theta)h$ be the unique solution of the linear system*

$$\dot{x} = A(\theta, t)x + \Pi(\theta)h. \tag{3.3.9}$$

Then $K(\theta) : L^2 \to C_p$ is a compact linear operator depending continuously on θ.

Proof. We derive a parameterized Green formula for $K(\theta)$ [110]. The orthogonal projection $\Pi(\theta) : L^2 \to L^2$ is given by the formula

$$\Pi(\theta)h = h - \sum_{i=1}^{d} c_i(\theta) v_i(\theta, \cdot), \tag{3.3.10}$$

where $c_i(\theta)$ solve the linear system

$$\sum_{i=1}^{d} c_i(\theta) \int\limits_0^1 \langle v_i(\theta, s), v_j(\theta, s) \rangle \, ds = \int\limits_0^1 \langle h(s), v_j(\theta, s) \rangle \, ds, \quad j = 1, 2, \cdots, d.$$

$$\tag{3.3.11}$$

The Gram matrix

$$G(\theta) = \left(\int\limits_0^1 \langle v_i(\theta, s), v_j(\theta, s) \rangle \, ds \right)_{i,j=1}^d$$

is invertible, so (3.3.11) implies

$$(c_1(\theta), \cdots, c_d(\theta))^* = G^{-1}(\theta) \left(\int\limits_0^1 \langle h(s), v_1(\theta, s) \rangle \, ds, \cdots, \int\limits_0^1 \langle h(s), v_d(\theta, s) \rangle \, ds \right)^*$$

$$= \int\limits_0^1 G^{-1}(\theta) \left(\langle h(s), v_1(\theta, s) \rangle, \cdots, \langle h(s), v_d(\theta, s) \rangle \right)^* \, ds.$$

This gives by (3.3.10) the formula of $\Pi(\theta)$:

$$(\Pi(\theta)h)(t) = h(t) - \int_0^1 M_1(\theta, t, s)h(s) \, ds, \tag{3.3.12}$$

where

$$M_1(\theta, t, s)h := (v_1(\theta, t)^*, \cdots, v_d(\theta, t)^*) \, G^{-1}(\theta) \left(\langle h, v_1(\theta, s) \rangle, \cdots, \langle h, v_d(\theta, s) \rangle \right)^*$$

is a continuous $n \times n$-matrix on $\mathcal{O} \times [0,1]^2$.

Next, a general solution of (3.3.9) is given by

$$x(t) = U(\theta, t)x_0 + \int\limits_0^t U(\theta, t)U^{-1}(\theta, s)(\Pi(\theta)h)(s) \, ds \tag{3.3.13}$$

and it satisfies $x(1) = x_0$ if and only if

$$(\mathbb{I} - U(\theta, 1)) x_0 = \int\limits_0^1 U(\theta, 1)U^{-1}(\theta, s)(\Pi(\theta)h)(s) \, ds. \tag{3.3.14}$$

Note

$$\left(\mathcal{R}(\mathbb{I} - U(\theta, 1)) \right)^{\perp} = \text{span} \, \{ v_i(\theta, 0) \mid i = 1, 2, \cdots, d \} \,.$$

So (3.3.14) has a solution x_0 if and only if

$$\left\langle \int\limits_0^1 U(\theta, 1)U^{-1}(\theta, s)(\Pi(\theta)h)(s) \, ds, v_i(\theta, 0) \right\rangle = 0, \quad \forall i = 1, 2, \cdots, d. \tag{3.3.15}$$

But using (3.3.8) we derive

$$\left\langle \int\limits_0^1 U(\theta, 1)U^{-1}(\theta, s)(\Pi(\theta)h)(s) \, ds, v_i(\theta, 0) \right\rangle$$

$$= \int\limits_0^1 \langle (\Pi(\theta)h)(s), U^{-1}(\theta, s)^* U(\theta, 1)^* v_i(\theta, 0) \rangle = \int\limits_0^1 \langle (\Pi(\theta)h)(s), v_i(\theta, s) \rangle = 0$$

according to the definition of $\Pi(\theta)$. So (3.3.14) has a solution x_0. We show its uniqueness. Put for simplicity

$$w_i(\theta, s) := \frac{\partial}{\partial \theta_i} \gamma(\theta, s), \quad i = 1, 2, \cdots, d-1, \quad w_d(\theta, s) = \dot{\gamma}(\theta, s).$$

Let

$$\text{span}\ \{e_{d+1}(\theta), \cdots, e_n(\theta)\} = \{w_i(\theta, 0) \mid i = 1, 2, \cdots, d\}^\perp$$

with $e_k(\theta)$ continuous. Then

$$x_0 = \sum_{j=1}^d \eta_j(\theta) w_j(\theta, 0) + \sum_{j=d+1}^n \eta_j(\theta) e_j(\theta) \tag{3.3.16}$$

and so

$$(\mathbb{I} - U(\theta, 1)) x_0 = \sum_{j=1}^d \eta_j(\theta) (\mathbb{I} - U(\theta, 1)) w_j(\theta, 0)$$

$$+ \sum_{j=d+1}^n \eta_j(\theta) (\mathbb{I} - U(\theta, 1)) e_j(\theta) = \sum_{j=d+1}^n \eta_j(\theta) (\mathbb{I} - U(\theta, 1)) e_j(\theta).$$

Introducing an $n \times (n-d)$-matrix

$$M(\theta) := \left([(\mathbb{I} - U(\theta, 1)) e_{d+1}(\theta)]^*, \cdots, [(\mathbb{I} - U(\theta, 1)) e_n(\theta)]^* \right),$$

(3.3.14) has the form

$$M(\theta) (\eta_{d+1}(\theta), \cdots, \eta_n(\theta))^* = \int_0^1 U(\theta, 1) U^{-1}(\theta, s) (\Pi(\theta) h)(s) \, ds. \tag{3.3.17}$$

The $n \times (n-d)$-matrix $M(\theta)$ has the rank $n-d$, so the $(n-d) \times (n-d)$-matrix $M(\theta)^* M(\theta)$ is invertible. Hence (3.3.12), (3.3.17) and the Fubiny theorem [171] imply

$$(\eta_{d+1}(\theta), \cdots, \eta_n(\theta))^* = \int_0^1 M_2(\theta, s) h(s) \, ds, \tag{3.3.18}$$

where

$$M_2(\theta, s) :=$$

$$[M(\theta)^* M(\theta)]^{-1} M(\theta)^* U(\theta, 1) \left[U^{-1}(\theta, s) - \int_0^1 U^{-1}(\theta, z) M_1(\theta, z, s) \, dz \right]$$

is a continuous $(n-d) \times n$-matrix on $\mathcal{O} \times [0, 1]$.

Now we compute the rest η_i. Using the assumption of Lemma 3.3.1:

$$\int_0^1 \langle x(s), w_j(\theta, s) \rangle \, ds = 0, \quad j = 1, 2, \cdots, d,$$

together with (3.3.13), we derive

$$\int_0^1 \left\langle \sum_{i=1}^d \eta_i(\theta) U(\theta, s) w_i(\theta, 0), w_j(\theta, s) \right\rangle ds = \widetilde{H}_j(\theta), \quad j = 1, 2, \cdots, d, \quad (3.3.19)$$

where

$$\widetilde{H}_j(\theta) := -\int_0^1 \left\langle \sum_{i=d+1}^n \eta_i(\theta) U(\theta, s) e_i(\theta), w_j(\theta, s) \right\rangle ds$$

$$- \int_0^1 \left\langle \int_0^s U(\theta, s) U^{-1}(\theta, z)(\Pi(\theta)h)(z) \, dz, w_j(\theta, s) \right\rangle ds$$

for $j = 1, 2, \cdots, d$. Using the Fubiny theorem we see that

$$\left(\widetilde{H}_1(\theta), \cdots, \widetilde{H}_d(\theta) \right)^* = \int_0^1 M_3(\theta, s) h(s) \, ds$$

for a continuous $d \times n$-matrix $M_3(\theta, s)$ on $\mathcal{O} \times [0, 1]$. Since

$$\int_0^1 \left\langle \sum_{i=1}^d \eta_i(\theta) U(\theta, s) w_i(\theta, 0), w_j(\theta, s) \right\rangle ds = \sum_{i=1}^d \eta_i(\theta) \int_0^1 \langle w_i(\theta, s), w_j(\theta, s) \rangle \, ds$$

$$= \widetilde{G}(\theta) \left(\eta_1(\theta), \cdots, \eta_d(\theta) \right)^*$$

and the Gram matrix

$$\widetilde{G}(\theta) := \left(\int_0^1 \langle w_i(\theta, s), w_j(\theta, s) \rangle \, ds \right)_{i,j=1}^d$$

is invertible, from (3.3.19) we derive

$$(\eta_1(\theta), \cdots, \eta_d(\theta))^* = \int_0^1 \widetilde{G}(\theta)^{-1} M_3(\theta, s) h(s) \, ds. \quad (3.3.20)$$

Inserting formulas (3.3.18) and (3.3.20) into (3.3.16), we get

$$x_0 = \int_0^1 M_4(\theta, s) h(s) \, ds \quad (3.3.21)$$

for a continuous $n \times n$-matrix $M_4(\theta, s)$ on $\mathcal{O} \times [0, 1]$. Next plugging (3.3.12) and (3.3.21) into (3.3.13), and using the Fubiny theorem, we obtain a formula of $K(\theta)h$ of the form

$$(K(\theta)h)(t) = \int_0^1 K_1(\theta, t, s)h(s)\, ds + \int_0^t K_2(\theta, t, s)h(s)\, ds$$

for continuous $n \times n$-matrix functions $K_1(\theta, t, s)$ and $K_2(\theta, t, s)$ on $\mathcal{O} \times [0, 1]^2$. Using standard method [199] based on the uniform continuity of continuous functions on compact intervals along with the Arzela-Ascolli theorem [59], we see that $K(\theta) : L^2 \to C_p$ is a compact linear operator which is continuous with respect to θ. Finally, introducing the parameterized Green function $G(\theta, t, s)$ as follows

$$G(\theta, t, s) := \begin{cases} K_1(\theta, t, s) + K_2(\theta, t, s) & \text{for } 0 \le s \le t \le 1, \\ K_1(\theta, t, s) & \text{for } 0 \le t \le s \le 1, \end{cases}$$

we see that

$$(K(\theta)h)(t) = \int_0^1 G(\theta, t, s)h(s)\, ds\,.$$

The proof is finished. □

3.3.3   Bifurcation of Periodics in Nonautonomous Systems

In this subsection, we study bifurcation of periodic solutions to (3.3.3) by using Lemma 3.3.1 together with topological degree arguments. For this reason, we first scale $\mu \to s\mu$ for $s \in \mathbb{R} \setminus \{0\}$ and take the change of variables [75]

$$x(t + \alpha) = \gamma(\theta, t) + sz(t), \quad \alpha \in \mathbb{R}, \quad s \in \mathbb{R} \setminus \{0\}, \tag{3.3.22}$$

so then (3.3.3) possesses the form

$$\dot{z}(t) - A(\theta, t)z(t) \in g(z(t), \theta, \alpha, \mu, s, t) \quad \text{a.e. on} \quad [0, 1], \tag{3.3.23}$$

where

$$g(x, \theta, \alpha, \mu, s, t) = \Bigg\{ v \in \mathbb{R}^n : v \in \frac{1}{s}\Big[f(sx + \gamma(\theta, t)) - f(\gamma(\theta, t)) - A(\theta, t)sx\Big]$$
$$+ \sum_{j=1}^k \mu_j f_j(sx + \gamma(\theta, t), s\mu, t + \alpha) \Bigg\}$$

and g is extended to $s = 0$ by putting the term in the square brackets equal to zero. Then $g : \mathbb{R}^n \times \mathcal{O} \times \mathbb{R} \times \mathbb{R}^k \times \mathbb{R} \times \mathbb{R} \to 2^{\mathbb{R}^n} \setminus \{\emptyset\}$ is upper–semicontinuous with compact and convex values. Introducing a multivalued mapping $G : C_p \times \mathcal{O} \times \mathbb{R} \times \mathbb{R}^k \times \mathbb{R} \to 2^{L^2}$ by the formula

$$G(z, \theta, \alpha, \mu, s) = \Big\{ h \in L^2 : h(t) \in g(z(t), \theta, \alpha, \mu, s, t) \quad \text{a.e. on} \quad [0, 1] \Big\},$$

(3.3.23) is considered as a multivalued equation for z of the form

$$\dot{z} - A(\theta, \cdot)z \in G(z, \theta, \alpha, \mu, s). \tag{3.3.24}$$

Like for (3.1.17), $G(z, \theta, \alpha, \mu, s)$ are non–empty, closed, convex and bounded in L^2, and they are also weakly compact in L^2. In order to solve (3.3.24), we consider the homotopy

$$\dot{z} - A(\theta, \cdot)z \in F(z, \theta, \alpha, \mu, s, \lambda), \quad \lambda \in [0, 1], \tag{3.3.25}$$

for

$$F(z, \theta, \alpha, \mu, s, \lambda) = \left\{ h \in L^2 \ : \ h(t) \in p(z(t), \theta, \alpha, \mu, s, \lambda, t) \text{ a.e. on } [0, 1] \right\},$$

with

$$p(x, \theta, \alpha, \mu, s, \lambda, t) = \left\{ v \in \mathbb{R}^n \ : \ v \in \tfrac{1}{s} \Big[f(sx + \gamma(\theta, t)) - f(\gamma(\theta, t)) - A(\theta, t)sx \Big] \right.$$
$$\left. + \lambda \sum_{j=1}^{k} \mu_j f_j(sx + \gamma(\theta, t), s\mu, t + \alpha) + (1 - \lambda) \sum_{j=1}^{k} \mu_j f_j(\gamma(\theta, t), 0, t + \alpha) \right\}.$$

Again p is extended to $s = 0$ like g above. Now we use the Fredholm alternative result of Lemma 3.3.1 to (3.3.25) by introducing a projection $L(\theta) : L^2 \to \mathbb{R}^d$ given by

$$L(\theta)h = \left(\int_0^1 \langle h(s), v_1(\theta, s) \rangle \, ds, \cdots, \int_0^1 \langle h(s), v_d(\theta, s) \rangle \, ds \right),$$

and then decomposing (3.3.25) as follows:

$$\begin{cases} 0 \in H(z, \theta, \alpha, \mu, s, \lambda) \\ H(z, \theta, \alpha, \mu, s, \lambda) = \left\{ (z - \lambda K(\theta)h, L(\theta)h) \ : \ h \in F(z, \theta, \alpha, \mu, s, \lambda) \right\}. \end{cases} \tag{3.3.26}$$

Like in Subsection 3.1.3, $H : C_p \times \mathcal{O} \times \mathbb{R} \times \mathbb{R}^k \times \mathbb{R} \times [0, 1] \to 2^{C_p \times \mathbb{R}^d} \setminus \{\emptyset\}$ is upper–semicontinuous and $\mathbb{I}_{C_p \times \mathbb{R}^d} - H$ has compact convex values and maps bounded sets into relatively compact ones. Finally for stating Theorem 3.3.2 below, we introduce a multivalued mapping $M_\mu : \mathcal{O} \times \mathbb{R} \to 2^{\mathbb{R}^d} \setminus \{\emptyset\}$ by

$$M_\mu(\theta, \alpha) = \left\{ L(\theta)h \ : \ h \in L^2, \ h(t) \in \sum_{j=1}^{k} \mu_j f_j(\gamma(\theta, t), 0, t + \alpha) \text{ a.e. on } [0, 1] \right\}. \tag{3.3.27}$$

Arguing as for (3.1.21), the mapping M_μ is upper–semicontinuous with compact convex values and maps bounded sets into bounded ones. Now we are ready to prove the main results of this section.

Theorem 3.3.2. *Let $d > 1$. If there is a non–empty open bounded set $\mathcal{B} \subset \mathcal{O} \times \mathbb{R}$ and $\mu_0 \in S^{k-1}$ such that*

(i) $0 \notin M_{\mu_0}(\partial \mathcal{B})$

(ii) $\deg(M_{\mu_0}, \mathcal{B}, 0) \neq 0$

Then there is a constant $K > 0$ and a region in $\mathbb{R}^k$ for μ of the form

$$\mathcal{R} = \Big\{ \ s\tilde{\mu} : s \text{ and } \tilde{\mu} \text{ are from open small connected neighborhoods}$$
$$U_1 \text{ and } U_2 \subset S^{k-1} \text{ of } 0 \in \mathbb{R} \text{ and of } \mu_0, \text{ respectively} \Big\}$$

such that for any $\mu \in \mathcal{R}$ of the form $\mu = s\tilde{\mu}$, $s \in U_1$, $\tilde{\mu} \in U_2$, the differential inclusion (3.3.3) possesses a 1–periodic solution x_μ satisfying, according to (3.3.22),

$$\sup_{0 \leq t \leq 1} \left| x_\mu(t) - \gamma(\theta_\mu, t - \alpha_\mu) \right| \leq Ks,$$

where $\alpha_\mu \in \mathbb{R}$ and $\theta_\mu \in \mathcal{O}$.

Proof. We have to solve (3.3.24) which is inserted into the homotopy (3.3.26). In order to handle (3.3.26), we need the following results.

Lemma 3.3.3. *Under condition (i) of Theorem 3.3.2, for any $A > 0$ there are constants $\delta > 0$, $1 > s_0 > 0$ such that $|L(\theta)h| \geq \delta$ for all $h \in F(z, \theta, \alpha, \mu, s, \lambda)$ and for any $|s| < s_0$, $\lambda \in [0, 1]$, $|z| \leq A$, $(\theta, \alpha) \in \partial \mathcal{B}$, $|\mu - \mu_0| \leq \delta$.*

Proof. We prove Lemma 3.3.3 by the contrary. So assume that there is an $A > 0$ and

$$s_p \to 0, \quad \lambda_p \to \lambda_0, \quad |z_p| \leq A, \quad \mu_p \to \mu_0, \quad p \in \mathbb{N},$$
$$\partial \mathcal{B} \ni (\theta_p, \alpha_p) \to (\theta_0, \alpha_0) \in \partial \mathcal{B}, \quad h_p \in F(z_p, \theta_p, \alpha_p, \mu_p, s_p, \lambda_p)$$

such that $L(\theta_p)h_p \to 0$ as $p \to \infty$. Using the boundedness of $\{h_p\}_1^p$ in L^2 and Theorem 2.1.1, we can assume that h_p tends weakly to some $h_0 \in L^2$. Then by Mazur's Theorem 2.1.2 like in the proof of Lemma 3.1.7, $h_0(t) \in \sum_{j=1}^k \mu_{0j} f_j(\gamma(\theta_0, t), 0, t + \alpha_0)$ a.e. on $[0, 1]$, and $L(\theta_0)h_0 = 0$. This contradicts to (i) of this theorem. The Lemma 3.3.3 is proved. $\qquad\square$

From Lemma 3.3.3 and the construction of H it follows:

Lemma 3.3.4. *There are open small connected neighborhoods $U_1 \subset \mathbb{R}$, $U_2 \subset S^{k-1}$ of 0 and μ_0, respectively, and a constant $K_1 > 0$ such that $0 \notin H(\partial \Omega, \mu, s, \lambda)$ for any $s \in U_1$, $\mu \in U_2$, $\lambda \in [0, 1]$ with*

$$\Omega = \Big\{ (z, \theta, \alpha) \in C_p \times \mathbb{R}^d : |z| < K_1, \ (\theta, \alpha) \in \mathcal{B} \Big\}.$$

Finally applying Lemma 3.3.4 for any $s \in U_1$, $\mu \in U_2$ we derive

$$\deg \left(H(\cdot, \mu, s, 1), \Omega, 0 \right) = \deg \left(H(\cdot, \mu_0, 0, 0), \Omega, 0 \right) = \deg(M_{\mu_0}, \mathcal{B}, 0) \neq 0.$$

Hence (3.3.26) has a solution in Ω for any $s \in U_1$, $\mu \in U_2$ and $\lambda = 1$. This solution gives a solution of (3.3.24) according to the definition of (3.3.26). The proof of Theorem 3.3.2 is finished. $\qquad\square$

Applying Theorem 3.3.2 with $\mathcal{B} = (a, b)$ together with the proof of Theorem 3.1.9 we get

Corollary 3.3.5. *Let $d = 1$. If there are constants $a < b$ and $\mu_0 \in S^{k-1}$ such that $M_{\mu_0}(a)M_{\mu_0}(b) \subset (-\infty, 0)$, then the conclusion of Theorem 3.3.2 is applicable.*

Remark 3.3.6. The restriction $|\mu_0| = 1$ is not essential, since M_μ in (3.3.27) is homogeneous with respect to the variable μ.

Remark 3.3.7. Let $f(x) = Bx$ for a matrix B and let $\gamma_1, \cdots, \gamma_d$ be the maximum number of linearly independent 1–periodic solutions of $\dot{x} = Bx$. Then we take $\gamma(\theta, t) = \theta_1\gamma_1 + \cdots + \theta_d\gamma_d$ and $\frac{\partial \gamma}{\partial \theta_i}$, $i = 1, \cdots, d$ represent the maximum number of linearly independent 1–periodic solutions of $\dot{x} = Bx$. Considering now (3.3.22) with $\alpha = 0$ and repeating the above procedure, we get $\mathcal{O} = \mathbb{R}^d$ and $M_\mu : \mathbb{R}^d \to 2^{\mathbb{R}^d} \setminus \{\emptyset\}$ of the form

$$M_\mu(\theta) = \Big\{ \quad Lh : h \in L^2,$$
$$h(t) \in \sum_{j=1}^k \mu_j f_j(\theta_1\gamma_1(t) + \cdots + \theta_d\gamma_d(t), 0, t) \text{ a.e. on } [0, 1] \Big\},$$

(3.3.28)

where $Lh = \left(\int_0^1 \langle h(s), v_1(s) \rangle \, ds, \cdots, \int_0^1 \langle h(s), v_d(s) \rangle \, ds \right)$ and v_i, $i = 1, \cdots, d$ are linearly independent 1-periodic solutions of $\dot{v} = -B^*v$. Then Theorem 3.3.2 is valid with such M_μ (3.3.28).

3.3.4 Bifurcation of Periodics in Autonomous Systems

When all f_j are independent of t in (3.3.3) then we need to modify the arguments of Subsection 3.3.3. So we start from the differential inclusion

$$\dot{x}(t) \in f(x(t)) + \sum_{j=1}^k \mu_j f_j(x(t), \mu) \quad \text{a.c. on } \mathbb{R} \qquad (3.3.29)$$

with $x \in \mathbb{R}^n$, $\mu \in \mathbb{R}^k$, $\mu = (\mu_1, \cdots, \mu_k)$ and f, f_j satisfying the assumptions (i), (ii) and (iv). Scaling $\mu \to s\mu$ for $s \in \mathbb{R} \setminus \{0\}$ and taking now, instead of (3.3.22), the change of variables

$$x((1 + s\alpha)t) = \gamma(\theta, t) + sz(t), \quad \alpha \in \mathbb{R}, \quad s \in \mathbb{R} \setminus \{0\}, \qquad (3.3.30)$$

(3.3.29) is rewritten in the form

$$\dot{z}(t) - A(\theta, t)z(t) \in \tilde{g}(z(t), \theta, \alpha, \mu, s, t) \quad \text{a.e. on } [0, 1] \qquad (3.3.31)$$

with

$$\tilde{g}(x, \theta, \alpha, \mu, s, t) = \Big\{ v \in \mathbb{R}^n : v \in \frac{1}{s}\Big[f(sx + \gamma(\theta, t)) - f(\gamma(\theta, t)) - A(\theta, t)sx \Big]$$
$$+\alpha f(\gamma(\theta, t) + sx) + (1 + s\alpha) \sum_{j=1}^k \mu_j f_j(sx + \gamma(\theta, t), s\mu) \Big\}.$$

Because of (3.3.31) has a similar form like (3.3.23), we do repeat the approach of Subsection 3.3.3 to (3.3.31) and we arrive like for (3.3.27) at a multivalued mapping $N_\mu : \mathcal{O} \times \mathbb{R} \to 2^{\mathbb{R}^d} \setminus \{\emptyset\}$ with

$$N_\mu(\theta, \alpha) = \left\{ L(\theta)h : h \in L^2, h(t) \in \alpha\dot\gamma(\theta, t) + \sum_{j=1}^{k} \mu_j f_j(\gamma(\theta, t), 0) \text{ a.e. on } [0,1] \right\}.$$

(3.3.32)

N_μ is again upper–semicontinuous with compact convex values and maps bounded sets into bounded ones. Following the proof of Theorem 3.3.2, we get the following result.

Theorem 3.3.8. *Let $d > 1$. Let there exist a non–empty open bounded set $\mathcal{B} \subset \mathcal{O} \times \mathbb{R}$ and $\mu_0 \in S^{k-1}$ such that $0 \notin N_{\mu_0}(\partial\mathcal{B})$ and $\deg(N_{\mu_0}, \mathcal{B}, 0) \neq 0$. Then there is a constant $K > 0$ and a region in $\mathbb{R}^k$ for μ of the form*

$$\mathcal{R} = \Big\{ \quad s\tilde\mu : s \text{ and } \tilde\mu \text{ are from open small connected neighborhoods}$$
$$U_1 \text{ and } U_2 \subset S^{k-1} \text{ of } 0 \in \mathbb{R} \text{ and of } \mu_0, \text{ respectively} \Big\}$$

such that for any $\mu \in \mathcal{R}$ of the form $\mu = s\tilde\mu$, $s \in U_1$, $\tilde\mu \in U_2$, the differential inclusion (3.3.29) possesses a $(1 + s\alpha_\mu)$–periodic solution x_μ satisfying, according to (3.3.30),

$$\sup_{0 \le t \le 1 + s\alpha_\mu} |x_\mu(t) - \gamma(\theta_\mu, t/(1 + s\alpha_\mu))| \le Ks,$$

where $\alpha_\mu \in \mathbb{R}$, $|\alpha_\mu| \le K$ and $\theta_\mu \in \mathcal{O}$.

We proceed with the case $d = 1$. Then $\gamma(\theta, t) = \gamma(t)$ and the adjoint variational equation (3.3.7) has a unique (up to scalar multiples) 1–periodic solution $v(t)$.

Corollary 3.3.9. *If*

$$\int_0^1 \langle \dot\gamma(s), v(s) \rangle \, ds \neq 0,$$

(3.3.33)

then there is a constant $K > 0$ such that for any sufficiently small μ, the differential inclusion (3.3.29) possesses a periodic solution x_μ with the properties of Theorem 3.3.8.

Proof. There is a $\widetilde{K} > 0$ such that

$$\text{dist}\left\{ N_\mu(\alpha), \alpha \int_0^1 \langle \dot\gamma(s), v(s) \rangle \, ds \right\} < \widetilde{K}$$

for any $\mu \in S^{k-1}$ and $\alpha \in \mathbb{R}$. Then there is a $\bar{K} > 0$ such that $N_\mu(-\bar{K}) N_\mu(\bar{K}) \subset (-\infty, 0)$, which gives $|\deg(N_\mu, (-\bar{K}, \bar{K}), 0)| = 1$ uniformly for $\mu \in S^{k-1}$. Applying Theorem 3.3.8 with $\mathcal{B} = (-\bar{K}, \bar{K})$, the proof is finished. $\square$

Remark 3.3.10. Corollary 3.3.9 is an extension of a classical bifurcation result of [110, p. 416, Theorem 2.4] to the multivalued case (3.3.29), since when (3.3.33) is satisfied and (3.3.29) is smooth then we can apply the implicit function theorem in order to solve the variable α near 0 from a bifurcation equation as a function of a small parameter μ. Note (3.3.33) is a transversality assumption for α. As a consequence, for any small μ, there is a unique 1-periodic solution near γ in the smooth case.

To complete the subject of Corollary 3.3.9, we suppose

$$\int_0^1 \langle \dot{\gamma}(s), v(s) \rangle \, ds = 0 \,. \tag{3.3.34}$$

Then we loose the variable α in N_μ, and we deal with a second-order singularity of (3.3.29) of a fold-type [56]. We need to modify the above approach. To this end, we scale $\mu \to s^2\mu$ for $s > 0$ and change the variables

$$x((1+s\alpha)t) = \gamma(t) + s\alpha u(t) + s^2 z(t), \quad \alpha \in \mathbb{R}, \quad s > 0, \tag{3.3.35}$$

where $u \in C_p$ is the unique solution of $\dot{u} - A(t)u = \dot{\gamma}, \int_0^1 \langle u(s), \dot{\gamma}(s) \rangle \, ds = 0$ and $A(t) = D_x f(\gamma(t))$. Then (3.3.31) has the form

$$\dot{z}(t) - A(t)z(t) \in \bar{g}(z(t), \alpha, \mu, s, t) \quad \text{a.e. on} \quad [0,1] \,, \tag{3.3.36}$$

where

$$\bar{g}(x, \alpha, \mu, s, t) = \Big\{ v \in \mathbb{R}^n : v \in \tfrac{1}{s^2} \big[f(s\alpha u(t) + s^2 x + \gamma(t)) - f(\gamma(t)) \big.$$
$$\big. - A(t)(s\alpha u(t) + s^2 x) \big] + \tfrac{\alpha}{s} \big[f(\gamma(t) + s\alpha u(t) + s^2 x) - f(\gamma(t)) \big]$$
$$+ (1+s\alpha) \sum_{j=1}^k \mu_j f_j(s\alpha u(t) + s^2 x + \gamma(t), s^2 \mu) \Big\} \,.$$

Repeating the procedure of Subsection 3.3.3 for (3.3.36), the resulting multivalued function is (like for (3.3.27)) as follows:

$$P_\mu(\alpha) = \Big\{ \int_0^1 \langle h(s), v(s) \rangle \, ds \, : \, h \in L^2 \text{ satisfying a.e. on } [0,1] \text{ the relation}$$

$$h(t) \in \alpha^2 \big[\tfrac{1}{2} D_x^2 f(\gamma(t))(u(t), u(t)) + D_x f(\gamma(t))u(t) \big] + \sum_{j=1}^k \mu_j f_j(\gamma(t), 0) \Big\} \,.$$

$$\tag{3.3.37}$$

We have the following extension of Corollary 3.3.9.

Theorem 3.3.11. *Let $d = 1$ and* (3.3.34) *hold. Suppose*

$$A_0 := \int_0^1 \langle \tfrac{1}{2} D_x^2 f(\gamma(s))(u(s), u(s)) + D_x f(\gamma(s))u(s), v(s) \rangle \, ds > 0$$

and there exists $\mu_0 \in S^{k-1}$ such that $\int_0^1 \langle h(s), v(s) \rangle \, ds < 0$ for any $h \in L^2$ satisfy-

ing $h(t) \in \sum_{j=1}^k \mu_{0j} f_j(\gamma(t), 0)$ a.e. on $[0,1]$. Then there is a constant $K > 0$ and a wedge–shaped region in $\mathbb{R}^k$ for μ of the form

$$\mathcal{R} = \Big\{ \; s^2 \tilde{\mu} : s > 0 \text{ and } \tilde{\mu} \text{ are from open small connected neighborhoods}$$
$$U_1 \text{ and } U_2 \subset S^{k-1} \text{ of } 0 \in \mathbb{R} \text{ and of } \mu_0, \text{ respectively} \Big\}$$

such that for any $\mu \in \mathcal{R}$ of the form $\mu = s^2 \tilde{\mu}$, $0 < s \in U_1$, $\tilde{\mu} \in U_2$, the differential inclusion (3.3.29) possesses two $(1 + s\alpha_{\pm,\mu})$–periodic solutions $x_{\pm,\mu}$ satisfying, according to (3.3.35),

$$\sup_{0 \leq t \leq 1 + s\alpha_{\pm,\mu}} \big| x_{\pm,\mu}(t) - \gamma(t/(1 + s\alpha_{\pm,\mu})) - s\alpha_{\pm,\mu} u(t/(1 + s\alpha_{\pm,\mu})) \big| \leq Ks^2,$$

where $\alpha_{+,\mu} > 0$, $\alpha_{-,\mu} < 0$ and $|\alpha_{\pm,\mu}| \leq K$.

Proof. Since $P_{\mu_0}(\alpha) = \alpha^2 A_0 + S$, where $A_0 > 0$ and

$$S = \left\{ \int_0^1 \langle h(s), v(s) \rangle \, ds : h \in L^2, \; h(t) \in \sum_{j=1}^k \mu_{0j} f_j(\gamma(t), 0) \text{ a.e. on } [0,1] \right\}$$

is a set of negative numbers, we have that $P_{\mu_0}(0) = S$ contains only negative numbers, while there is an $\alpha_0 > 0$ such that $P_{\mu_0}(\pm \alpha_0)$ contains only
positive numbers. Then according to the proof of Theorem 3.1.9, we derive
$\deg(P_{\mu_0}, (-\alpha_0, 0), 0) = -1$ and $\deg(P_{\mu_0}, (0, \alpha_0), 0) = 1$. Applying Theorem 3.3.8
with $\mathcal{B} = (-\alpha_0, 0)$ and $\mathcal{B} = (0, \alpha_0)$, respectively, the proof is finished. $\square$

Remark 3.3.12. Theorem 3.3.11 is an extension of a *saddle-node bifurcation* of
periodic solutions [108, p. 197], [151] to the multivalued case (3.3.29). Indeed,
if (3.3.29) is smooth then we can use the implicit function theorem for solving
a bifurcation equation, and we get the singularity fold. This means that when
$k = 1$, for $\mu > 0$, there are only two 1-periodic solutions near γ, for $\mu = 0$ the
only one is γ near it, while for $\mu < 0$ there are no 1-periodic solutions near γ.

In the rest of this subsection, we deal with planar differential inclusions under
the condition (i), and (ii), (iv) are replaced by

(v) There are numbers $0 < c < e$ and a C^2–mapping $\gamma : (c, e) \times \mathbb{R} \to \mathbb{R}^2$
such that $\gamma(\theta, t)$ has the minimal period θ in t and $\gamma(\theta, \cdot)$ is a solution of
$\dot{x} = f(x)$.

Condition (v) usually holds when $\dot{x} = f(x)$ is a nonlinear conservative
second-order equation, for instance like the Duffing equation $\ddot{z} - z + z^3 = 0$
possessing a 1-parametric family of periodic solutions (see Fig. 1.2) with explicit
formulas expressed in terms of the Jacobi elliptic functions [108, p. 198]. Next,
differentiating $\dot{\gamma}(\theta, t) = f(\gamma(\theta, t))$ with respect to t and θ, we see that the linear

variational equation $\dot{u} = D_x f(\gamma(\theta, t))u$ has two solutions: $\dot{\gamma}(\theta, t)$ and $\frac{\partial}{\partial \theta}\gamma(\theta, t)$. The first one is θ–periodic, but the second one is not, since $\gamma(\theta, t + \theta) = \gamma(\theta, t)$ implies $\frac{\partial}{\partial \theta}\gamma(\theta, t + \theta) + \dot{\gamma}(\theta, t) = \frac{\partial}{\partial \theta}\gamma(\theta, t)$. Consequently, $\dot{u} = D_x f(\gamma(\theta, t))u$ has the single (up to scalar multiples) θ–periodic solution $\dot{\gamma}(\theta, t)$. Then according to Subsection 3.3.2, (v) implies the existence of a non-zero, continuous mapping $v : (c, e) \to \mathbb{R}^2$ such that $v(\theta, t)$ is a θ–periodic solution of the adjoint equation $\dot{v} = -A(\theta, t)^* v$. Then $w(\theta, t) = v(\theta, \theta t)$ is a 1–periodic solution (up to scalar multiples) of $\dot{w} = -\theta A(\theta, \theta t)^* w$.

To study bifurcation of θ-periodic solutions in (3.3.29), we scale $\mu \to s\mu$ like above for $s \in \mathbb{R} \setminus \{0\}$ and change variables as follows

$$x(\theta t) = \gamma(\theta, \theta t) + sz(t), \quad s \in \mathbb{R} \setminus \{0\}. \tag{3.3.38}$$

Then (3.3.29) has the form

$$\dot{z}(t) - \theta A(\theta, \theta t)z(t) \in \bar{p}(z(t), \theta, \mu, s, t) \quad \text{a.e. on} \quad [0, 1], \tag{3.3.39}$$

where

$$\bar{p}(x, \theta, \mu, s, t) = \left\{ v \in \mathbb{R}^2 : v \in \frac{\theta}{s}\left[f(sx + \gamma(\theta, \theta t)) - f(\gamma(\theta, \theta t)) - A(\theta, \theta t)sx \right] \right.$$
$$\left. + \theta \sum_{j=1}^{k} \mu_j f_j(sx + \gamma(\theta, \theta t), s\mu) \right\}.$$

Applying the procedure of Subsection 3.3.3, we can solve (3.3.39) in $z \in C_p$ considering $\theta \in (c, e)$ as a parameter. The corresponding multivalued mapping has the form

$$Q_\mu(\theta) = \left\{ \int_0^1 \langle h(s), w(\theta, s) \rangle \, ds : h \in L^2, \right.$$
$$\left. h(t) \in \theta \sum_{j=1}^{k} \mu_j f_j(\gamma(\theta, \theta t), 0) \quad \text{a.e. on} \quad [0, 1] \right\}$$
$$= \left\{ \int_0^\theta \langle h(s), v(\theta, s) \rangle \, ds : h \in L^2([0, \theta], \mathbb{R}^2), \right. \tag{3.3.40}$$
$$\left. h(t) \in \sum_{j=1}^{k} \mu_j f_j(\gamma(\theta, t), 0) \quad \text{a.e. on} \quad [0, \theta] \right\}.$$

Summarizing, we obtain the following result.

Theorem 3.3.13. *Let (i) and (v) be satisfied. If there are constants $c < a < b < e$ and $\mu_0 \in S^{k-1}$ such that $Q_{\mu_0}(a)Q_{\mu_0}(b) \subset (-\infty, 0)$, then there is a constant $K > 0$ and a region in $\mathbb{R}^k$ for μ of the form*

$$\mathcal{R} = \left\{ s\tilde{\mu} : s \text{ and } \tilde{\mu} \text{ are from open small connected neighborhoods} \right.$$
$$\left. U_1 \text{ and } U_2 \subset S^{k-1} \text{ of } 0 \in \mathbb{R} \text{ and of } \mu_0, \text{ respectively} \right\}$$

such that for any $\mu \in \mathcal{R}$ *of the form* $\mu = s\tilde{\mu}$, $s \in U_1$, $\tilde{\mu} \in U_2$, *the differential inclusion* (3.3.29) *possesses a* θ_μ*-periodic solution* x_μ *satisfying, according to* (3.3.38),

$$\sup_{0 \leq t \leq \theta_\mu} \left| x_\mu(t) - \gamma(\theta_\mu, t) \right| \leq Ks,$$

where $\theta_\mu \in (c, e)$.

Remark 3.3.14. If $\dot{x} = f(x)$ is a Hamiltonian system and $\gamma(\theta, t) = (\gamma_1(\theta, t), \gamma_2(\theta, t))$ then $v(\theta, t) = (\dot{\gamma}_2(\theta, t), -\dot{\gamma}_1(\theta, t))$ and

$$\int_0^\theta \langle h(s), v(\theta, s) \rangle \, ds = \int_0^\theta h(s) \wedge \dot{\gamma}(\theta, s) \, ds = \int_0^\theta h(s) \wedge f(\gamma(\theta, s)) \, ds,$$

where $\wedge$ is the wedge product given as $a \wedge b := a_1 b_2 - a_2 b_1$ for $a = (a_1, a_2)$ and $b = (b_1, b_2)$ [108, p. 187].

Remark 3.3.15. $Q_\mu(\theta)$ in (3.3.40) is an extension of a classical Melnikov function [55], [108, p. 195] to the multivalued case (3.3.29). Hence Theorem 3.3.13 is a generalization of the *Poincaré-Andronov bifurcation theorem*.

3.3.5   Applications to Discontinuous O.D.Eqns

First we consider the following simple version of (3.3.1):

$$\ddot{x} + \mu_1 \tau(x) + \mu_2 \operatorname{sgn}(\dot{x} + v_0 \beta(t)) = 0, \tag{3.3.41}$$

with $\tau \in C(\mathbb{R}, \mathbb{R})$, $\beta \in C(\mathbb{R}, \mathbb{R})$ is 1–periodic and $\mu_{1,2} > 0$, $v_0 > 0$ are constants. Rewriting (3.3.41) in the form of (3.3.3)

$$\dot{z} = y, \quad \dot{y} \in -\mu_1 \tau(z) - \mu_2 \operatorname{Sgn}(y + v_0 \beta(t)), \tag{3.3.42}$$

we get $f = 0$, so we apply Remark 3.3.7. 1–periodic solutions of the variational equation $\dot{z} = y$, $\dot{y} = 0$ and adjoint one $\dot{u} = 0$, $\dot{v} = -u$ are constant ones $\gamma_1(t) = (1, 0)$ and $v_1(t) = (0, 1)$, respectively.

Theorem 3.3.16. *Let* $0 \leq t_1 < t_2 < \cdots < t_{2j} < 1$, $j \geq 1$ *be the only zero points of* β. *Let* $\inf \tau = \Gamma_1 < 0 < \Gamma_2 = \sup \tau$. *If* $\mu_{1,2} > 0$ *are sufficiently small satisfying*

$$\mu_1 / \mu_2 > \max \left\{ -\frac{\Gamma_3}{\Gamma_1}, -\frac{\Gamma_3}{\Gamma_2} \right\}, \quad \text{where}$$

$$\Gamma_3 = \left(-1 + 2 \sum_{i=1}^{2j} (-1)^i t_i \right) \operatorname{sgn} \beta \left(\frac{t_1 + t_2}{2} \right), \tag{3.3.43}$$

then (3.3.41) *has a 1–periodic solution.*

Proof. Multivalued function (3.3.28) has now the form

$$M_\mu(\theta) = \left\{ \int_0^1 h(s)\, ds \; : \; h \in L^2, \right.$$

$$\left. h(t) \in -\mu_1 \tau(\theta) - \mu_2 \operatorname{Sgn}(v_0 \beta(t)) \quad \text{a.e. on} \quad [0,1] \right\}$$

which, by the assumptions of this theorem, has the form

$$M_\mu(\theta) = -\mu_1 \tau(\theta) - \mu_2 \left(-1 + 2 \sum_{i=1}^{2j} (-1)^i t_i \right) \operatorname{sgn} \beta \left(\frac{t_1 + t_2}{2} \right).$$

From (3.3.43) there are $a < b$ such that $M_\mu(a) M_\mu(b) < 0$. The proof is completed by Corollary 3.3.5. □

The next example is a multivalued van der Pol oscillator (see [55]) of the form

$$\dot{x} = y, \quad \dot{y} \in -x + \mu \phi(x) y, \tag{3.3.44}$$

where $\phi(x) = [\phi_1(x), \phi_2(x)]$, $\mu > 0$ and $\phi_{1,2} \in C(\mathbb{R}, \mathbb{R})$, $\phi_1(x) \leq \phi_2(x)$. Note (3.3.42) and (3.3.44) are multivalued perturbations of linear second-order o.d.eqns.

Theorem 3.3.17. *If there are constants $0 < \theta_1$, $0 < \theta_2$ such that*

$$0 < \int_0^{2\pi} \phi_1(\theta_1 \cos s) \sin^2 s\, ds, \quad \int_0^{2\pi} \phi_2(\theta_2 \cos t) \sin^2 s\, ds < 0,$$

then (3.3.44) has a periodic solution for any sufficiently small $\mu > 0$.

Proof. We apply Theorem 3.3.8. The unperturbed equation in (3.3.44) is just the harmonic oscillator, so we have

$$\gamma(\theta, t) = \theta(\cos t, -\sin t), \quad \dot{\gamma}(\theta, t) = -\theta(\sin t, \cos t), \quad \theta > 0$$
$$v_1(\theta, t) = (\cos t, -\sin t), \quad v_2(\theta, t) = (\sin t, \cos t),$$

and (3.3.32) now possesses the form

$$N_1(\theta, \alpha) = \left\{ \left(-\int_0^{2\pi} h(s) \sin s\, ds, -2\theta \alpha \pi + \int_0^{2\pi} h(s) \cos s\, ds \right) : \right.$$

$$\left. h \in L^2([0, 2\pi], \mathbb{R}), \quad h(t) \in [\phi_1(\theta \cos t), \phi_2(\theta \cos t)](-\theta \sin t) \text{ a.e. on } [0, 2\pi] \right\}.$$

To compute $\deg(N_1, \mathcal{B}, 0)$ with $\mathcal{B} = (\theta_1, \theta_2) \times (-\alpha_0, \alpha_0)$ for $\alpha_0 > 0$ large, we consider the homotopy

$$N_{1,\lambda}(\theta, \alpha) =$$

$$\left\{ \left((\lambda\theta + 1 - \lambda) \int_0^{2\pi} h(s) \sin^2 s \, ds, -2(\lambda\theta + 1 - \lambda)\alpha\pi - \lambda\theta \int_0^{2\pi} h(s) \sin s \cos s \, ds \right) : \right.$$

$$\left. h \in L^2([0, 2\pi], \mathbb{R}), \quad h(t) \in [\phi_1(\theta\cos t), \phi_2(\theta\cos t)] \text{ a.e. on } [0, 2\pi] \right\}.$$

If $\alpha_0 > 0$ is sufficiently large then $0 \notin N_{1,\lambda}(\partial\mathcal{B})$, and so $\deg(N_1, \mathcal{B}, 0) = \deg(N_{1,0}, \mathcal{B}, 0) = \pm 1$. Theorem 3.3.8 gives the result. □

Now we study (3.3.2) by assuming:

(vi) There are numbers $0 < c < e$ and a C^2–mapping $\gamma : (c, e) \times \mathbb{R} \to \mathbb{R}$ such that $\gamma(\theta, t)$ has the minimal period θ in t, $\dot\gamma(\theta, 0) = 0$ and $\gamma(\theta, \cdot)$ is a solution of $\ddot x + q(x) = 0$.

Fixing $\theta = 2\pi/\omega$, provided that $c < 2\pi/\omega < e$, condition (iv) holds with $d = 1$. We need the following well-known property of γ.

Lemma 3.3.18. $\dot\gamma(2\pi/\omega, \pi/\omega) = 0$ and $\dot\gamma(2\pi/\omega, t) \neq 0$ for any $\pi/\omega \neq t \in (0, 2\pi/\omega)$.

Proof. Since $\dot\gamma(2\pi/\omega, t)$ is periodic, there is a smallest t_1, $0 < t_1 \leq 2\pi/\omega$ such that $\dot\gamma(2\pi/\omega, t_1) = 0$. Then $y(t) = \gamma(2\pi/\omega, 2t_1 - t)$ is also a solution of $\ddot x + q(x) = 0$ such that $y(t_1) = \gamma(2\pi/\omega, t_1)$, $\dot y(t_1) = \dot\gamma(2\pi/\omega, t_1)$. Hence $\gamma(2\pi/\omega, t) = \gamma(2\pi/\omega, 2t_1 - t)$. Similarly we have $\gamma(2\pi/\omega, -t) = \gamma(2\pi/\omega, t)$. So $\gamma(2\pi/\omega, t)$ is $2t_1$ periodic, and this gives $2t_1 = 2\pi/\omega$. □

Now we can prove the following

Theorem 3.3.19. *Let $v_0 > 0$ be sufficiently large. If (vi) holds and $c < 2\pi/\omega < e$, then for any sufficiently small $\mu > 0$, (3.3.2) has a $2\pi/\omega$–periodic solution near the family $\gamma(\theta, t)$, $\theta \in (c, e)$ from (vi).*

Proof. We apply Corollary 3.3.5. Like in Remark 3.3.14, we have $v_1(t) = (\ddot\gamma(2\pi/\omega, t), -\dot\gamma(2\pi/\omega, t))$. Next, for $v_0 > 0$ sufficiently large the equation

$$\dot\gamma(2\pi/\omega, \alpha) + v_0 \sin\omega(t + \alpha) = 0, \quad 0 \leq \alpha \leq 2\pi/\omega$$

has as precisely the solutions $t_1(\alpha) + 2\pi j/\omega$, $t_2(\alpha) + 2\pi j/\omega$, $j \in \mathbb{Z}$, where t_1, t_2 are continuous functions such that $t_1(\alpha) < t_2(\alpha)$, $t_1(0) = \pi/\omega$, $t_2(0) = 2\pi/\omega$, $t_1(\pi/\omega) = 0$, $t_2(\pi/\omega) = \pi/\omega$, $t_1(\alpha)$ is near $\pi/\omega - \alpha$ and $t_2(\alpha)$ is near $2\pi/\omega - \alpha$. Moreover, for $v_0 > 0$ sufficiently large, $\dot\gamma(2\pi/\omega, t) + v_0 \sin\omega(t + \alpha)$

is positive on $(t_2(\alpha), t_1(\alpha) + 2\pi/\omega)$ and negative on $(t_1(\alpha), t_2(\alpha))$, respectively. Using this, (3.3.27) possesses the form

$$
\begin{aligned}
M_1(\alpha) &= \left\{ - \int\limits_0^{2\pi/\omega} \langle h(s), \dot\gamma(2\pi/\omega, s)\rangle \, ds \; : \; h \in L^2, \right. \\
&\qquad \left. h(t) \in - \operatorname{Sgn}\big(\dot\gamma(2\pi/\omega, t) + v_0 \sin\omega(t+\alpha)\big) \text{ a.e. on } [0, 2\pi/\omega] \right\} \\
&= \int\limits_{t_2(\alpha)}^{t_1(\alpha)+2\pi/\omega} \dot\gamma(2\pi/\omega, s) \, ds - \int\limits_{t_1(\alpha)}^{t_2(\alpha)} \dot\gamma(2\pi/\omega, s) \, ds \\
&= 2\big(\gamma(2\pi/\omega, t_1(\alpha)) - \gamma(2\pi/\omega, t_2(\alpha))\big) \, .
\end{aligned}
$$

We have

$$
M_1(0) = 2\big(\gamma(2\pi/\omega, \pi/\omega) - \gamma(2\pi/\omega, 2\pi/\omega)\big) = -M_1(\pi/\omega) \, .
$$

According to Lemma 3.3.18, $\gamma(2\pi/\omega, t)$ has the global extrema on $[0, 2\pi/\omega]$ at $t = 0, \pi/\omega$, and so $\gamma(2\pi/\omega, \pi/\omega) \neq \gamma(2\pi/\omega, 2\pi/\omega)$, i.e. $M_1(\pi/\omega) \neq 0$. This gives that $M_1(\pi/\omega)M_1(0) < 0$. Taking $a = 0$, $b = \pi/\omega$ in Corollary 3.3.5, the proof is finished. $\qquad\square$

The following example is motivated by a clock–pendulum [123].

Theorem 3.3.20. *Consider the equation*

$$
\ddot{x} + q(x) + \mu \operatorname{sgn} x \cdot \operatorname{sgn} \dot{x} = 0, \tag{3.3.45}
$$

where (vi) holds. Let us assume in addition

(vii) $\gamma(\theta, 0) > 0$ *and* $\gamma(\theta, \theta/2) < 0$.

If there are constants $c < a < b < e$ *such that*

$$
\big(\gamma(a, 0) + \gamma(a, a/2)\big)\big(\gamma(b, 0) + \gamma(b, b/2)\big) < 0 \, , \tag{3.3.46}
$$

then (3.3.45) has a periodic solution for any sufficiently small $\mu > 0$.

Proof. We apply Theorem 3.3.13 to (3.3.45). According to the proof of Lemma 3.3.18 and (vii), on $[0, \theta]$, $\gamma(\theta, t)$ has a global positive maximum at $t = 0$ and a global negative minimum at $t = \theta/2$. Then there are continuous mappings $\tilde{t}_1(\cdot), \tilde{t}_2(\cdot) : (c, e) \to (0, \infty)$ such that $0 < \tilde{t}_1(\theta) < \theta/2 < \tilde{t}_2(\theta) < \theta$ and $\gamma(\theta, \tilde{t}_1(\theta)) = \gamma(\theta, \tilde{t}_2(\theta)) = 0$. Clearly, $\gamma(\theta, t)$, $\dot\gamma(\theta, t)$ have the signs $(+, -)$, $(-, -)$, $(-, +)$, $(+, +)$ on the intervals $(0, \tilde{t}_1(\theta))$, $(\tilde{t}_1(\theta), \theta/2)$, $(\theta/2, \tilde{t}_2(\theta))$, $(\tilde{t}_2(\theta), \theta)$, respectively. Again $v(\theta, t) = \big(\ddot\gamma(\theta, t), -\dot\gamma(\theta, t)\big)$, and then (3.3.40) for (3.3.45) has the form

$$
\begin{aligned}
Q_1(\theta) &= - \int\limits_0^{\tilde{t}_1(\theta)} \dot\gamma(\theta, s) \, ds + \int\limits_{\tilde{t}_1(\theta)}^{\theta/2} \dot\gamma(\theta, s) \, ds - \int\limits_{\theta/2}^{\tilde{t}_2(\theta)} \dot\gamma(\theta, s) \, ds + \int\limits_{\tilde{t}_2(\theta)}^{\theta} \dot\gamma(\theta, s) \, ds \\
&= 2\big(\gamma(\theta, 0) + \gamma(\theta, \theta/2)\big) \, .
\end{aligned}
$$

Since the assumptions of this theorem imply $Q_1(a)Q_1(b) < 0$, the proof is completed by Theorem 3.3.13. $\square$

Remark 3.3.21. Since $\bar{q}(\gamma(\theta, 0)) = \bar{q}(\gamma(\theta, \theta/2))$ for $\bar{q}(x) = \int\limits_0^x q(s)\, ds$, inequality (3.3.46) could be verified from the graph of $\bar{q}$.

We conclude this subsection by considering coupled oscillators

$$\begin{aligned}
\ddot{x}_1 + q_1(x_1) + \mu_1 \operatorname{sgn}(\dot{x}_1 - \dot{x}_2) &= 0, \\
\ddot{x}_2 + q_2(x_2) + \mu_2 \operatorname{sgn}(\dot{x}_2 - \dot{x}_1) &= \mu_3 \operatorname{sw} t,
\end{aligned} \tag{3.3.47}$$

where $q_{1,2} \in C^2(\mathbb{R}, \mathbb{R})$, $\mu_{1,2,3} > 0$ are parameters and

$$\operatorname{sw} t = \begin{cases} 1 & \text{for } [2t] \text{ even} \\ 0 & \text{for } [2t] \text{ odd}. \end{cases}$$

Here $[t]$ is the integer part of t. (3.3.47) represents a movement of two masses on a ribbon coupled with an interference of the masses given by the relative velocity $\dot{x}_1 - \dot{x}_2$. The term $\operatorname{sw} t$ is a switching. In this interpretation of (3.3.47), the coupling is given by the dry friction. We assume that there are constants $0 < c < 1 < e$ and mappings $\rho_{1,2} : (c, e) \times \mathbb{R} \to \mathbb{R}$ such that ρ_i and $\ddot{x}_i + q_i(x_i) = 0$, $i = 1, 2$ satisfy the conditions (vi) and (vii).

Theorem 3.3.22. *Let us assume*

(viii) There are continuous functions $\bar{t}_1, \bar{t}_2 : [0, 1] \to \mathbb{R}$ such that $\bar{t}_1(\theta) < \bar{t}_2(\theta)$, $\bar{t}_1(0) = 1/2$, $\bar{t}_2(0) = 1$, $\bar{t}_1(1/2) = 0$, $\bar{t}_2(1/2) = 1/2$ and $\dot{\rho}_1(1, t) - \dot{\rho}_2(1, t+\theta)$ is positive or negative on $(\bar{t}_1(\theta), \bar{t}_2(\theta))$, $(\bar{t}_2(\theta), \bar{t}_1(\theta) + 1)$, respectively.

If $\mu_{1,2,3} > 0$ are sufficiently small such that $\mu_3 > 2\mu_2$, then (3.3.47) has a 1–periodic solution.

Proof. First, defining a multivalued mapping by

$$\operatorname{Sw} t = \begin{cases} 1 & \text{for } [2t] \text{ even and } 2t \notin \mathbb{Z} \\ [0, 1] & \text{for } 2t \in \mathbb{Z} \\ 0 & \text{for } [2t] \text{ odd and } 2t \notin \mathbb{Z}, \end{cases}$$

(3.3.47) is rewritten in the form of (3.3.3):

$$\begin{aligned}
\dot{x}_1 = y_1, \quad \dot{y}_1 &\in -q(x_1) - \mu_1 \operatorname{Sgn}(y_1 - y_2), \\
\dot{x}_2 = y_2, \quad \dot{y}_2 &\in -q(x_2) - \mu_2 \operatorname{Sgn}(y_2 - y_1) + \mu_3 \operatorname{Sw} t.
\end{aligned} \tag{3.3.48}$$

According to Remark 3.3.14, (ii) (iv) hold with

$$\begin{aligned}
\gamma(\theta, t) &= \left(\rho_1(1, t), \dot{\rho}_1(1, t), \rho_2(1, t+\theta), \dot{\rho}_2(1, t+\theta)\right), \\
v_1(\theta, t) &= \left(\ddot{\rho}_1(1, t), -\dot{\rho}_1(1, t), 0, 0\right), \\
v_2(\theta, t) &= \left(0, 0, \ddot{\rho}_2(1, t+\theta), -\dot{\rho}_2(1, t+\theta)\right).
\end{aligned}$$

We intend to apply Theorem 3.3.2 to (3.3.48). In this case, (3.3.27) reads

$$M_\mu(\theta, \alpha) = \Big\{ \Big(-\int_0^1 h_1(s)\dot\rho_1(1, s)\, ds, -\int_0^1 h_2(s)\dot\rho_2(1, s+\theta)\, ds \Big) : h_1 \in L^2,$$
$$h_1(t) \in -\mu_1 \operatorname{Sgn}\big(\dot\rho_1(1, t) - \dot\rho_2(1, t+\theta)\big) \quad \text{a.e. on} \quad [0, 1], \quad h_2 \in L^2,$$
$$h_2(t) \in -\mu_2 \operatorname{Sgn}\big(\dot\rho_2(1, t+\theta) - \dot\rho_1(1, t)\big) + \mu_3 \operatorname{Sw}(t+\alpha) \quad \text{a.e. on} \quad [0, 1] \Big\}.$$

Using (viii), we can simplify $M_\mu = (M_{1\mu}, M_{2\mu})$ as follows:

$$M_{1\mu}(\theta) = \mu_1 \Big(\int_{\bar t_1(\theta)}^{\bar t_2(\theta)} \dot\rho_1(1, s)\, ds - \int_{\bar t_2(\theta)}^{\bar t_1(\theta)+1} \dot\rho_1(1, s)\, ds \Big)$$
$$= 2\mu_1 \big(\rho_1(1, \bar t_2(\theta)) - \rho_1(1, \bar t_1(\theta))\big),$$

$$M_{2\mu}(\theta, \alpha) = \mu_2 \Big(\int_{\bar t_2(\theta)}^{\bar t_1(\theta)+1} \dot\rho_2(1, s+\theta)\, ds - \int_{\bar t_1(\theta)}^{\bar t_2(\theta)} \dot\rho_2(1, s+\theta)\, ds \Big)$$
$$- \mu_3 \big((\rho_2(1, \tfrac{1}{2} + \theta - \alpha) - \rho_2(1, \theta - \alpha)\big)$$
$$= 2\mu_2 \big(\rho_2(1, \bar t_1(\theta) + \theta) - \rho_2(1, \bar t_2(\theta) + \theta)\big)$$
$$- \mu_3 \big((\rho_2(1, \tfrac{1}{2} + \theta - \alpha) - \rho_2(1, \theta - \alpha)\big).$$

We have
$$M_{1\mu}(0) = 2\mu_1\big(\rho_1(1, 1) - \rho_1(1, 1/2)\big) = -M_{1\mu}(1/2).$$

Like at the end of the proof of Theorem 3.3.19, we derive $M_{1\mu}(0) > 0$ and $M_{1\mu}(1/2) < 0$ provided that $\mu_1 > 0$. Finally we take

$$\mathcal{B} = \Big\{ (\theta, \alpha) : \theta \in (0, 1/2), \quad \theta < \alpha < \tfrac{1}{2} + \theta \Big\}$$

and put M_μ in the homotopy $M_{\mu, \lambda} = (M_{1\mu, \lambda}, M_{2\mu, \lambda})$ given by

$$M_{1\mu, \lambda}(\theta) = \lambda M_{1\mu}(\theta) + (1 - \lambda)\big(\tfrac{1}{4} - \theta\big),$$

$$M_{2\mu, \lambda} = \lambda M_{2\mu}(\theta, \alpha) + (1 - \lambda)\big(\theta - \alpha + \tfrac{1}{4}\big).$$

Since $\max_{[0,1]} \rho_2(1, t) = \rho_2(1, 0)$ and $\min_{[0,1]} \rho_2(1, t) = \rho_2(1, 1/2)$, we see that $\mu_1 > 0$, $\mu_3 > 2\mu_2 > 0$ implies $0 \notin M_{\mu, \lambda}(\partial\mathcal{B})$, $\lambda \in [0, 1]$. Hence

$$\deg(M_\mu, \mathcal{B}, 0) = \deg(M_{\mu, 0}, \mathcal{B}, 0) = 1.$$

Now the result follows from Theorem 3.3.2. $\qquad\square$

3.3.6   Concluding Remarks

Remark 3.3.23. Coulomb's law for the dry friction [64, 123] includes a statistic coefficient of friction μ_s and a dynamic coefficient of friction μ_d. If $\mu_s = \mu_d = \mu$, then the friction law may be written as $\dot{x} \to \mu \operatorname{sgn} \dot{x}$. This is done in this section. On the other hand, since usually $\mu_s > \mu_d$, we can apply Remark 3.1.29.

Remark 3.3.24. Our method is clearly applied to piecewise smoothly perturbed problems like the problem

$$\ddot{x} + \mu_1(x^+)^2 + \mu_2 x^- = \mu_3 \beta(t), \tag{3.3.49}$$

where $\mu_{1,2,3} \in \mathbb{R}$ are small parameters and $\beta \in C(\mathbb{R}, \mathbb{R})$ is 1–periodic. Then we have similarly as for (3.3.41) that (3.3.27) now assumes the form $M_\mu(\theta) = -\mu_1(\theta^+)^2 - \mu_2\theta^- + \mu_3 \int\limits_0^1 \beta(s)\,ds$. By estimating the number of simple roots of M_μ, we obtain from Corollary 3.3.5 that (3.3.49) has a 1–periodic solution for any sufficiently small $\mu_{1,2,3}$ satisfying one of the following conditions:

(a) $\mu_3 \int\limits_0^1 \beta(s)\,ds < 0$ and either $\mu_1 < 0$ or $\mu_2 > 0$

(b) $\mu_3 \int\limits_0^1 \beta(s)\,ds > 0$ and either $\mu_1 > 0$ or $\mu_2 < 0$

Moreover, (3.3.49) has at least two 1–periodic solutions for any sufficiently small $\mu_{1,2,3}$ satisfying one of the following conditions:

(c) $\mu_3 \int\limits_0^1 \beta(s)\,ds < 0$ and $\mu_1 < 0$, $\mu_2 > 0$

(d) $\mu_3 \int\limits_0^1 \beta(s)\,ds > 0$ and $\mu_1 > 0$, $\mu_2 < 0$

Remark 3.3.25. Finally we note that by combining the method of this section with that of Section 3.2, we can straightforwardly extend the results of this section to singularly perturbed differential inclusions of the form

$$\begin{aligned}
\dot{x}(t) &\in f(x(t), y(t)) + \varepsilon h_1(x(t), y(t), t) \quad \text{a.e. on} \quad \mathbb{R}, \\
\varepsilon \dot{y}(t) &\in g(x(t), y(t)) + \varepsilon h_2(x(t), y(t), t) \quad \text{a.e. on} \quad \mathbb{R}
\end{aligned} \tag{3.3.50}$$

with $x \in \mathbb{R}^n$, $y \in \mathbb{R}^k$, $\varepsilon > 0$ is small, all $h_i(x, y, t)$ are 1-periodic in t and assumptions (i) and (iii) of Subsection 3.2.1 are satisfied along with the following one that the reduced equation $\dot{x} = f(x, 0)$ of (3.3.50) has a nondegenerate manifold of 1–periodic solutions, i.e. $\dot{x} = f(x, 0)$ satisfies assumptions (ii) and (iv) of this section.

By assuming in addition the validity of the condition **(H)** of Subsection 3.2.3, multivalued mappings like (3.3.27), (3.3.32) and (3.3.40) can be derived for both the non–autonomous and autonomous versions of (3.3.50). For instance, a multivalued mapping $M : \mathbb{R}^d \to 2^{\mathbb{R}^d} \setminus \{\emptyset\}$ corresponding to (3.3.27) has the form

$$M(\theta, \alpha) = \left\{ L(\theta)h\, ds \; : \; h \in L^2 \text{ satisfying a.e. on } [0, 1] \right.$$

$$\left. \text{the relation} \quad h(t) \in D_y f(\gamma(\theta, t), 0)(C(\theta, t, \alpha)) + h_1(\gamma(\theta, t), 0, t + \alpha) \right\},$$

where $C : \mathcal{O} \times [0, 1] \times \mathbb{R} \to 2^{\mathbb{R}^k} \setminus \{\emptyset\}$ is the upper–semicontinuous mapping from the condition (H) of Subsection 3.2.3.

3.4 Bifurcation of Periodics in Relay Systems

3.4.1 Systems with Relay Hysteresis

Oscillations in systems with relay hysteresis are extensively studied in literature [42, 153, 192] using several approaches ranging from harmonic balance methods [143, 144] to analysis of Poincaré maps [39, 192]. Such oscillators model a variety of phenomena from electrical circuitry to circadian biological clocks, chemical oscillators and ecological systems as well (see [192] for more references). For instance, electrical engineers are interested in the periodic behavior of circuits with hysteresis which could be modeled by $L_m y = f(y)$, where L_m is an mth–order differential operator and f is a relay hysteresis operator. In this section, we continue with this study. To deal with much more general equations, we are interested in the periodic oscillations of systems given by

$$\dot{x} = Ax + \mu f(x_1)b, \qquad (3.4.1)$$

where A is a constant $n \times n$ matrix, x_1 is the first component of $x \in \mathbb{R}^n$, $b \in \mathbb{R}^n$ is a constant vector and $\mu \in \mathbb{R}$ is a small parameter.

A relay hysteresis operator f is defined as follows: there is given a pair of real numbers $\alpha < \beta$ (thresholds) and a pair of real–valued continuous functions $h_o \in C([\alpha, \infty), \mathbb{R})$, $h_c \in C((-\infty, \beta], \mathbb{R})$ such that $h_o(u) \geq h_c(u)\, \forall\, u \in [\alpha, \beta]$. Moreover, we suppose that h_o, h_c are bounded on $[\alpha, \infty)$, $(-\infty, \beta]$, respectively (see Fig. 3.6). For a given continuous input $u(t)$, $t \geq t_0$, one defines the output $v(t) = f(u)(t)$ of the relay hysteresis operator as follows

$$f(u)(t) = \begin{cases} h_o(u(t)) & \text{if} \quad u(t) \geq \beta, \\ h_c(u(t)) & \text{if} \quad u(t) \leq \alpha, \\ h_o(u(t)) & \text{if} \quad u(t) \in (\alpha, \beta) \quad \text{and} \quad u(\tau(t)) = \beta, \\ h_c(u(t)) & \text{if} \quad u(t) \in (\alpha, \beta) \quad \text{and} \quad u(\tau(t)) = \alpha, \end{cases}$$

where $\tau(t) = \sup\{s \; : \; s \in [t_0, t], \, u(s) = \alpha \text{ or } u(s) = \beta\}$. If $\tau(t)$ does not exist (i.e. $u(\sigma) \in (\alpha, \beta)$ for $\sigma \in [t_0, t]$), then $f(u)(\sigma)$ is undefined and we have to initially set the relay open or closed when $u(t_0) \in (\alpha, \beta)$. Of course, when either $h_o(\beta) > h_c(\beta)$ or $h_o(\alpha) > h_c(\alpha)$ then $f(u)$ is generally discontinuous.

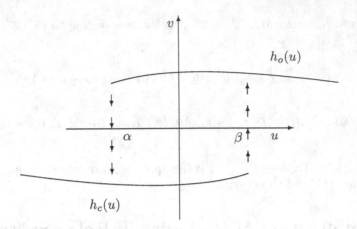

Figure 3.6: A relay hysteresis

Contrary to the above-mentioned papers, we assume that (3.4.1) is at resonance, i.e. $\dot{x} = Ax$ has a nonzero periodic solution. Since (3.4.1) is generally discontinuous, we consider it as a differential inclusion.

3.4.2   Bifurcation of Periodics

First we study the linear equation $\dot{x} = Ax$ with its adjoint one $\dot{x} = -A^*x$. We suppose that the following condition holds

(i) There is an $x_0 \in \mathcal{N}\left(\mathbb{I} - e^A\right)$ such that $Ax_0 \neq 0$.

Then $\dim \mathcal{N}\left(\mathbb{I} - e^{-A^*}\right) = \dim \mathcal{N}\left(\mathbb{I} - e^A\right) > 1$. Let $\langle \cdot, \cdot \rangle$ be the inner product on $\mathbb{R}^n$. By Lemma 3.3.1, we know that the linear equation

$$\dot{x} = Ax + h(t), \quad h \in L^2 := L^2([0,1], \mathbb{R}^n)$$

has a solution $x \in W^{1,\infty} := W^{1,\infty}([0,1], \mathbb{R}^n)$ satisfying $x(0) = x(1)$ if and only if $\int_0^1 \langle h(s), e^{-A^*s}w \rangle \, ds = 0 \ \forall w \in \mathcal{N}\left(\mathbb{I} - e^{-A^*}\right)$. The norm on $W^{1,\infty}$ is denoted by $\|\cdot\|$. This solution is unique if it satisfies $\int_0^1 \langle x(s), e^{As}z \rangle \, ds = 0 \ \forall z \in \mathcal{N}\left(\mathbb{I} - e^A\right)$. Let $x = \mathcal{K}h$ be this solution. We put

$$X = \left\{ x \in W^{1,\infty} : \int_0^1 \langle x(s), e^{As}z \rangle \, ds = 0 \quad \forall z \in \mathcal{N}\left(\mathbb{I} - e^A\right) \right\}.$$

Let

$$\Pi : L^2 \to \left\{ h \in L^2 : \int_0^1 \langle h(s), e^{-A^*s}w \rangle \, ds = 0 \quad \forall w \in \mathcal{N}\left(\mathbb{I} - e^{-A^*}\right) \right\}$$

be the orthogonal projection. Then $\mathcal{K} : \mathcal{R}\Pi \to X$ is linear and bounded.

According to assumption (i) there is a basis $\{w_1, \cdots, w_d\}$ of $\mathcal{N}\left(\mathbb{I} - e^A\right)$ such that any 1-periodic solution of $\dot{x} = Ax$ has a form $\gamma(\theta, t + \omega)$ for some $\omega \in \mathbb{R}$ and $\theta \in \mathbb{R}^{d-1}$, where

$$\gamma(\theta, t) = \sum_{i=1}^{d-1} \theta_i\, e^{At} w_i, \quad \theta_i \in \mathbb{R}.$$

Let $\gamma_1(\theta, t)$ be the first coordinate of $\gamma(\theta, t)$. Now we suppose the following conditions:

(ii) There is an open bounded subset $\emptyset \neq \mathcal{O} \subset \mathbb{R}^{d-1}$ such that $\forall \theta \in \mathcal{O}$ and $\forall t_0 \in \mathbb{R}$ it holds

$$\gamma_1(\theta, t_0) = \alpha, \beta \implies \dot{\gamma}_1(\theta, t_0) \neq 0.$$

(iii) $\displaystyle\min_{t \in \mathbb{R}} \gamma_1(\theta, t) < \alpha, \quad \max_{t \in \mathbb{R}} \gamma_1(\theta, t) > \beta \quad \forall \theta \in \mathcal{O}.$

In order to state the next theorem, we introduce a mapping given by

$$M : \mathbb{R} \times \mathcal{O} \to \mathbb{R}^d, \quad M(\omega, \theta) = \mathcal{L}h$$
$$h(t) = f(\gamma_1(\theta, \cdot))(t)b + \omega A\gamma(\theta, t) \quad \text{a.e. on} \quad [0, 1] \tag{3.4.2}$$

where $\mathcal{L} : L^2 \to \mathbb{R}^d$ is defined by

$$\mathcal{L}h := \left(\int_0^1 \langle h(s),\, e^{-A^* s}\tilde{w}_1 \rangle\, ds, \cdots, \int_0^1 \langle h(s),\, e^{-A^* s}\tilde{w}_d \rangle\, ds \right)$$

for a basis $\{\tilde{w}_1, \cdots, \tilde{w}_d\}$ of $\mathcal{N}\left(\mathbb{I} - e^{-A^*}\right)$. From (ii) and (iii) we see that M is well-defined and continuous.

Theorem 3.4.1. *Assume that (i–iii) hold. If there is a non–empty open bounded set $\mathcal{B}$ such that $\overline{\mathcal{B}} \subset \mathbb{R} \times \mathcal{O}$ and*

(i) $0 \notin M(\partial\mathcal{B})$

(ii) $\deg(M, \mathcal{B}, 0) \neq 0$

where M is given by (3.4.2). Then there are constants $K_1 > 0$ and $\mu_0 > 0$ such that for any $|\mu| < \mu_0$, there are $(\omega_\mu, \theta_\mu) \in \mathcal{B}$ and an $(1 + \mu\omega_\mu)$–periodic solution x_μ of (3.4.1) satisfying

$$\sup_{t \in \mathbb{R}} \left| x_\mu(t) - \gamma(\theta_\mu, t/(1 + \mu\omega_\mu)) \right| \leq K_1 |\mu|.$$

Proof. Following arguments from Subsection 3.3.4, first we make in (3.4.1) the change of variables

$$x\big((1+\mu\omega)t\big) = \mu z(t) + \gamma(\theta, t), \quad \omega \in \mathbb{R}$$

and consider (3.4.1) as a differential inclusion of the form

$$\dot{x} - Ax \in \mu F(x_1)b, \tag{3.4.3}$$

where F is a multivalued mapping defined as follows

$$F(u)(t) = \begin{cases} f(u)(t) & \text{if } u(t) \neq \alpha, \beta, \\ h_c(\alpha) & \text{if } u(t) = \alpha \text{ and } u(\tau(s)) = \alpha \text{ for any } s < t \text{ near } t, \\ h_o(\beta) & \text{if } u(t) = \beta \text{ and } u(\tau(s)) = \beta \text{ for any } s < t \text{ near } t, \\ [h_c(\alpha), h_o(\alpha)] & \text{if } u(t) = \alpha \text{ and } u(\tau(s)) = \beta \text{ for any } s < t \text{ near } t, \\ [h_c(\beta), h_o(\beta)] & \text{if } u(t) = \beta \text{ and } u(\tau(s)) = \alpha \text{ for any } s < t \text{ near } t. \end{cases}$$

The conditions (ii) and (iii) imply that for any $\tilde{K} > 0$ there is an $\mu_0 > 0$ such that if $z \in X$ satisfies $\|z\| \leq \tilde{K}$ and $|\mu| \leq \mu_0$ then $u(t) = \mu z_1(t) + \gamma_1(\theta, t)$ strictly monotonically crosses α and β for any $\theta \in \mathcal{O}$. Consequently, $F(u)$ is well–defined.

In variable $z(t)$, (3.4.3) has the form

$$\dot{z}(t) - Az(t) \in (1+\mu\omega)F\big(\mu z_1 + \gamma_1(\theta, \cdot)\big)(t)b + \omega A\big(\mu z(t) + \gamma(\theta, t)\big). \tag{3.4.4}$$

Since we intend to use functional-analytical method, we take the mapping

$$G(z, \omega, \theta, \mu, \lambda) = \Big\{ h \in L^2 : \text{ satisfying a.e. on } [0, 1] \text{ the relation}$$

$$h(t) \in (1+\lambda\mu\omega)F\big(\lambda\mu z_1 + \gamma_1(\theta, \cdot)\big)(t)b + \omega A\big(\lambda\mu z(t) + \gamma(\theta, t)\big) \Big\}$$

and consider (3.4.4) in the form

$$\dot{z} - Az \in G(z, \omega, \theta, \mu, 1). \tag{3.4.5}$$

To solve (3.4.5), using Π and $\mathcal{K}$, we rewrite (3.4.5) as follows

$$\begin{cases} 0 \in H(z, \omega, \theta, \mu, 1) \\ H(z, \omega, \theta, \mu, \lambda) = \Big\{ (z - \lambda\mathcal{K}\Pi h, \mathcal{L}h) : h \in G(z, \omega, \theta, \mu, \lambda) \Big\}. \end{cases} \tag{3.4.6}$$

Since f is bounded in (3.4.1), there are $\mu_0 > 0$ and $K > 0$ such that $\|\mathcal{K}\Pi h\| \leq K$ for any $h \in G(z, \omega, \theta, \mu, \lambda)$, $(z, \omega, \theta) \in \Omega$, $|\mu| \leq \mu_0$ and $\lambda \in [0, 1]$ where

$$\Omega = \Big\{ (z, \omega, \theta) \in X \times \mathbb{R}^d : \|z\| < K + 1, \quad (\omega, \theta) \in \mathcal{B} \Big\}.$$

Moreover, if μ_0 is sufficiently small then by (ii) and (iii), the mapping

$$H : \Omega \times [-\mu_0, \mu_0] \times [0, 1] \to 2^{X \times \mathbb{R}^d} \tag{3.4.7}$$

is well–defined and singlevalued. It is easy to show that $H : \Omega \times [-\mu_0, \mu_0] \times [0, 1] \to X \times \mathbb{R}^d$ is continuous and $\mathbb{I}_{X \times \mathbb{R}^d} - H$ is compact as well. Next we show

$$0 \notin H\big(\partial\Omega \times [-\mu_0, \mu_0] \times [0, 1]\big)$$

for any $\mu_0 > 0$ sufficiently small. Assume the contrary. So there are

$$[0, 1] \ni \lambda_i \to \lambda_0, \quad \|z_i\| \le K + 1, \quad \mu_i \to 0, \quad i \in \mathbb{N}$$
$$\partial\mathcal{B} \ni (\omega_i, \theta_i) \to (\omega_0, \theta_0) \in \partial\mathcal{B}, \quad h_i \in G(z_i, \omega_i, \theta_i, \mu_i, \lambda_i)$$

such that $\mathcal{L}h_i = 0$. We can assume that $z_i \to z$ in $C([0, 1], \mathbb{R}^n)$ and h_i tends weakly to some $h_0 \in L^2$. Then by applying Mazur's Theorem 2.1.2 like for (3.1.21), we obtain

$$h \in G(z, \omega_0, \theta_0, 0, \lambda_0) \quad \text{and} \quad \mathcal{L}h_0 = 0,$$

i.e. $0 = M(\omega_0, \theta_0)$ for some $(\omega_0, \theta_0) \in \partial\mathcal{B}$. This contradicts to (i) of this theorem. Consequently, we compute for μ sufficiently small

$$\deg\big(H(\cdot, \cdot, \cdot, \mu, 1), \Omega, 0\big) = \deg\big(H(\cdot, \cdot, \cdot, \mu, 0), \Omega, 0\big) = \deg(M, \mathcal{B}, 0) \ne 0.$$

In this way, (3.4.6) has a solution $(z, \omega, \theta) \in \Omega$ for any μ sufficiently small. The proof is finished. □

Now we return to the differential equation of Subsection 3.4.1

$$L_m y = \sum_{i=0}^{m} a_i y^{(i)} = \mu f(y), \tag{3.4.8}$$

where $a_i \in \mathbb{R}$, $a_m = 1$ and $y^{(i)} = \frac{d^i}{dt^i} y$. Of course, (3.4.8) can be rewritten in the form of (3.4.1). We put

$$L_m^* y = \sum_{i=0}^{m} (-1)^i a_i y^{(i)}.$$

Let $\phi_1, \cdots, \phi_d$, respectively $\psi_1, \cdots, \psi_d$, be a basis of the space of all 1–periodic solutions of $L_m y = 0$, respectively $L_m^* y = 0$. Supposing that ϕ_d is non–constant, it could be of the form $\sin 2\pi k_d t$ for some $k_d \in \mathbb{N}$. Then we could take $\phi_{d-1}(t) = \cos 2\pi k_d t$, and as a result of this we see that any 1-periodic solution of $L_m y = 0$ has a form $\eta(\theta, t + \omega)$ for some θ and ω where $\eta(\theta, t) := \sum_{i=1}^{d-1} \theta_i \phi_i(t)$.

Theorem 3.4.2. *Assume that ϕ_d is non–constant and the following conditions hold:*

(a) *There is an open bounded subset $\emptyset \ne \mathcal{O} \subset \mathbb{R}^{d-1}$ such that $\forall \theta \in \mathcal{O}$ and $\forall t_0 \in \mathbb{R}$ it holds*

$$\eta(\theta, t_0) = \alpha, \beta \implies \dot{\eta}(\theta, t_0) \ne 0.$$

(b) $\min_{t \in \mathbb{R}} \eta(\theta, t) < \alpha, \quad \max_{t \in \mathbb{R}} \eta(\theta, t) > \beta \quad \forall \theta \in \mathcal{O}.$

If there is a non–empty open bounded set $\mathcal{B}$ such that $\overline{\mathcal{B}} \subset \mathbb{R} \times \mathcal{O}$ and

(i) $0 \notin M(\partial\mathcal{B})$

(ii) $\deg(M, \mathcal{B}, 0) \neq 0$

where $M : \mathbb{R} \times \mathcal{O} \to \mathbb{R}^d$ is given by

$$M(\omega, \theta) = \left(\int_0^1 h(s)\psi_1(s)\,ds, \cdots, \int_0^1 h(s)\psi_d(s)\,ds \right) \tag{3.4.9}$$

$$h(t) = f(\eta(\theta, \cdot))(t) + \omega \sum_{i=1}^m ia_i\eta^{(i)}(\theta, t) \quad \text{a.e. on} \quad [0,1].$$

Then there are constants $K_1 > 0$ and $\mu_0 > 0$ such that for any $|\mu| < \mu_0$, there are $(\omega_\mu, \theta_\mu) \in \mathcal{B}$ and an $(1 + \mu\omega_\mu)$–periodic solution y_μ of (3.4.8) satisfying

$$\sup_{t \in \mathbb{R}} \left| y_\mu(t) - \eta(\theta_\mu, t/(1 + \mu\omega_\mu)) \right| \leq K_1|\mu|.$$

Proof. We follow the proof of Theorem 3.4.1 by taking in (3.4.8) the change of variables

$$y\big((1 + \mu\omega)t\big) = \mu z(t) + \eta(\theta, t), \quad \omega \in \mathbb{R}.$$

Conditions (a) and (b) are analogies of (i–iii). Since computations are the same as for Theorem 3.4.1, we omit details. $\qquad\square$

Results of Remark 3.3.7 can be directly modified to existence results of subharmonic solutions of nonautonomous periodic versions of (3.4.1) expressed in the following theorems [84].

Theorem 3.4.3. *Consider*

$$\dot{x} = Ax + \mu\big(f(x_1)b + q(t)\big), \tag{3.4.10}$$

where $q \in C(\mathbb{R}, \mathbb{R}^n)$ is 1–periodic and A, f, b are given in (3.4.1). Assume that (i–iii) hold. If there is a non–empty open bounded set $\mathcal{B}$ such that $\overline{\mathcal{B}} \subset \mathbb{R} \times \mathcal{O}$, $0 \notin M(\partial\mathcal{B})$ and $\deg(M, \mathcal{B}, 0) \neq 0$, where M is given by

$$M : \mathbb{R} \times \mathcal{O} \to \mathbb{R}^d, \quad M(\omega, \theta) = \mathcal{L}h$$
$$h(t) = f(\gamma_1(\theta, \cdot))(t)b + q(t + \omega) \quad \text{a.e. on} \quad [0,1]. \tag{3.4.11}$$

Then there are constants $K_1 > 0$ and $\mu_0 > 0$ such that for any $|\mu| < \mu_0$, there are $(\omega_\mu, \theta_\mu) \in \mathcal{B}$ and an 1–periodic solution x_μ of (3.4.10) satisfying

$$\sup_{t \in \mathbb{R}} \left| x_\mu(t) - \gamma(\theta_\mu, t - \omega_\mu) \right| \leq K_1|\mu|.$$

Theorem 3.4.4. *Consider*

$$L_m y = \mu(f(y) + q(t)), \tag{3.4.12}$$

where L_m, f are given in (3.4.8) and $q \in C(\mathbb{R}, \mathbb{R})$ is 1–periodic. Assume that ϕ_d is non–constant, and (a) and (b) of Theorem 3.4.2 are valid. If there is a non–empty open bounded set $\mathcal{B}$ such that $\overline{\mathcal{B}} \subset \mathbb{R} \times \mathcal{O}$, $0 \notin M(\partial \mathcal{B})$ and $\deg(M, \mathcal{B}, 0) \neq 0$, where $M : \mathbb{R} \times \mathcal{O} \to \mathbb{R}^d$ is given by

$$M(\omega, \theta) = \left(\int_0^1 h(s) \psi_1(s)\, ds, \cdots, \int_0^1 h(s) \psi_d(s)\, ds \right) \tag{3.4.13}$$

$$h(t) = f(\eta(\theta, \cdot))(t) + q(t + \omega) \quad \text{a.e. on} \quad [0, 1].$$

Then there are constants $K_1 > 0$ and $\mu_0 > 0$ such that for any $|\mu| < \mu_0$, there are $(\omega_\mu, \theta_\mu) \in \mathcal{B}$ and an 1–periodic solution y_μ of (3.4.12) satisfying

$$\sup_{t \in \mathbb{R}} |y_\mu(t) - \eta(\theta_\mu, t - \omega_\mu)| \leq K_1 |\mu|.$$

Remark 3.4.5. The boundedness of h_o and h_c on $[\alpha, \infty)$, respectively $(-\infty, \beta]$, is not essential in our considerations.

Remark 3.4.6. The smallness of μ_0 in Theorems 3.4.1–3.4.4 can be estimated.

3.4.3   Third-Order O.D.Eqns with Small Relay Hysteresis

First, we consider the following autonomous problem

$$\dddot{y} + \ddot{y} + \dot{y} + y = \mu f(y), \tag{3.4.14}$$

where f is of the form

$$\alpha = -\delta, \quad \beta = \delta, \quad \delta > 0, \quad h_o = g + p, \quad h_c = g - p$$

with $p > 0$ constant and $g \in C(\mathbb{R}, \mathbb{R})$.

Theorem 3.4.7. *If there are numbers $\delta < a_1 < a_2$ such that the numbers*

$$4p\left(\frac{\delta}{a_i} + \sqrt{1 - \frac{\delta^2}{a_i^2}}\right) + \int_0^{2\pi} g(a_i \sin t) \sin t\, dt, \quad i = 1, 2 \tag{3.4.15}$$

have opposite signs, then there is a constant $K > 0$ such that for any μ sufficiently small there are $\theta_\mu \in (a_1, a_2)$, $\omega_\mu \in \left(-\frac{3\delta p}{\pi a_1^2}, -\frac{\delta p}{\pi a_2^2}\right)$ and a $2\pi(1 + \mu\omega_\mu)$–periodic solution y_μ of (3.4.14) satisfying

$$\sup_{t \in \mathbb{R}} \left|y_\mu(t) - \theta_\mu \sin \frac{t}{1 + \mu\omega_\mu}\right| \leq K|\mu|.$$

Proof. We apply Theorem 3.4.2 with

$$\phi_1(t) = \psi_1(t) = \sin t, \quad \phi_2(t) = \psi_2(t) = \cos t, \quad \eta(\theta, t) = \theta \sin t.$$

By taking $\mathcal{O} = (\delta, \infty)$, the conditions (a) and (b) of Theorem 3.4.2 are satisfied. Let $t_0 = \arcsin \frac{\delta}{\theta}$ for $\theta \in \mathcal{O}$. Computing (3.4.9) for this case, we derive

$$M(\omega, \theta) = \big(M_1(\omega, \theta), M_2(\omega, \theta)\big), \tag{3.4.16}$$

where

$$M_1(\omega, \theta) = \int_0^{2\pi} \omega(\theta \cos t - 2\theta \sin t - 3\theta \cos t) \sin t \, dt + \int_{t_0}^{t_0+\pi} (g(\theta \sin t) + p) \sin t \, dt$$

$$+ \int_{t_0+\pi}^{t_0+2\pi} (g(\theta \sin t) - p) \sin t \, dt = -2\pi\theta\omega + 4p\sqrt{1 - \frac{\delta^2}{\theta^2}} + \int_0^{2\pi} g(\theta \sin t) \sin t \, dt,$$

$$M_2(\omega, \theta) = \int_0^{2\pi} \omega(\theta \cos t - 2\theta \sin t - 3\theta \cos t) \cos t \, dt + \int_{t_0}^{t_0+\pi} (g(\theta \sin t) + p) \cos t \, dt$$

$$+ \int_{t_0+\pi}^{t_0+2\pi} (g(\theta \sin t) - p) \cos t \, dt = -2\pi\theta\omega - 4\frac{\delta p}{\theta}.$$

Now we verify (i) and (ii) of Theorem 3.4.2 when M is given by (3.4.16) and $\mathcal{B} = \left(-\frac{3\delta p}{\pi a_1^2}, -\frac{\delta p}{\pi a_2^2}\right) \times (a_1, a_2)$. For this reason, we put (3.4.16) in the homotopy

$$M(\omega, \theta, \lambda) = \big(M_1(\omega, \theta, \lambda), M_2(\omega, \theta, \lambda)\big), \quad \lambda \in [0, 1],$$

where

$$M_1(\omega, \theta, \lambda) = -2\pi\left(\lambda\theta + (1-\lambda)\frac{a_1 + a_2}{2}\right)\left(\omega - (1-\lambda)D\right) + 4p\sqrt{1 - \frac{\delta^2}{\theta^2}}$$

$$+ \int_0^{2\pi} g(\theta \sin t) \sin t \, dt + 4\frac{\delta p}{\theta} - \frac{4\lambda\delta p}{\left(\lambda\theta + (1-\lambda)\frac{a_1+a_2}{2}\right)},$$

$$M_2(\omega, \theta, \lambda) = -2\pi\left(\lambda\theta + (1-\lambda)\frac{a_1 + a_2}{2}\right)\left(\omega - (1-\lambda)D\right) - \frac{4\lambda\delta p}{\left(\lambda\theta + (1-\lambda)\frac{a_1+a_2}{2}\right)}$$

and $D := -\frac{\delta p}{\pi a_1^2} - \frac{\delta p}{\pi a_2^2}$.

It is elementary to see that $0 \notin M(\partial\mathcal{B}, \lambda)$, $\forall \lambda \in [0, 1]$. Indeed, for $\theta \in \{a_1, a_2\}$, this follows from (3.4.15). Next, let $\omega \in \left\{-\frac{3\delta p}{\pi a_1^2}, -\frac{\delta p}{\pi a_2^2}\right\}$ and $\theta \in (a_1, a_2)$. If $M_2(\omega, \theta, \lambda) = 0$, we have $\lambda\theta + (1-\lambda)\frac{a_1+a_2}{2} \in (a_1, a_2)$ and

$$\frac{\omega - D}{\lambda} + D = \frac{-2\delta p}{\pi\left(\lambda\theta + (1-\lambda)\frac{a_1+a_2}{2}\right)^2} \in \left(-\frac{2\delta p}{\pi a_1^2}, -\frac{2\delta p}{\pi a_2^2}\right) \tag{3.4.17}$$

for $\lambda \in (0, 1]$. If $\omega = -\frac{3\delta p}{\pi a_1^2}$ then $\frac{\omega - D}{\lambda} + D \leq \omega < -\frac{2\delta p}{\pi a_1^2}$, and if $\omega = -\frac{\delta p}{\pi a_2^2}$ then $\frac{\omega - D}{\lambda} + D \geq \omega > -\frac{2\delta p}{\pi a_2^2}$. This contradicts to (3.4.17). So $M_2(\omega, \theta, \lambda) \neq 0$. If $\lambda = 0$ then $M_2(\omega, \theta, 0) = -\pi (a_1 + a_2)(\omega - D) \neq 0$.

Consequently, we obtain

$$\deg\left(M(\cdot, \cdot, 1), \mathcal{B}, 0\right) = \deg\left(M(\cdot, \cdot, 0), \mathcal{B}, 0\right) = -\deg\left(M_1(D, \cdot, 0), (a_1, a_2), 0\right) \neq 0.$$

The proof is finished by Theorem 3.4.2. $\qquad\square$

Corollary 3.4.8. *If $g(x) = c_1 x + c_2$ in (3.4.14) with constant $c_{1,2}$ such that $c_1 < 0$ and $4p > -c_1 \delta \pi$, then the conclusion of Theorem 3.4.7 is applicable.*

Proof. In Theorem 3.4.7 now we compute

$$4p\left(\frac{\delta}{\theta} + \sqrt{1 - \frac{\delta^2}{a\theta^2}}\right) + \int_0^{2\pi} (c_1 \theta \sin t + c_2) \sin t \, dt = 4p\left(\frac{\delta}{\theta} + \sqrt{1 - \frac{\delta^2}{\theta^2}}\right) + c_1 \theta \pi .$$

Taking $a_1 > \delta$ near to δ and $a_2 > a_1$ sufficiently large, the proof is finished. $\qquad\square$

Next we consider a forced problem of (3.4.14)

$$\dddot{y} + \ddot{y} + \dot{y} + y = \mu\big(f(y) + \sin t\big), \tag{3.4.18}$$

where f is given in (3.4.14).

Theorem 3.4.9. *Assume that $4p = \pi$ and $g \in C^1(\mathbb{R}, \mathbb{R})$. If the function*

$$\rho \to \int_0^{2\pi} g(\delta \rho \sin t) \sin t \, dt$$

has a simple root $\rho_0 > 1$, then by putting $1/\rho_0 = \sin \omega_0$, $\pi/2 < \omega_0 < \pi$, there is a constant $K > 0$ such that for any μ sufficiently small there are (ω_μ, θ_μ) near to $(\omega_0, \delta\rho_0)$ and a 2π–periodic solution y_μ of (3.4.18) satisfying

$$\sup_{t \in \mathbb{R}} \left| y_\mu(t) - \theta_\mu \sin(t - \omega_\mu)\right| \leq K|\mu| .$$

Proof. We apply Theorem 3.4.4. The mapping (3.4.13) for (3.4.18) has the form

$$M(\omega, \theta) = \big(M_1(\omega, \theta), M_2(\omega, \theta)\big), \tag{3.4.19}$$

where

$$M_1(\omega, \theta) = \pi\sqrt{1 - \frac{\delta^2}{\theta^2}} + \int_0^{2\pi} g(\theta \sin t) \sin t \, dt + \int_0^{2\pi} \sin(t + \omega) \sin t \, dt$$

$$= \pi\sqrt{1 - \frac{\delta^2}{\theta^2}} + \int_0^{2\pi} g(\theta \sin t) \sin t \, dt + \pi \cos \omega ,$$

$$M_2(\omega, \theta) = -\frac{\delta\pi}{\theta} + \int_0^{2\pi} \sin(t + \omega) \cos t \, dt = -\frac{\delta\pi}{\theta} + \pi \sin \omega .$$

For $\pi/2 < \omega < \pi$, $M_1 = 0$, $M_2 = 0$ are equivalent to $\int_0^{2\pi} g\left(\frac{\sin t}{\sin \omega}\delta\right)\sin t\, dt = 0$.

Since $\rho_0 > 1$ is a simple root of $\rho \to \int_0^{2\pi} g(\delta\rho\sin t)\sin t\, dt$, then we can easily verify (see Lemma 3.5.5) that (ω_0, θ_0) defined as $\theta_0 = \delta\rho_0$, $1/\rho_0 = \sin\omega_0$, $\pi/2 < \omega_0 < \pi$ is a simple zero of $M = 0$ given by (3.4.19), i.e. $M(\omega_0, \theta_0) = 0$ and $DM(\omega_0, \theta_0)$ is invertible. So the Brouwer index of M at (ω_0, θ_0) is nonzero. Consequently, the proof is finished by Theorem 3.4.4 when $\mathcal{B}$ is taken as a small open neighborhood of (ω_0, θ_0). □

Corollary 3.4.10. *Assume that $4p = \pi$. If $g(x) = c_1 x^3 + c_2 x$ in (3.4.14) with constant $c_{1,2}$ such that $c_1 c_2 < \frac{-3}{4}c_1^2\delta^2$, then the conclusion of Theorem 3.4.9 is applicable.*

Proof. We apply Theorem 3.4.9. We have $\int_0^{2\pi} g(\delta\rho\sin t)\sin t\, dt = \frac{3}{4}\pi c_1\delta^3\rho^3 + \pi\delta c_2\rho$ for $g(x) = c_1 x^3 + c_2 x$. Under assumption $c_1 c_2 < \frac{-3}{4}c_1^2\delta^2$, equation $\frac{3}{4}\pi c_1\delta^3\rho^3 + \pi\delta c_2\rho = 0$ has a simple root $\rho_0 > 1$. The proof is finished. □

We finish this subsection with the case $c_2 = 0$.

Theorem 3.4.11. *Assume that $g(x) = c_1 x$ with a constant $c_1 > 0$ such that $\pi^2 - 16p^2 > c_1^2\delta^2\pi^2$. Then there is a constant $K > 0$ such that for any μ sufficiently small there are (ω_μ, θ_μ) near to (ω_0, θ_0) given by*

$$\theta_0 = \sqrt{\left(\frac{-4p + \pi\sqrt{1 - \delta^2 c_1^2}}{c_1\pi}\right)^2 + \delta^2}, \quad \sin\omega_0 = \frac{4\delta p}{\pi\theta_0}, \; \omega_0 \in (\pi/2, \pi), \quad (3.4.20)$$

and a 2π–periodic solution y_μ of (3.4.18) satisfying

$$\sup_{t\in\mathbb{R}} \left|y_\mu(t) - \theta_\mu\sin(t - \omega_\mu)\right| \le K|\mu|.$$

Proof. Now (3.4.19) has the form

$$M_1(\omega, \theta) = 4p\sqrt{1 - \frac{\delta^2}{\theta^2}} + \pi\cos\omega + c_1\theta\pi, \quad M_2(\omega, \theta) = -4\frac{\delta p}{\theta} + \pi\sin\omega.$$

Since $\pi > 4p$ and $\theta > \delta$, we can solve ω with $\pi/2 < \omega < \pi$ from $M_2(\omega, \theta) = 0$, and inserting it into $M_1(\omega, \theta) = 0$ we see that equation $M(\omega, \theta) = 0$ is equivalent to $4p\sqrt{1 - \frac{\delta^2}{\theta^2}} - \pi\sqrt{1 - \frac{16\delta^2 p^2}{\theta^2\pi^2}} + c_1\theta\pi = 0$, i.e.

$$8\pi c_1 p\sqrt{\theta^2 - \delta^2} + c_1^2\theta^2\pi^2 = \pi^2 - 16p^2. \quad (3.4.21)$$

Since $\pi^2 - 16p^2 > c_1^2\delta^2\pi^2$ then (3.4.21) has a unique simple root $\theta_0 > \delta$ given by (3.4.20) . Then (θ_0, ω_0) is a simple root of $M(\omega, \theta) = 0$. The proof is finished by Theorem 3.4.4 when $\mathcal{B}$ is taken as a small open neighborhood of (ω_0, θ_0). □

Similarly we have

Theorem 3.4.12. *Assume that $g(x) = c_1 x$ with a constant $c_1 < 0$ such that $16p^2(1 - c_1^2\delta^2) > \pi^2$. Then there is a constant $K > 0$ such that for any μ sufficiently small there are (ω_μ, θ_μ) near to (ω_0, θ_0) given by*

$$\theta_0 = \frac{1}{\pi}\sqrt{\left(\frac{\pi - 4p\sqrt{1 - \delta^2 c_1^2}}{c_1}\right)^2 + 16\delta^2 p^2}, \quad \sin\omega_0 = \frac{4\delta p}{\pi\theta_0}, \quad \omega_0 \in (\pi/2, \pi),$$

and a 2π–periodic solution y_μ of (3.4.18) satisfying

$$\sup_{t\in\mathbb{R}} \left|y_\mu(t) - \theta_\mu \sin(t - \omega_\mu)\right| \le K|\mu|.$$

Proof. We can directly follow the proof of Theorem 3.4.11 when now (3.4.21) has a form $16p^2 = c_1^2\theta^2\pi^2 - 2c_1\pi\sqrt{\theta^2\pi^2 - 16\delta^2 p^2} + \pi^2$. $\quad\square$

3.5 Nonlinear Oscillators with Weak Couplings

3.5.1 Weakly Coupled Systems

Systems with slowly varying coefficients often arise in applications, like as an Einstein pendulum

$$\ddot{x} + \omega(\varepsilon t)^2 x = 0$$

with slowly varying frequency or a pendulum

$$\frac{d}{dt}\left(l(\varepsilon t)^2 \dot{x}\right) + l(\varepsilon t) = g(\varepsilon t, x, \dot{x})$$

with slowly varying length and some other perturbations. We have a *Duffing*-type perturbation for $g(\tau, x, \dot{x}) = \mu l(\tau)x^3 - \sigma l(\tau)\dot{x}$ [173]. Parameter $\varepsilon > 0$ is assumed to be small in the both equations. Averaging method is usually applied to study the dynamics of such systems on the intervals $\left[0, O\left(\varepsilon^{-1}\right)\right]$. We also study some systems with slowly varying coefficients in Sections 3.1 and 4.2 (see (3.1.27) and (3.1.30)) by using different methods, since averaging method is not applicable to these equations. Similarly we can formulate nonlinear oscillators when some their coefficients are governed by weakly nonlinear equations like the following Duffing-type one

$$\ddot{x} + \omega^2 x + x^3 = \varepsilon\widetilde{f}(x, \dot{x}, t), \quad \dot{\omega} = \varepsilon\widetilde{g}(x, \dot{x}), \tag{3.5.1}$$

with slowly varying frequency. In this section, we investigate more general weakly coupled equations than (3.5.1) of the form

$$\begin{aligned} x' &= \varepsilon f(x, y, t, \varepsilon) \\ y' &= g(x, y) + \varepsilon h(x, y, t, \varepsilon), \end{aligned} \tag{3.5.2}$$

where $x \in \mathbb{R}^n$, $y \in \mathbb{R}^m$, f, g, h are sufficiently smooth, f, h are 1-periodic in t and $\varepsilon > 0$ is a small parameter. For $\varepsilon = 0$, we get the *unperturbed equation*

$$\dot{y} = g(x, y), \tag{3.5.3}$$

where x is considered as a parameter. We suppose that (3.5.3) has for any x in some open subset either a single 1-periodic solution or a nondegenerate family of 1-periodic solutions. By using the averaging method which is a combination of the Lyapunov-Schmidt method together with the Brouwer degree theory, we find conditions for bifurcation of 1-periodic solutions of (3.5.2) from the above 1-periodic solutions of (3.5.3). Finally, the averaging method is also used, for instance, in [44, 48, 98, 111, 135, 141] with many interesting applications. Other nonlinear boundary value problems are investigated in [105].

3.5.2 Forced Oscillations from Single Periodics

We start with the following condition:

(H1) (3.5.3) has a 1-periodic smooth solution $y = \varphi(t, x)$ for any $x \in \Omega$, where $\Omega \subset \mathbb{R}^n$ is an open subset.

Since we are looking for 1-periodic solutions of (3.5.2) bifurcating from $\varphi(t, x)$, $x \in \Omega$, we shift the time $t \leftrightarrow t + \alpha$, $\alpha \in \mathbb{R}$ and change the variable $y \leftrightarrow y + \varphi(t, x)$ to get the equation

$$
\begin{aligned}
x' &= \varepsilon f(x, y + \varphi(t, x), t + \alpha, \varepsilon) \\
y' &= g(x, y + \varphi(t, x)) - g(x, \varphi(t, x)) + \\
&\quad \varepsilon \Big(h(x, y + \varphi(t, x), t + \alpha, \varepsilon) - \varphi_x(t, x) f(x, y + \varphi(t, x), t + \alpha, \varepsilon) \Big).
\end{aligned}
\tag{3.5.4}
$$

Like in the previous sections, first we investigate the linearization of (3.5.3) along $\varphi(t, x)$, i.e. the variational equation

$$
v' = g_y(x, \varphi(t, x))v,
\tag{3.5.5}
$$

and its dual variational system

$$
w' = -g_y^*(x, \varphi(t, x))w.
\tag{3.5.6}
$$

Certainly the function $\varphi'(t, x)$ satisfies (3.5.5). We suppose

(H2) There are smooth basis $\{v_i(t, x)\}_{i=0}^r$ and $\{w_i(t, x)\}_{i=0}^r$ of 1-periodic solutions of (3.5.5) and (3.5.6), respectively, for any $x \in \Omega$. We assume $v_0(t, x) = \varphi'(t, x)$.

In order to apply the Lyapunov-Schmidt decomposition, we introduce the Banach spaces

$$
\begin{aligned}
X &= \Big\{ x \in C(\mathbb{R}, \mathbb{R}^n) \mid x(t + 1) = x(t) \, \forall t \in \mathbb{R} \Big\} \\
Y &= \Big\{ y \in C(\mathbb{R}, \mathbb{R}^m) \mid y(t + 1) = y(t) \, \forall t \in \mathbb{R} \Big\}
\end{aligned}
$$

and then the projections $P_1 : X \to X$, $P_x : Y \to Y$ defined by

$$P_1 x := x(t) - \int_0^1 x(s)\, ds$$

$$P_x y := y(t) - q_0 w_0(t, x) - q_1 w_1(t, x) - \cdots - q_r w_r(t, x),$$

$$(q_0, q_1, \cdots, q_r)^* := A(x)^{-1} \left(\int_0^1 (y(t), w_0(t, x))\, dt, \cdots, \int_0^1 (y(t), w_r(t, x))\, dt \right)^*$$

where $(\cdot, \cdot)$ is the scalar product on $\mathbb{R}^m$ and $A(x) : \mathbb{R}^r \to \mathbb{R}^r$ is the Gram matrix given by

$$A(x) := \left(\int_0^1 (w_i(t, x), w_j(t, x))\, dt \right)_{i,j=1}^r .$$

The meaning of these projections is the following: The nonhomogeneous variational equation of (3.5.2) along $\varphi(t, x)$ is given by

$$z' = h, \quad v' = g_y(x, \varphi(t, x))v + y, \quad h \in X, \quad y \in Y. \tag{3.5.7}$$

From Lemma 3.3.1 we know that (3.5.7) has a 1-periodic solution if and only if $P_1 h = h$ and $P_x y = y$. Moreover this solution is unique if $P_1 z = z$ and $\int_0^1 (v(t), v_i(t, x))\, dt = 0$, $i = 0, 1, \cdots, r$. Let us denote it by $z := \mathcal{K} h$ and $v := \mathcal{K}_x y$. Using these projections and operators $\mathcal{K}$, $\mathcal{K}_x$, like in Section 3.3, we take in (3.5.4) the changes

$$\varepsilon \leftrightarrow \varepsilon^2, \quad x = u + x_1, \quad u \in X, \quad P_1 u = u, \quad x_1 \in \mathbb{R}^n$$

$$y = v + \varepsilon \sum_{i=1}^r \beta_i v_i(t, u + x_1), \quad \int_0^1 (v(t), v_i(t))\, dt = 0, \quad i = 0, 1, \cdots, r$$

to obtain the decomposition of (3.5.4) on

$$u = \varepsilon^2 \mathcal{K} P_1 f \left(u + x_1, v + \varepsilon \sum_{i=1}^r \beta_i v_i(t, u + x_1) + \varphi(t, u + x_1), t + \alpha, \varepsilon^2 \right)$$

$$v = \mathcal{K}_{x_1} P_{x_1} H(u, v, x_1, \varepsilon, \alpha, \beta, t),$$

$$\tag{3.5.8}$$

and

$$\int_0^1 f \left(u + x_1, v + \varepsilon \sum_{i=1}^r \beta_i v_i(t, u + x_1) + \varphi(t, u + x_1), t + \alpha, \varepsilon^2 \right) dt = 0$$

$$\tag{3.5.9}$$

$$\frac{1}{\varepsilon^2} \int_0^1 (H(u, v, x_1, \varepsilon, \alpha, \beta, t), w_j(t, u + x_1))\, dt = 0 \quad j = 0, 1, \cdots, r,$$

where $\beta := (\beta_1, \beta_2, \cdots, \beta_r)$ and

$$
H(u, v, x_1, \varepsilon, \alpha, \beta, t) := g\left(u + x_1, v + \varepsilon \sum_{i=1}^{r} \beta_i v_i(t, u + x_1) + \varphi(t, u + x_1)\right) -
$$
$$
g(u + x_1, \varphi(t, u + x_1)) -
$$
$$
g_y(u + x_1, \varphi(t, u + x_1))\left(v + \varepsilon \sum_{i=1}^{r} \beta_i v_i(t, u + x_1)\right) +
$$
$$
+\left(g_y(u + x_1, \varphi(t, u + x_1)) - g_y(x_1, \varphi(t, x_1))\right)v - \varepsilon^3 \sum_{i=1}^{r} \beta_i v_{ix}(t, u + x_1) \times
$$
$$
P_1 f\left(u + x_1, v + \varepsilon \sum_{i=1}^{r} \beta_i v_i(t, u + x_1) + \varphi(t, u + x_1), t + \alpha, \varepsilon^2\right) +
$$
$$
\varepsilon^2 h\left(u + x_1, v + \varepsilon \sum_{i=1}^{r} \beta_i v_i(t, u + x_1) + \varphi(t, u + x_1), t + \alpha, \varepsilon^2\right) -
$$
$$
\varepsilon^2 \varphi_x(t, u + x_1) \times f\left(u + x_1, v + \varepsilon \sum_{i=1}^{r} \beta_i v_i(t, u + x_1) + \varphi(t, u + x_1), t + \alpha, \varepsilon^2\right).
$$

Since

$$
H(u, v, x_1, \varepsilon, \alpha, \beta, t) = \frac{1}{2} g_{yy}(u + x_1, \varphi(t, u + x_1))\left(v + \varepsilon \sum_{i=1}^{r} \beta_i v_i(t, u + x_1)\right)^2 +
$$
$$
\varepsilon^2 h\left(u + x_1, v + \varepsilon \sum_{i=1}^{r} \beta_i v_i(t, u + x_1) + \varphi(t, u + x_1), t + \alpha, \varepsilon^2\right) -
$$
$$
\varepsilon^2 \varphi_x(t, u + x_1) \times f\left(u + x_1, v + \varepsilon \sum_{i=1}^{r} \beta_i v_i(t, u + x_1) + \varphi(t, u + x_1), t + \alpha, \varepsilon^2\right)
$$
$$
+ O\left(\left(v + \varepsilon \sum_{i=1}^{r} \beta_i v_i(t, u + x_1)\right)^3\right) + O(|u|)|v| + O(\varepsilon^3)
$$
$$
= O(|v|^2) + O(|u|)|v| + O(\varepsilon),
$$

we can uniquely solve (3.5.8) in u, v small by means of the implicit function theorem and moreover, $u = O(\varepsilon^2)$ and $v = O(\varepsilon^2)$. Then we insert these solutions to (3.5.9) to get the bifurcation equation

$$
0 = G(x_1, \beta, \alpha, \varepsilon) = G_0(x_1, \beta, \alpha) + G_1(x_1, \beta, \alpha)\varepsilon + \cdots +
$$
$$
G_p(x_1, \beta, \alpha)\varepsilon^p + Q_p(x_1, \beta, \alpha, \varepsilon)\varepsilon^{p+1} := \tag{3.5.10}
$$
$$
q_p(x_1, \beta, \alpha, \varepsilon) + Q_p(x_1, \beta, \alpha, \varepsilon)\varepsilon^{p+1},
$$

where $G(x_1, \beta, \alpha, \varepsilon)$ is the left-hand side of (3.5.9). Summarizing, we arrive at the following result.

Theorem 3.5.1. *Suppose* (H1) *and* (H2) *hold. If there is an open bounded subset* $\mathcal{O} \subset \Omega \times \mathbb{R}^r \times \mathbb{R}$ *and a constant* $c_p > 0$ *such that*

$$|q_p(x_1, \beta, \alpha, \varepsilon)| \geq c_p \varepsilon^p$$

on the boundary $\partial \mathcal{O}$ *for any* $\varepsilon > 0$ *small, and* $\deg \Big(q_p(\cdot, \cdot, \cdot, \varepsilon), \mathcal{O}, 0 \Big) \neq 0$. *Then* (3.5.2) *has a 1-periodic solution for* $\varepsilon > 0$ *small.*

Proof. To solve (3.5.10), we put it in the homotopy

$$G(x_1, \beta, \alpha, \varepsilon, \lambda) := q_p(x_1, \beta, \alpha, \varepsilon) + \lambda Q_p(x_1, \beta, \alpha, \varepsilon) \varepsilon^{p+1}$$

for $\lambda \in [0, 1]$. Then the assumptions of this theorem imply that $G(\cdot, \cdot, \cdot, \varepsilon, \lambda) \neq 0$ on $\partial \mathcal{O}$ for any $\varepsilon > 0$ small and $\lambda \in [0, 1]$. Hence

$$\deg \Big(G(\cdot, \cdot, \cdot, \varepsilon), \mathcal{O}, 0 \Big) = \deg \Big(q_p(\cdot, \cdot, \cdot, \varepsilon), \mathcal{O}, 0 \Big) \neq 0.$$

So (3.5.10) is solvable for any $\varepsilon > 0$ small. The proof is finished. $\qquad\square$

For $p = 0$, we can immediately derive from (3.5.8) and (3.5.9) that

$$G_0(x_1, \beta, \alpha) = \left(\int_0^1 f(x_1, \varphi(t, x_1), t + \alpha, 0) \, dt, \sum_{i,j=1}^r \beta_i \beta_j a_{ijk}(x_1) \right.$$

$$\left. + \int_0^1 \Big(h(x_1, \varphi(t, x_1), t + \alpha, 0) - \varphi_x(t, x_1) f(x_1, \varphi(t, x_1), t + \alpha, 0), w_k(t, x_1) \Big) \, dt \right),$$

$$(3.5.11)$$

where $k = 0, 1, 2, \cdots, r$ and

$$a_{ijk}(x_1) := \frac{1}{2} \int_0^1 (g_{yy}(x_1, \varphi(t, x_1))(v_i(t, x_1), v_j(t, x_1)), w_k(t, x_1)) \, dt.$$

Of course, higher-order terms $G_i(x_1, \beta, \alpha)$, $i \geq 1$ are much more complicated, for this reason, we do not derive their general forms.

3.5.3 Forced Oscillations from Families of Periodics

In the case that unperturbed equation (3.5.3) possesses some symmetries then very often instead of condition (H1) the following one holds (see Subsection 3.3.2)

(C1) Equation(3.5.3) has a smooth family $\varphi(t, x, \theta)$ of 1-periodic solutions for any $x \in \Omega$ and $\theta \in \Gamma$, where $\Omega \subset \mathbb{R}^n$, $\Gamma \subset \mathbb{R}^r$ are open bounded subsets.

So the symmetry causes that (3.5.3) has not only a single 1-periodic solution for any $x \in \Omega$, like in (H1), but a family parameterized by $\theta \in \Gamma$ (see arguments below (3.5.24) for a concrete problem). Then we can repeat the above procedure to (3.5.2) with the next modifications: First, (3.5.5) is changed to

$$v' = g_y(x, \varphi(t, x, \theta))v. \tag{3.5.12}$$

Clearly $\varphi'(t, x, \theta)$, $\varphi_{\theta_i}(t, x, \theta)$, $i = 1, 2, \cdots, r$, $\theta = (\theta_1, \theta_2, \cdots, \theta_r)$ are 1-periodic solutions of (3.5.12). We suppose

(C2) The family $\varphi(t, x, \theta)$ is *non-degenerate*, i.e. the functions $\widetilde{v}_0(t, x, \theta) := \varphi'(t, x, \theta)$, $\widetilde{v}_i(t, x, \theta) := \varphi_{\theta_i}(t, x, \theta)$, $i = 1, 2, \cdots, r$ form a basis of the space of 1-periodic solutions of (3.5.12).

From Subsection 3.3.2 we know that condition (C2) implies the existence of a smooth basis $\widetilde{w}_j(t, x, \theta)$, $j = 0, 1, \cdots, r$ of the space of 1-periodic solutions of the adjoint system $w' = -g_y^*(x, \varphi(t, x, \theta))w$ to (3.5.12).

Now, in the above procedure, we keep the projection P_1, but we replace P_x with $P_{x,\theta} : Y \to Y$ defined by

$$P_{x,\theta}y := y(t) - \widetilde{q}_0\widetilde{w}_0(t, x, \theta) - \widetilde{q}_1\widetilde{w}_1(t, x, \theta) - \cdots - \widetilde{q}_r\widetilde{w}_r(t, x, \theta),$$

$$(\widetilde{q}_0, \widetilde{q}_1, \cdots, \widetilde{q}_r)^* :=$$

$$\widetilde{A}(x, \theta)^{-1} \left(\int_0^1 (y(t), \widetilde{w}_0(t, x, \theta)) \, dt, \cdots, \int_0^1 (y(t), \widetilde{w}_r(t, x, \theta)) \, dt \right)^*$$

where $\widetilde{A}(x, \theta) := \left(\int_0^1 (\widetilde{w}_i(t, x, \theta), \widetilde{w}_j(t, x, \theta)) \, dt \right)_{i,j=1}^r$ is an $r \times r$-matrix. Then changing

$$x = u + x_1, \quad u \in X, \quad P_1 u = u, \quad x_1 \in \mathbb{R}^n$$

$$y = v + \varphi(t, u + x_1, \theta), \quad \int_0^1 (v(t), \widetilde{v}_i(t)) \, dt = 0, \quad i = 0, 1, \cdots, r$$

in (3.5.2) and using projections P_1, $P_{x,\theta}$, we derive like above

$$\begin{aligned} u' &= \varepsilon P_1 f(u + x_1, v + \varphi(t, u + x_1, \theta), t + \alpha, \varepsilon) \\ v' - g_y(x_1, \varphi(t, x_1, \theta))v &= P_{x_1,\theta}\widetilde{H}(u, v, \varepsilon, \alpha, \theta, t), \end{aligned} \tag{3.5.13}$$

and

$$\int_0^1 f(u + x_1, v + \varphi(t, u + x_1, \theta), t + \alpha, \varepsilon) \, dt = 0 \tag{3.5.14}$$

$$\frac{1}{\varepsilon} \int_0^1 (\widetilde{H}(u, v, \varepsilon, \alpha, \theta, t), \widetilde{w}_j(t, u + x_1, \theta)) \, dt = 0 \quad j = 0, 1, \cdots, r,$$

where

$$\widetilde{H}(u, v, \varepsilon, \alpha, \theta, t) := g(u + x_1, v + \varphi(t, u + x_1, \theta)) - $$
$$g(u + x_1, \varphi(t, u + x_1, \theta)) - g_y(x_1, \varphi(t, u + x_1, \theta))v + $$
$$\varepsilon \Big(h\left(u + x_1, v + \varphi(t, u + x_1, \theta), t + \alpha, \varepsilon\right) - $$
$$\varphi_x(t, u + x_1, \theta) f\left(u + x_1, v + \varphi(t, u + x_1, \theta), t + \alpha, \varepsilon\right) \Big).$$

Finally, using again the implicit function theorem we get solutions $u = O(\varepsilon)$ and $v = O(\varepsilon)$ of (3.5.13), and then inserting them into (3.5.14), we obtain the bifurcation equation

$$0 = \widetilde{G}(x_1, \theta, \alpha, \varepsilon) = \widetilde{G}_0(x_1, \theta, \alpha) + \widetilde{G}_1(x_1, \theta, \alpha)\varepsilon + \cdots +$$
$$\widetilde{G}_p(x_1, \theta, \alpha)\varepsilon^p + \widetilde{Q}_p(x_1, \theta, \alpha, \varepsilon)\varepsilon^{p+1} := \widetilde{q}_p(x_1, \theta, \alpha, \varepsilon) + \widetilde{Q}_p(x_1, \theta, \alpha, \varepsilon)\varepsilon^{p+1},$$

where $\widetilde{G}(x_1, \beta, \alpha, \varepsilon)$ is the left-hand side of (3.5.14). Consequently, the Brouwer degree method again gives the following result.

Theorem 3.5.2. *Suppose* (C1) *and* (C2). *If there is an open bounded subset* $\mathcal{O} \subset \Omega \times \Gamma \times \mathbb{R}$ *and a constant* $\widetilde{c}_p > 0$ *such that*

$$|\widetilde{q}_p(x_1, \beta, \alpha, \varepsilon)| \geq \widetilde{c}_p \varepsilon^p$$

on the boundary $\partial \mathcal{O}$ *for any* $\varepsilon > 0$ *small, and* $\deg\left(\widetilde{q}_p(\cdot, \cdot, \cdot, \varepsilon), \mathcal{O}, 0\right) \neq 0$. *Then* (3.5.2) *has a 1-periodic solution for* $\varepsilon > 0$ *small.*

Since again, the higher-order terms $\widetilde{G}_i(x_1, \theta, \alpha)$, $i \geq 1$ are still rather complicated, we do not derive them. For $p = 0$ from (3.5.13) and (3.5.14) we derive

$$\widetilde{G}_0(x_1, \theta, \alpha) = \left(\int_0^1 f(x_1, \varphi(t, x_1, \theta), t + \alpha, 0)\, dt, \int_0^1 \Big(h(x_1, \varphi(t, x_1, \theta), t + \alpha, 0) \right.$$

$$\left. -\varphi_x(t, x_1, \theta) f(x_1, \varphi(t, x_1, \theta), t + \alpha, 0), \widetilde{w}_k(t, x_1, \theta) \Big)\, dt \right)$$

(3.5.15)

for $k = 0, 1, 2, \cdots, r$.

3.5.4 Applications to Weakly Coupled Nonlinear Oscillators

We present in this part two examples, where we apply Theorems 3.5.1 and 3.5.2.

Example 3.5.3. We first apply Theorem 3.5.1 to the system

$$y_1' = (x^2 + 1)(y_1^2 + y_2^2)y_2 + \varepsilon y_1 \qquad (3.5.16)$$
$$y_2' = -(x^2 + 1)(y_1^2 + y_2^2)y_1 - \varepsilon y_2^3$$
$$x' = \varepsilon(y_1^3 \sin 2\pi t + y_2 \cos 2\pi t).$$

We need to verify conditions (H1) and (H2) for the unperturbed system (3.5.3) of the form

$$y_1' = (x^2 + 1)(y_1^2 + y_2^2)y_2, \quad y_2' = -(x^2 + 1)(y_1^2 + y_2^2)y_1 \qquad (3.5.17)$$

possessing a smooth family of 1-periodic solutions

$$\varphi(t, x) = \sqrt{\frac{2\pi k}{x^2 + 1}}\left(\sin 2\pi kt, \cos 2\pi kt\right) \qquad (3.5.18)$$

for $k \in \mathbb{Z} \setminus \{0\}$. The linearization of (3.5.17) along (3.5.18) is

$$v_1' = 2\pi k\left(\sin 4\pi ktv_1 + (2 + \cos 4\pi kt)v_2\right)$$
$$v_2' = -2\pi k\left((2 - \cos 4\pi kt)v_1 + \sin 4\pi ktv_2\right) \qquad (3.5.19)$$

and the adjoint system is

$$w_1' = 2\pi k\left(-\sin 4\pi ktw_1 + (2 - \cos 4\pi kt)w_2\right)$$
$$w_2' = 2\pi k\left(-(2 + \cos 4\pi kt)w_1 + \sin 4\pi ktw_2\right). \qquad (3.5.20)$$

Clearly (3.5.19) has solutions

$$v_0(t, x) = (\cos 2\pi kt, -\sin 2\pi kt)$$

$$\widetilde{v}(x, t) = \left(\sin 2\pi kt + 4\pi kt \cos 2\pi kt, \cos 2\pi kt - 4\pi kt \sin 2\pi kt\right).$$

Hence $v_0(x, t)$ is a basis of 1-periodic solutions of (3.5.19). Furthermore, the function

$$w_0(t, x) = (\sin 2\pi kt, \cos 2\pi kt)$$

is a basis of 1-periodic solutions of (3.5.20). Now we do not have parameters β. For simplicity we take $k = 1$. After some computations, the function G_0 of (3.5.11) for this case (3.5.16) has the form

$$G_0(x_1, \alpha) = \frac{\sqrt{\pi}}{2\sqrt{2}(x_1^2 + 1)^{3/2}} \times \left((2 + 3\pi + 2x_1^2)\cos 2\alpha\pi,\right.$$
$$\left. 2 - 3\pi + 2x_1^2 + \frac{x\sqrt{2\pi}}{(x_1^2 + 1)^{3/2}}(2 + 3\pi + 2x_1^2)\cos 2\alpha\pi\right). \qquad (3.5.21)$$

We immediately see that (3.5.21) has a simple root $x_1 = \sqrt{\frac{3\pi - 2}{2}}$ and $\alpha = \frac{1}{4}$. Taking a small neighborhood $\mathcal{O}$ of $\left(\sqrt{\frac{3\pi - 2}{2}}, \frac{1}{4}\right)$, Theorem 3.5.1 gives an 1-periodic solution of (3.5.16) for any $\varepsilon > 0$ small which is in an $O(\varepsilon)$-neighborhood of $\left(-\frac{2\sqrt{3}}{3}\cos 2\pi t, \frac{2\sqrt{3}}{3}\sin 2\pi t, \sqrt{\frac{3\pi - 2}{2}}\right)$.

Example 3.5.4. Finally, we consider the system

$$y_1' = y_2, \quad y_2' = -y_1 - (x^2 + 1)(y_1^2 + y_3^2)y_1 - \varepsilon\delta y_2 - \varepsilon\mu_1(y_1 - y_3) - \varepsilon\mu_2 \cos 2\pi t$$
$$y_3' = y_4, \quad y_4' = -y_3 - (x^2 + 1)(y_1^2 + y_3^2)y_3 - \varepsilon\delta y_4 - \varepsilon\mu_1(y_3 - y_1) - \varepsilon\mu_2 \cos 2\pi t$$
$$x' = \varepsilon(y_1 \sin 2\pi t + y_3 \cos 2\pi t),$$

$$(3.5.22)$$

where δ, μ_1, μ_2 are positive parameters. We verify assumptions (C1) and (C2) for its unperturbed system

$$\begin{aligned} y_1' &= y_2, \quad y_2' = -y_1 - (x^2 + 1)(y_1^2 + y_3^2)y_1 \\ y_3' &= y_4, \quad y_4' = -y_3 - (x^2 + 1)(y_1^2 + y_3^2)y_3 \,. \end{aligned} \tag{3.5.23}$$

We note that (3.5.23) has the form

$$\ddot{w} + (1 + (x^2 + 1)\|w\|^2)w = 0 \tag{3.5.24}$$

for $w = (y_1, y_3)$ and $\|w\| = \sqrt{y_1^2 + y_3^2}$. For $\Gamma(\theta) = \begin{pmatrix} \cos\theta & -\sin\theta \\ \sin\theta & \cos\theta \end{pmatrix}$ we see that if $w(t)$ solves (3.5.24) then $\Gamma(\theta)w(t)$ is also its solution. We know [132] that

$$y_1(t) = v(t, x, k) = \frac{\sqrt{2}k}{\sqrt{(1 - 2k^2)(x^2 + 1)}} \, \text{cn} \, \frac{t}{\sqrt{1 - 2k^2}}$$

solves $y_1' = y_2, y_2' = -y_1 - (x^2 + 1)y_1^3$, where cn is the Jacobi elliptic function and k is the elliptic modulus. Consequently, (3.5.23) has a smooth family of periodic solutions

$$y(t, x, \theta, k) = \Big(\cos\theta \, v(t, x, k), \cos\theta \, v(t, x, k)', \sin\theta \, v(t, x, k), \sin\theta \, v(t, x, k)' \Big)$$

$$(3.5.25)$$

The function $y(t, x, \theta, k)$ has the period $T(k) = 4K(k)\sqrt{1 - 2k^2}$ for the complete elliptic integral $K(k)$ of the first kind. We note $T(0) = 2\pi$ and $T(\sqrt{2}/2) = 0$. By numerically solving the equation $T(k) = 1$, we find its unique solution $k_0 \doteq 0.700595$ with $T(k_0)' \neq 0$. So we fix $k = k_0$ and take

$$\varphi(t, x, \theta) = y(t, x, \theta, k_0)$$

to satisfy condition (C1). Next we show the nondegeneracy of $\varphi(t, x, \theta)$ from condition (C2). The linearization of (3.5.24) at w has the form

$$\ddot{z} + (1 + (x^2 + 1)\|w\|^2)z + 2(x^2 + 1) < w, z > w = 0, \tag{3.5.26}$$

where $< \cdot, \cdot >$ is the usual scalar product on $\mathbb{R}^2$. Furthermore, we can easily check that if $z = \Gamma(\theta)z_1$ and $w = \Gamma(\theta)w_1$, then

$$\ddot{z}_1 + (1 + (x^2 + 1)\|w_1\|^2)z_1 + 2(x^2 + 1) < w_1, z_1 > w = 0. \tag{3.5.27}$$

Consequently, in order to study the linearization of (3.5.23) (or (3.5.26)) at $\varphi(t,x,\theta)$, we study the linearization of (3.5.27) at $w_1(t) = (v(t,x),0)$, $v(t,x) = v(t,x,k_0)$ which has the form

$$v_1' = v_2, \quad v_2' = -(1 + 3w(t,k_0))v_1 \tag{3.5.28}$$

$$v_3' = v_4, \quad v_4' = -(1 + w(t,k_0))v_3 \tag{3.5.29}$$

for $w(t,k) = \frac{2k^2}{1-2k^2}\,\mathrm{cn}^2\frac{t}{\sqrt{1-2k^2}}$. Equation (3.5.28) has an 1-periodic solution $v(t,x)'$ and a non-1-periodic solution $\frac{\partial}{\partial k}v(t,x,k_0)$. Equation (3.5.29) has a 1-periodic solution $v(t,x)$ and by solving numerically (3.5.29) with initial value conditions $v_3(0) = 0$, $v_4(0) = 1$, we see that the second solution of (3.5.29) is non-1-periodic. Consequently, condition (C2) is satisfied with

$$\tilde{v}_0(t,x,\theta) = \Big(\cos\theta\, v(t,x)', \cos\theta\, v(t,x)'', \sin\theta\, v(t,x)', \sin\theta\, v(t,x)'' \Big)$$
$$\tilde{v}_1(t,x,\theta) = \Big(-\sin\theta\, v(t,x), -\sin\theta\, v(t,x)', \cos\theta\, v(t,x), \cos\theta\, v(t,x)' \Big).$$

Similarly we derive

$$\tilde{w}_0(t,x,\theta) = \Big(-\cos\theta\, v(t,x)'', \cos\theta\, v(t,x)', -\sin\theta\, v(t,x)'', \sin\theta\, v(t,x)' \Big)$$
$$\tilde{w}_1(t,x,\theta) = \Big(\sin\theta\, v(t,x)', -\sin\theta\, v(t,x), -\cos\theta\, v(t,x)', \cos\theta\, v(t,x) \Big).$$

Now we insert the above formulas to (3.5.15) and by using the evenness of function cn, after some computations we get the first-order bifurcation function

$$\tilde{G}_0(x,\theta,\alpha) = \Big(\tilde{G}_{01}(x,\theta,\alpha), \tilde{G}_{02}(x,\theta,\alpha), \tilde{G}_{03}(x,\theta,\alpha) \Big),$$

where

$$\tilde{G}_{01}(x,\theta,\alpha) = \frac{\sin(2\pi\alpha + \theta)}{\sqrt{x^2+1}} \int_0^1 w(t)\cos 2\pi t\, dt,$$

$$\tilde{G}_{02}(x,\theta,\alpha) =$$

$$\frac{1}{(x^2+1)} \int_0^1 \Big\{ -\delta\dot{w}(t)^2 + \mu_2(\cos\theta + \sin\theta)\sqrt{x^2+1}\sin 2\pi t \sin 2\pi\alpha \dot{w}(t)$$

$$+ \frac{x}{(x^2+1)^{3/2}} \big(\dot{w}(t)^2 - \ddot{w}(t)w(t)\big) w(t)\cos 2\pi t \sin(2\pi\alpha + \theta) \Big\}\, dt,$$

$$\tilde{G}_{0,3}(x,\theta,\alpha) =$$

$$\frac{1}{x^2+1} \int_0^1 \Big\{ \mu_1\cos 2\theta\, w(t)^2 - \sqrt{x^2+1}\,\mu_2(\cos\theta - \sin\theta)\cos 2\pi t \cos 2\pi\alpha w(t) \Big\}\, dt,$$

and $w(t) = \frac{\sqrt{2}k_0}{\sqrt{1-2k_0^2}}\,\mathrm{cn}\frac{t}{\sqrt{1-2k_0^2}}$. In order to prove the next theorem, we need the following obvious result.

Lemma 3.5.5. *Let $F_1 \in C^1(\Omega_1 \times \Omega_2, \mathbb{R}^n)$, $F_2 \in C^1(\Omega_1 \times \Omega_2, \mathbb{R}^m)$, $\Omega_1 \subset \mathbb{R}^n$, $\Omega_2 \subset \mathbb{R}^m$ be open subsets. Suppose that for any $y \in \Omega_2$ there is a $x := f(y) \in \Omega_1$ such that $F_1(f(y), y) = 0$ and $D_x F_1(f(y), y) : \mathbb{R}^n \to \mathbb{R}^n$ is regular, i.e. $F_1(x, y) = 0$ has a simple root $x = f(y)$ in Ω_1 for any $y \in \Omega_2$. Assume that $G(y) := F_2(f(y), y) = 0$ has a simple root $y_0 \in \Omega_2$, i.e. $G(y_0) = 0$ and $DG(y_0)$ is regular. Then (x_0, y_0), $x_0 := f(y_0)$ is a simple root of $F = (F_1, F_2)^*$, i.e. $F(x_0, y_0) = 0$ and $DF(x_0, y_0)$ is regular. Note a local uniqueness of simple roots and their smooth dependence on parameters follow from the implicit function theorem, so we suppose that $f \in C^1(\Omega_2, \Omega_1)$.*

Proof. From $DG(y) = D_y F_2(f(y), y) - D_x F_2(f(y), y) D_x F_1(f(y), y)^{-1} D_y F_1(f(y), y)$ and $DF = \begin{pmatrix} D_x F_1 & D_y F_1 \\ D_x F_2 & D_y F_2 \end{pmatrix}$, we derive

$$\begin{pmatrix} D_x F_1^{-1} & 0_{n \times m} \\ -D_x F_2 D_x F_1^{-1} & \mathbb{I}_{m \times m} \end{pmatrix} \circ DF = \begin{pmatrix} \mathbb{I}_{n \times n} & D_x F_1^{-1} D_y F_1 \\ 0_{m \times n} & DG \end{pmatrix}.$$

Hence $DG(y_0)$ is regular if and only if $DF(x_0, y_0)$ is regular. The proof is finished. $\square$

We are ready to prove the following result [93].

Theorem 3.5.6. *If one of the following assumptions is satisfied*

$$\delta \neq 0.313471\mu_1, \quad \delta > 0.02236\mu_2, \tag{3.5.30}$$

$$F(6.38018\delta/\mu_1) < 44.7227\delta/\mu_2, \quad \delta \neq 0.313471\mu_1, \tag{3.5.31}$$

$$0.00463021\mu_2 < \delta < 0.0211411\mu_1, \tag{3.5.32}$$

where

$$F(A) = 3A\left(1 + A + \frac{A^2 - 4A + 1}{C^{1/3}} + C^{1/3}\right)^{-1},$$

$$C = 1 + 21A - 6A^2 + A^3 + 3\sqrt{6A + 42A^2 - 18A^3 + 3A^4}.$$

Then system (3.5.22) possesses an 1-periodic solution for any $\varepsilon > 0$ small.

Proof. In order to apply Theorem 3.5.2, we search for a simple root of $\widetilde{G}_0(x, \theta, \alpha)$ by using Lemma 3.5.5. Then its Brouwer index is nonzero. Since $\int_0^1 w(t)$ $\cos 2\pi t \, dt \doteq 3.49859$, we see that $\widetilde{G}_{01}(x, \theta, \alpha) = 0$ gives two possibilities: either $2\pi\alpha + \theta = 0$ or $2\pi\alpha + \theta = \pi$. Then $\widetilde{G}_{02}(x, \theta, \alpha) = 0$ implies

$$\sqrt{x^2 + 1} = \frac{\delta \int_0^1 \dot{w}(t)^2 \, dt}{\mu_2 (\cos\theta + \sin\theta) \sin 2\pi\alpha \int_0^1 \dot{w}(t) \sin 2\pi t \, dt} > 1 \tag{3.5.33}$$

By inserting (3.5.33) into $\widetilde{G}_{03}(x, \theta, \alpha) = 0$ we get the following equivalent equation

$$\delta(\cos\theta - \sin\theta)\cos 2\pi\alpha \int_0^1 \dot{w}(t)^2\, dt \int_0^1 w(t)\cos 2\pi t\, dt -$$

$$\mu_1 \cos 2\theta(\cos\theta + \sin\theta)\sin 2\pi\alpha \int_0^1 w(t)^2\, dt \int_0^1 \dot{w}(t)\sin 2\pi t\, dt = 0 \qquad (3.5.34)$$

First we consider the case $2\pi\alpha = -\theta$. Then from (3.5.33) we obtain

$$\frac{\delta}{\mu_2(\cos\theta + \sin\theta)\sin\theta} > 0.02236\,, \qquad (3.5.35)$$

while (3.5.34) gives

$$(\cos\theta - \sin\theta)\left\{6.38018\delta\cos\theta - \mu_1(\sin\theta + \cos\theta)^2\sin\theta\right\} = 0 \qquad (3.5.36)$$

To solve (3.5.36), we first consider either $\theta_0 = \pi/4$ or $\theta_0 = 5\pi/4$, which are simple roots of (3.5.36) if $\delta/\mu_1 \neq 0.313471$. Then by inserting $\theta = \theta_0$ into (3.5.35), we get $\delta/\mu_2 > 0.02236$. So Theorem 3.5.6 is proved when (3.5.30) holds, since $\left(\sqrt{2000.1216\frac{\delta^2}{\mu_2^2} - 1}, \theta_0, -\frac{\theta_0}{2\pi}\right)$ is a simple root of $\widetilde{G}_0(x, \theta, \alpha)$.

Now we consider that $\theta \neq \theta_0$ and still $2\pi\alpha = -\theta$, then (3.5.35) and (3.5.36) are equivalent to

$$A = 6.38018\delta/\mu_1 = \frac{(1 + \tan\theta)^2}{1 + \tan^2\theta}\tan\theta = \Psi_1(\tan\theta) \qquad (3.5.37)$$

and

$$44.7227\delta/\mu_2 > \frac{1 + \tan\theta}{1 + \tan^2\theta}\tan\theta = \Psi_2(\tan\theta) > 0\,. \qquad (3.5.38)$$

So we take $\theta \in (0, \pi/2) \setminus \{\pi/4\}$. Since Ψ_1 is increasing on $[0, \infty)$ and $\Psi_1(0) = 0$, we see that (3.5.37) is uniquely solvable in $\tan\theta$ as a function of $A \geq 0$. Then inserting this solution into the right hand side of (3.5.38), we obtain (3.5.31). Note condition $\delta/\mu_1 \neq 0.313471$ comes from $6.38018\delta/\mu_1 = \Psi_1(\tan\theta) \neq \Psi_1(\tan\pi/4) = 2$. So Theorem 3.5.6 is proved also for this assumption, since

$$\left(\sqrt{\frac{2000.1216\delta^2}{\mu_2^2\Psi_2(\tan\theta_1)^2} - 1}, \theta_1, -\frac{\theta_1}{2\pi}\right)$$

with $\theta_1 = \arctan[\Psi_1^{-1}(6.38018\delta/\mu_1)] \in (0, \pi/2) \setminus \{\pi/4\}$ is a simple root of $\widetilde{G}_0(x, \theta, \alpha)$.

Finally, we consider the case $2\pi\alpha = \pi - \theta$. Then (3.5.37) and (3.5.38) are changing to

$$-6.38018\delta/\mu_1 = \Psi_1(\tan\theta) \qquad (3.5.39)$$

and

$$-44.7227\delta/\mu_2 < \Psi_2(\tan\theta) < 0\,, \qquad (3.5.40)$$

respectively. So we take $\theta \in (-\pi/4, 0)$. Now the situation is different: functions $\Psi_{1,2}$ are not invertible on interval $\mathcal{I} = [-1, 0]$. They are both non-positive on $\mathcal{I}$. Function Ψ_1 has the minimum -0.134884 on $\mathcal{I}$ at -0.295598, while function Ψ_2 has the minimum -0.207107 on $\mathcal{I}$ at -0.414214. So in order to solve (3.5.39) we suppose $-6.38018\delta/\mu_1 > -0.134884$, while the condition $-44.7227\delta/\mu_2 < -0.207107$ is sufficient for holding (3.5.40). We can put these two inequalities into one (3.5.32). So Theorem 3.5.6 is proved also for the last assumption, since $\widetilde{G}_0(x, \theta, \alpha)$ has a simple root at

$$\left(\sqrt{\frac{2000.1216\delta^2}{\mu_2^2 \Psi_2 (\tan\theta_2)^2} - 1}, \theta_2, \frac{\pi - \theta_2}{2\pi} \right)$$

for $\theta_2 = \arctan[\Psi_1^{-1} (6.38018\delta/\mu_1)] \in (-\pi/4, 0)$. $\qquad \square$

Remark 3.5.7. When several conditions of (3.5.30–3.5.32) hold simultaneously, then we get multiple 1-periodic solutions. For instance, we can numerically check that function $F(A)$ has a global maximum $F(A_0) \doteq 1.20711$ on $[0, \infty)$ at $A_0 \doteq 4.12132$. Hence for $\delta/\mu_2 > 0.02699$ and $\delta/\mu_1 \neq 0.313471$ both (3.5.30) and (3.5.31) are satisfied, and we get 3 different 1-periodic solutions of (3.5.22): 2 solutions bifurcating for $\theta_{0,1} = \pi/4$, $\theta_{0,2} = 5\pi/4$, and the 3rd one for $\theta_1 \in (0, \pi/2) \setminus \{\pi/4\}$ in (3.5.25). Moreover, for $0.02699\mu_2 < \delta < 0.0211411\mu_1$ we have 4 different 1-periodic solutions of (3.5.22), the 4th one bifurcating for $\theta_2 \in (-\pi/4, 0)$ in (3.5.25).

The case $\delta = 0$ is a different situation. Theorem 3.5.2 gives the following result.

Theorem 3.5.8. *If $\delta = 0$ and $\mu_1 > 0.14264\mu_2$ then system (3.5.22) possesses an 1-periodic solution for any $\varepsilon > 0$ small.*

Proof. We again search for a simple root of $\widetilde{G}_0(x, \theta, \alpha)$. The form of bifurcation function $\widetilde{G}_0(x, \theta, \alpha)$ remains and $\widetilde{G}_{01}(x, \theta, \alpha) = 0$ gives still that either $2\pi\alpha = -\theta$ or $2\pi\alpha = \pi - \theta$. But equations $\widetilde{G}_{0j}(x, \theta, \alpha) = 0$, $j = 2, 3$ now imply the following ones

$$(\cos\theta + \sin\theta)\sin\theta = 0 ,$$
$$(\cos\theta - \sin\theta)(7.0107\mu_1(\cos\theta + \sin\theta) - \mu_2\cos 2\pi\alpha\sqrt{x^2 + 1}) = 0 . \tag{3.5.41}$$

By analyzing system (3.5.41), we get a solution:

$$\theta = 0, \quad \alpha = 0, \quad 7.0107\mu_1/\mu_2 = \sqrt{x^2 + 1} .$$

For $\mu_1 > 0.14264\mu_2$, we get a simple root $\left(\sqrt{49.1499\mu_1^2/\mu_2^2 - 1}, 0, 0 \right)$ of $\widetilde{G}_0(x, \theta, \alpha)$. The proof is finished. $\qquad \square$

Of course, (3.5.1) could be similarly handled like (3.5.22).

Chapter 4

Bifurcation of Chaotic Solutions

4.1 Chaotic Differential Inclusions

4.1.1 Nonautonomous Discontinuous O.D.Eqns

Motivated by several coupled oscillators with small quasiperiodic forcing terms and with small dry friction effects like the following one

$$\ddot{x} = x - 2x(x^2 + y^2) + \varepsilon\mu_1 \cos\omega_1 t - \varepsilon\delta_1 \operatorname{sgn}\dot{x},$$
$$\ddot{y} = y - 2y(x^2 + y^2) + \varepsilon\mu_2 \cos\omega_2 t - \varepsilon\delta_2 \operatorname{sgn}\dot{y}, \tag{4.1.1}$$

where $\omega_1 > \omega_2$, $\delta_{1,2}$ and $\mu_{1,2}$ are positive constants and $\varepsilon > 0$ is a small parameter, in this section, we consider differential inclusions of the form

$$\dot{x}(t) \in f(x(t)) + \varepsilon g(x(t), \varepsilon, t) \quad \text{a.e. on} \quad \mathbb{R} \tag{4.1.2}$$

with $x \in \mathbb{R}^n$ and $\varepsilon \in \mathbb{R}$ small. Similar systems are studied in Sections 3.1 and 3.3 with periodic perturbations. In this section we proceed in this investigation to show more complicated solutions under the following assumptions about (4.1.2):

(i) $f \in C^2(\mathbb{R}^n, \mathbb{R}^n)$ and $g : \mathbb{R}^n \times \mathbb{R} \times \mathbb{R} \to 2^{\mathbb{R}^n} \setminus \{\emptyset\}$ has a form $g(x, \varepsilon, t) = F(x, \varepsilon) + G(x, \varepsilon, t)$, where mapping $F : \mathbb{R}^n \times \mathbb{R} \to 2^{\mathbb{R}^n} \setminus \{\emptyset\}$ is upper-semicontinuous with compact and convex values, and $G \in C(\mathbb{R}^{n+2}, \mathbb{R}^n)$. Moreover, for any bounded subset $\Omega \subset \mathbb{R}^{n+1}$, mapping $G(x, \varepsilon, t)$ is bounded and uniformly continuous on $\Omega \times \mathbb{R}$.

(ii) $f(0) = 0$ and the eigenvalues of $Df(0)$ lie off the imaginary axis.

(iii) The unperturbed equation has a homoclinic solution. That is, there exists a differentiable function $t \to \gamma(t) \neq 0$ such that $\lim_{t \to +\infty} \gamma(t) = \lim_{t \to -\infty} \gamma(t) = 0$ and $\dot{\gamma}(t) = f(\gamma(t))$.

M. Fečkan, *Topological Degree Approach to Bifurcation Problems*, 121–142.
© Springer Science + Business Media B.V., 2008

Hence we again suppose a homoclinic structure for the unperturbed equation $\dot{x} = f(x)$ of (4.1.2) like in Section 3.1, where bifurcation of infinitely many subharmonics from homoclinics is shown. Here we study bifurcation of more oscillatory solutions when perturbations are not necessary periodic. The most interesting case is when

$$g(x, \varepsilon, t) = q(x, \varepsilon, \omega_1 t, \cdots, \omega_m t) \tag{4.1.3}$$

for $\omega_1, \cdots, \omega_m \in \mathbb{R}$ and the multivalued mapping $q : \mathbb{R}^n \times \mathbb{R} \times \mathbb{R}^m \to 2^{\mathbb{R}^n} \setminus \{\emptyset\}$ is upper-semicontinuous with compact and convex values. Moreover, the multivalued mapping $q(x, \varepsilon, \theta_1, \cdots, \theta_m)$ is 1-periodic in each θ_i, $i = 1, 2, \cdots, m$. We note that our method of Subsection 4.1.5 is applicable to (4.1.3).

When $g(x, \varepsilon, t)$ is periodic in t, this problem is also solved in [10–12]. Almost periodic ordinary differential equations are investigated in [152, 160, 195] while partial differential equations are studied in [38]. Our main result on (4.1.2) is as follows: For multivalued almost periodic perturbations, we find conditions ensuring that for any sequence $E = \{e_j\}_{j \in \mathbb{Z}} \in \mathcal{E}$ and $\varepsilon > 0$ small, there is a bounded solution of (4.1.2) on $\mathbb{R}$. Moreover, different solutions correspond to different sequences and in addition, each sequence E characterizes (or counts) turnings of the corresponding solution around γ. These chaotic solutions are in a narrow layer around γ. Consequently, we extend the deterministic chaos of Section 2.5.3 to almost periodically perturbed problems of (4.1.2).

4.1.2　The Linearized Equation

To prove our main results, we extend the method of Section 3.1 to (4.1.2). Hence we first extend Theorem 3.1.4 to more general cases as follows. First, we fix $\tau > 0$ and define the following Banach spaces:

$$Z_\tau = C\left([-\tau, \tau], \mathbb{R}^n\right), \quad Y_\tau = L^\infty\left([-\tau, \tau], \mathbb{R}^n\right)$$

with supremum norms $\| \cdot \|_\tau$ and $| \cdot |_\tau$ for Z_τ, respectively for Y_τ. Then we consider the linear equation

$$\dot{v}(t) = Df(0)v(t), \tag{4.1.4}$$

and put

$$V(t) = e^{tDf(0)}, \quad Q_{us} = C \begin{pmatrix} \mathbb{I}_s & 0 \\ 0 & 0 \end{pmatrix} C^{-1}$$

$$Q_{ss} = 0, \quad Q_{uu} = 0, \quad Q_{su} = C \begin{pmatrix} 0 & 0 \\ 0 & \mathbb{I}_u \end{pmatrix} C^{-1},$$

where C is from Theorem 3.1.2. The next result follows directly from assumption (ii).

Theorem 4.1.1. *By considering in Theorems 3.1.2-3.1.3 the exchanges*

$$U(t) \leftrightarrow V(t), \quad P_{ss} \leftrightarrow Q_{ss}, \quad P_{su} \leftrightarrow Q_{su}, \quad P_{us} \leftrightarrow Q_{us}, \quad P_{uu} \leftrightarrow Q_{uu},$$

Theorems 3.1.2–3.1.3 are valid for (4.1.4).

Now for any finite sequence $E_p = \{e_j\}_{j=-p}^p \in \{0,1\}^{2p+1}$, $p \in \mathbb{N}$ we put:

if $e_j = 1$ then

$$A_j(t) = Df(\gamma(t)), \quad U_j = U, \quad P_{ss}^j = P_{ss}$$
$$P_{su}^j = P_{su}, \quad P_{us}^j = P_{us}, \quad P_{uu}^j = P_{uu};$$

if $e_j = 0$ then

$$A_j(t) = Df(0), \quad U_j = V, \quad P_{ss}^j = Q_{ss}$$
$$P_{su}^j = Q_{su}, \quad P_{us}^j = Q_{us}, \quad P_{uu}^j = Q_{uu}.$$

Let $I_p = \{-p, -p+1, \cdots, p\}$. Now we fix a sequence $\{t_i\}_{i\in\mathbb{Z}}$, $t_i < t_{i+1}$ such that $t_i \to \pm\infty$ as $i \to \pm\infty$. For any $\alpha \in A_{E_p}$, where the set A_{E_p} is defined by

$$A_{E_p} = \left\{ (\alpha_{-p}, \cdots, \alpha_p) \in \mathbb{R}^{2p+1} : \alpha_j \in \mathbb{R} \text{ if } e_j = 1 \text{ and } \alpha_j = 0 \text{ if } e_j = 0 \right\},$$

we consider the non-homogeneous coupled linear equations

$$\dot{z}_j = A_j(t - \alpha_j)z_j + h_j, \quad j \in I_p, \quad h_j \in Y_{\tau_j}$$
$$z_j(\tau_j) = z_{j+1}(-\tau_{j+1}) \quad \text{for} \quad -p \le j \le p-1 \qquad (4.1.5)$$
$$z_p(\tau_p) = z_{-p}(-\tau_{-p}),$$

for $\tau_j = (t_{j+1} - t_j)/2$, $j \in I_p$. We put

$$\mathcal{Y}_p = \times_{j \in I_p} Y_{\tau_j}$$

with the norm $|h|_p = \max_{j \in I_p} |h_j|_{\tau_j}$, $h = (h_{-p}, \cdots, h_p)$, $h_j \in Y_{\tau_j}$, $j \in I_p$. Repeating arguments of the proof of Theorem 3.1.4, we have a Fredholm-like alternative result for (4.1.5).

Theorem 4.1.2. *For any $K > 0$, there exist $m_0 > 0$, $A > 0$, $B > 0$ such that if $\tau_j > m_0$ for every $j \in I_p$ and $\alpha \in A_{E_p}$ such that $|\alpha| \le K$, then there exist linear functions $\mathcal{L}_{\alpha,j} : \mathcal{Y}_p \to \mathbb{R}^n$, $j \in I_p$ with $\|P_{uu}^j \mathcal{L}_{\alpha,j}\| \le A e^{-2M\tau_j}$ and with the property that if $h \in \mathcal{Y}_p$ satisfies*

$$\int_{-\tau_j}^{\tau_j} P_{uu}^j U_j(t - \alpha_j)^{-1} h_j(t)\, dt + P_{uu}^j \mathcal{L}_{\alpha,j} h = 0$$

for every $j \in I_p$, then (4.1.5) has solutions in $z_j \in Z_{\tau_j}$ satisfying

$$P_{ss}^j U_j(-\alpha_j)^{-1} z_j(0) = 0 \quad \text{and} \quad \max_{j \in I_p} \|z_j\|_{\tau_j} \le B \max_{j \in I_p} |h_j|_{\tau_j}.$$

Moreover, these solutions z_j depend linearly on h and continuously on α as well.

Proof. Putting $\tilde{U}_j(t) = U_j(t - \alpha_j)$, $j \in I_p$, we consider the following two solutions to (4.1.5):

$$z_{1,j}(t) = \tilde{U}_j(t)P_{su}^j\xi_{1,j} + \tilde{U}_j(t)(P_{us}^j + P_{uu}^j)\tilde{U}_j(-\tau_j)^{-1}\varphi_{1,j}$$

$$+\tilde{U}_j(t)\int_0^t (P_{ss}^j + P_{su}^j)\tilde{U}_j(s)^{-1}h_j(s)\,ds$$

$$+\tilde{U}_j(t)\int_{-\tau_j}^t (P_{us}^j + P_{uu}^j)\tilde{U}_j(s)^{-1}h_j(s)\,ds\,,$$

$$z_{2,j}(t) = \tilde{U}_j(t)P_{us}^j\xi_{2,j} + \tilde{U}_j(t)(P_{su}^j + P_{uu}^j)\tilde{U}_j(\tau_j)^{-1}\varphi_{2,j}$$

$$+\tilde{U}_j(t)\int_0^t (P_{ss}^j + P_{us}^j)\tilde{U}_j(s)^{-1}h_j(s)\,ds$$

$$-\tilde{U}_j(t)\int_t^{\tau_j} (P_{su}^j + P_{uu}^j)\tilde{U}_j(s)^{-1}h_j(s)\,ds$$

for arbitrary $\xi_{1,j}$, $\xi_{2,j}$, $\varphi_{1,j}$ and $\varphi_{2,j}$. They satisfy $P_{ss}^j\tilde{U}_j(0)^{-1}z_j(0) = 0$, while we consider $z_{1,j}(t)$ for $t \in [-\tau_j, 0]$ and $z_{2,j}(t)$ for $t \in [0, \tau_j]$. To find the desired solution, we consider the first matching conditions determined by $z_{1,j}(0) = z_{2,j}(0)$ and then the second ones given at end points determined by $z_{1,j+1}(-\tau_{j+1}) = z_{2,j}(\tau_j)$. Now we see that the proof of Theorem 3.1.4 can be directly modified, so we omit further details and refer the reader to [81] for a complete proof. $\square$

Following Subsection 3.1.2, we define closed linear subspaces $Y_{\alpha,j,E_p} \subset \mathcal{Y}_p$ by

$$Y_{\alpha,j,E_p} = \left\{ z \in \mathcal{Y}_p : \int_{-\tau_j}^{\tau_j} P_{uu}^j U_j(t - \alpha_j)^{-1}z_j(t)\,dt + P_{uu}^j\mathcal{L}_{\alpha,j}z = 0 \right\}$$

and a variation of constants map

$$K_{\alpha,E_p} : Y_{\alpha,E_p} = \bigcap_{j\in I_p} Y_{\alpha,j,E_p} \to \times_{j\in I_p} \mathcal{Z}_{\tau_j} = \mathcal{Z}_p$$

by taking $K_{\alpha,E_p}(h)$, $h = (h_{-p}, \cdots, h_p)$ to be the solution in $\mathcal{Z}_p$ to (4.1.5) from Theorem 4.1.2. Then K_{α,E_p} is a compact linear operator with the norm $\|K_{\alpha,E_p}\|$ uniformly bounded with respect to E_p and $\alpha \in \mathcal{A}_{E_p}$ bounded. We note that the norm on $\mathcal{Z}_p$ is defined as $\max_{j\in I_p} \|z_j\|_{\tau_j}$.

Remark 4.1.3. If $e_j = 0$ then $Y_{\alpha,j,E_p} = \mathcal{Y}_p$ and $P_{uu}^j = 0$.

Finally, Lemma 3.1.5 can be also simply adapted to closed linear subspaces Y_{α,j,E_p}.

4.1.3 Bifurcation of Chaotic Solutions

We find chaotic solutions of (4.1.2) in this subsection. First we set a mapping $M : \mathbb{R}^d \to 2^{\mathbb{R}^d} \setminus \{\emptyset\}$, $M = (M_1, \cdots, M_d)$ given by

$$M_l(\alpha, \beta) = \left\{ \int_{-\infty}^{\infty} \langle h(s), u_l^{\perp}(s) \rangle \, ds : h \in L^2_{\mathrm{loc}}(\mathbb{R}, \mathbb{R}^n) \text{ satisfying a.e. on } \mathbb{R} \right.$$

$$\left. h(t) \in \left(\frac{1}{2} \sum_{i,r=1}^{d-1} \beta_i \beta_r D^2 f(\gamma(t))(u_{d+i}(t), u_{d+r}(t)) + g(\gamma(t), 0, t + \alpha) \right) \right\}.$$

$$(4.1.6)$$

Like in Subsection 3.1.3 we know that M is upper–semicontinuous with compact convex values and maps bounded sets into bounded ones. Now we can prove the following result.

Theorem 4.1.4. *Let (i–iii) hold and $d > 1$. If there are non-empty open bounded sets $\mathcal{B}_j \subset \mathbb{R}^{d-1}$, $j \in \mathbb{Z}$ along with a sequence of intervals $\{(a_j, b_j)\}_{j \in \mathbb{Z}}$ such that*

(i) $a_j \to +\infty$ as $j \to +\infty$, and $b_j \to -\infty$ as $j \to -\infty$

(ii) $\bigcup_{j \in \mathbb{Z}} \mathcal{B}_j$ is bounded and $\sup_{j \in \mathbb{Z}}(b_j - a_j) < \infty$

(iii) $\inf_{j \in \mathbb{Z}} \mathrm{dist}\left(0, M\left(\partial((a_j, b_j) \times \mathcal{B}_j)\right)\right) > 0$

(iv) $\deg(M, (a_j, b_j) \times \mathcal{B}_j, 0) \neq 0$

Then there is a constant $K > 0$ and for any sufficiently small $s > 0$ there are increasing sequences $\{t_j\}_{j \in \mathbb{Z}}$, $t_{j+1} - t_j \geq 2/s$, $\forall j \in \mathbb{Z}$, and $\{k(j)\}_{j \in \mathbb{Z}}$, $k(j) \in \mathbb{Z}$, such that for $\varepsilon = s^2$ and any infinite sequence $E = \{e_j\}_{j \in \mathbb{Z}} \in \mathcal{E}$, the differential inclusion (4.1.2) possesses a solution $x_{E,s}$ satisfying

$$\sup_{t_j \leq t \leq t_{j+1}} |x_{E,s}(t) - \gamma(t - \bar{\alpha}_{j,E,s})| \leq Ks \quad when \quad e_j = 1,$$

$$\sup_{t_j \leq t \leq t_{j+1}} |x_{E,s}(t)| \leq Ks^2 \quad when \quad e_j = 0,$$

$$(4.1.7)$$

where $\bar{\alpha}_{j,E,s} \in (a_{k(j)}, b_{k(j)}) \subset (t_j, t_{j+1})$ for any $j \in \mathbb{Z}$ with $e_j = 1$. The mapping $E \to z_{E,s}$ is injective.

Proof. Let $E = \{e_j\}_{j \in \mathbb{Z}} \in \mathcal{E}$ be given. We first find solutions of (4.1.2) associated to any $E_p = \{e_j\}_{j \in I_p} \in \{0, 1\}^{2p+1}$, $p \in \mathbb{N}$ and then by passing to the limit $p \to \infty$, we show the desired solutions. We closely follow the method of Section 3.1. We start with a construction of sequence $\{t_i\}_{i \in \mathbb{Z}}$ as follows: By using assumptions (i), (ii) of this theorem, we choose step by step an increasing sequence $\{k(j)\}_{j \in \mathbb{Z}}$, $k(j) \in \mathbb{Z}$ such that $\bar{t}_j = \frac{a_{k(j)} + b_{k(j)}}{2}$, $j \in \mathbb{Z}$ satisfy

$$\bar{t}_j \geq \frac{1}{s} + 2\bar{t}_{j-1} - 2\bar{t}_{j-2} + \cdots + (-1)^{j+1} 2\bar{t}_0, \quad j \geq 1$$
$$\bar{t}_0 \geq \frac{1}{s}, \quad \bar{t}_{-1} \leq -\frac{1}{s},$$
$$\bar{t}_j \leq -\frac{1}{s} + 2\bar{t}_{j+1} - 2\bar{t}_{j+2} + \cdots + (-1)^j 2\bar{t}_{-1}, \quad j \leq -2.$$

Here we suppose $2/\left(\sup\limits_{j\in\mathbb{Z}}(b_j-a_j)\right) > s > 0$. Then we get by

$$t_j = 2\bar{t}_{j-1} - 2\bar{t}_{j-2} + \cdots + (-1)^{j+1}2\bar{t}_0, \quad j \geq 1$$
$$t_0 = 0$$
$$t_j = 2\bar{t}_j - 2\bar{t}_{j+1} + \cdots + (-1)^{j+1}2\bar{t}_{-1}, \quad j \leq -1$$

an increasing sequence $\{t_j\}_{j\in\mathbb{Z}}$ such that $t_{j+1} - t_j \geq 2/s$, $\bar{t}_j = \frac{t_{j+1}+t_j}{2} = \frac{a_{k(j)}+b_{k(j)}}{2}$ and $(a_{k(j)}, b_{k(j)}) \subset (t_j, t_{j+1})$.

Next we fix E_p. For any $\alpha \in \mathcal{A}_{E_p}$ we put:

if $e_j = 1$ then

$$\gamma_j(t) = \gamma(t - \alpha_j), \quad \beta_j = (\beta_{1,j}, \cdots, \beta_{d-1,j}) \in \mathbb{R}^{d-1}$$
$$u_{i+d,j}(t) = u_{i+d}(t - \alpha_j), \ i \in \{1, \cdots, d-1\};$$

if $e_j = 0$ then

$$\gamma_j(t) = 0, \quad \beta_j = (0, \cdots, 0) \in \mathbb{R}^{d-1},$$
$$u_{i+d,j} = 0, \ i \in \{1, \cdots, d-1\}.$$

We define the functions b_j, $j \in I_p$ by

$$b_j(\alpha_j, \alpha_{j+1}, \beta_j, \beta_{j+1}, r_1, r_2)$$
$$= \gamma_{j+1}(-r_2) - \gamma_j(r_1) + \sum_{i=1}^{d-1}\left(\beta_{i,j+1}u_{i+d,j+1}(-r_2) - \beta_{i,j}u_{i+d,j}(r_1)\right).$$

Clearly
$$|b_j(\alpha_j, \alpha_{j+1}, \beta_j, \beta_{j+1}, r_1, r_2)| = O(e^{-2Mr})$$

as $r = \min\{r_1, r_2\} \to \infty$ uniformly with respect to bounded $\alpha_j, \alpha_{j+1}, \beta_j, \beta_{j+1}$.

Then we consider (4.1.2) with $\varepsilon = s^2$ on the interval $[t_{-p}, t_{p+1}]$ and we make on each interval $[t_j, t_{j+1}]$, $j \in I_p$ the changes of variables

$$x(t + \bar{t}_j) = \gamma_j(t) + s^2 z_j(t) + \sum_{i=1}^{d-1} s\beta_{i,j}u_{i+d,j}(t)$$

$$+ \frac{1}{2\tau_j}b_j(\alpha_j, \alpha_{j+1}, s\beta_j, s\beta_{j+1}, \tau_j, \tau_{j+1})(t + \tau_j), \quad z_j \in Z_{\tau_j},$$

(4.1.8)

where $\tau_j = (t_{j+1} - t_j)/2$, $j \in I_p$. We consider in (4.1.8) for any j with $e_j = 1$ that $\beta_j \in \mathcal{B}_{k(j)}$ and $\alpha_j \in \left(-\frac{b_{k(j)}-a_{k(j)}}{2}, \frac{b_{k(j)}-a_{k(j)}}{2}\right)$. Assumption (ii) of this theorem implies that β_j, α_j are uniformly bounded with respect to $j \in \mathbb{Z}$. We set $\tau_{p+1} = \tau_{-p}$. Note if $z_j(\tau_j) = z_{j+1}(-\tau_{j+1})$ for $-p \leq j \leq p-1$ and $z_p(\tau_p) = z_{-p}(-\tau_{-p})$, and z_j are continuous then x in (4.1.8) is continuously extended on $[t_{-p}, t_{p+1}]$.

Inserting (4.1.8) into (4.1.2), the differential inclusions for z_j, $j \in I_p$ are

$$\dot{z}_j(t) - Df(\gamma_j(t))z_j(t) \in g_{s,j,E_p}(z_j(t), \alpha_j, \alpha_{j+1}, \beta_j, \beta_{j+1}, t)$$

(4.1.9)

a.e. on $[-\tau_j, \tau_j]$, $j \in I_p$, where

$$
g_{s,j,E_p}(x, \alpha_j, \alpha_{j+1}, \beta_j, \beta_{j+1}, t) = \left\{ v \in \mathbb{R}^n : v \in \tfrac{1}{s^2} \Big\{ f\Big(s^2 x + \gamma_j(t) \right.
$$
$$
+ s \sum_{i=1}^{d-1} \beta_{i,j} u_{i+d,j}(t) + \tfrac{1}{2\tau_j} b_j(\alpha_j, \alpha_{j+1}, s\beta_j, s\beta_{j+1}, \tau_j, \tau_{j+1})(t + \tau_j)\Big) - f(\gamma_j(t))
$$
$$
- s \sum_{i=1}^{d-1} \beta_{i,j} \dot{u}_{i+d,j}(t) - \tfrac{1}{2\tau_j} b_j(\alpha_j, \alpha_{j+1}, s\beta_j, s\beta_{j+1}, \tau_j, \tau_{j+1}) - Df(\gamma_j(t)) s^2 x \Big\}
$$
$$
+ g\Big(s^2 x + \gamma_j(t) + s \sum_{i=1}^{d-1} \beta_{i,j} u_{i+d,j}(t)
$$
$$
\left. + \tfrac{1}{2\tau_j} b_j(\alpha_j, \alpha_{j+1}, s\beta_j, s\beta_{j+1}, \tau_j, \tau_{j+1})(t + \tau_j), s^2, t + \bar{t}_j \Big) \right\}.
$$

We note that $Df(\gamma_j(t)) = A_j(t - \alpha_j)$ in the notations of Subsection 4.1.2. Now we can repeat with the help of Theorem 4.1.2 the arguments of Subsection 3.1.3 to solve (4.1.9) in $\mathcal{Z}_p$. We omit details, since we can directly modify the proofs without any changes to arrive at a similar inclusion like (3.1.20). So for $j \in I_p$ such that $e_j = 1$, according to Subsection 3.1.3 (see (3.1.20)), (4.1.9) is homotopically associated to a mapping $\tilde{M}_j : \mathbb{R}^d \to 2^{\mathbb{R}^d} \setminus \{\emptyset\}$ given by $\tilde{M}_j(\alpha_j, \beta_j) = M(\bar{t}_j + \alpha_j, \beta_j)$. While for $j \in I_p$ such that $e_j = 0$, according to Subsection 3.1.3 and Remark 4.1.3 (see again (3.1.20)), (4.1.9) is homotopically trivial. Consequently, like in the proof of Theorem 3.1.6, the solvability of (4.1.9) is homotopically reduced to the solvability of the system of p_{E_p} inclusions

$$
\underbrace{0 \in \tilde{M}_{j_1}, \quad \cdots, \quad 0 \in \tilde{M}_{j_{p_{E_p}}}}_{p_{E_p} - \text{times}} \tag{4.1.10}
$$

on the set

$$
\Omega_{E_p} := \times_{k \in p_{E_p}} \left[\left(-\frac{b_{k(j_k)} - a_{k(j_k)}}{2}, \frac{b_{k(j_k)} - a_{k(j_k)}}{2} \right) \times \mathcal{B}_{k(j_k)} \right].
$$

Here p_{E_p} is the number of 1 in E_p and each above $\tilde{M}_{j_k}$, $k = 1, 2, \cdots, p_{E_p}$ corresponds to such j_k that $e_{j_k} = 1$. By assumption (iv) of this theorem, system of inclusions (4.1.10) is solvable on Ω_{E_p}. Consequently we get a solution $x_{E_p,s}$ of (4.1.2) satisfying (4.1.7). Finally passing to the limit $p \to \infty$ for $x_{E_p,s}$ like in Theorem 3.1.33, we get the solution $x_{E,s}$ of this theorem. The proof is finished. $\qquad \square$

Similarly we get the next result.

Theorem 4.1.5. *Let* (i–iii) *hold and* $d = 1$. *If there is a sequence of intervals* $\{(a_j, b_j)\}_{j \in \mathbb{Z}}$ *along with a constant* $\delta > 0$ *such that*

(i) $a_j \to +\infty$ *as* $j \to +\infty$, $b_j \to -\infty$ *as* $j \to -\infty$, *and* $\sup_{j \in \mathbb{Z}}(b_j - a_j) < \infty$,

(ii) Either $M(a_j) \subset [\delta, \infty)$ and $M(b_j) \subset (-\infty, -\delta]$, or $M(a_j) \subset (-\infty, -\delta]$ and $M(b_j) \subset [\delta, \infty)$, for any $j \in \mathbb{Z}$,

where $M : \mathbb{R} \to 2^{\mathbb{R}} \setminus \{\emptyset\}$ is defined by

$$M(\alpha) = \left\{ \int_{-\infty}^{\infty} \langle h(s), u_1^{\perp}(s) \rangle \, ds : \right.$$
$$\left. h \in L_{loc}^2(\mathbb{R}, \mathbb{R}^n), h(t) \in g(\gamma(t), 0, t + \alpha) \text{ a.e. on } \mathbb{R} \right\}$$

Then there is a constant $K > 0$ and for any sufficiently small $s > 0$, there are increasing sequences $\{t_j\}_{j \in \mathbb{Z}}$, $t_{j+1} - t_j \geq 2/s$, and $\{k(j)\}_{j \in \mathbb{Z}}$, $k(j) \in \mathbb{Z}$, such that for $\varepsilon = \pm s^2$ and any infinite sequence $E = \{e_j\}_{j \in \mathbb{Z}} \in \mathcal{E}$, the differential inclusion (4.1.2) possesses a solution $x_{E,s}$ satisfying

$$\sup_{t_j \leq t \leq t_{j+1}} \left| x_{E,s}(t) - \gamma(t - \bar{\alpha}_{j,E,s}) \right| \leq K s^2 \quad \text{when} \quad e_j = 1,$$

$$\sup_{t_j \leq t \leq t_{j+1}} \left| x_{E,s}(t) \right| \leq K s^2 \quad \text{when} \quad e_j = 0, \tag{4.1.11}$$

where $\bar{\alpha}_{j,E,s} \in (a_{k(j)}, b_{k(j)}) \subset (t_j, t_{j+1})$ for any $j \in \mathbb{Z}$ with $e_j = 1$. The mapping $E \to z_{E,s}$ is injective.

Hence for any $\varepsilon > 0$ sufficiently small and $E \in \mathcal{E}$, (4.1.2) possesses a solution $z_{E,s}$ satisfying either (4.1.7) or (4.1.11). These estimates (4.1.7) and (4.1.11) give the chaotic behavior of solutions $z_{E,s}$ for any $E \in \mathcal{E}$. This is more discussed in the next section.

4.1.4   Chaos from Homoclinic Manifolds

In many cases, the assumption (iii) of Subsection 4.1.1 is replaced by

(iv) The unperturbed equation $\dot{x} = f(x)$ has a manifold of homoclinic solutions. That is, there exists a C^2-smooth mapping $\gamma : \mathbb{R}^{d-1} \times \mathbb{R} \to \mathbb{R}^n$ such that $t \to \gamma(\theta, t)$ is a homoclinic solution of $\dot{x} = f(x)$ to 0 and $\left\{ \frac{\partial \gamma}{\partial \theta_i}(\theta, t), \dot{\gamma}(\theta, t) \right\}_{i=1}^{d-1}$, $\theta = (\theta_1, \cdots, \theta_{d-1})$ form a basis of bounded solutions on $\mathbb{R}$ of the variational equation $\dot{u}(t) = Df(\gamma(\theta, t))u(t)$.

This usually happens when $\dot{x} = f(x)$ has some symmetry. Bifurcations from family of periodics are studied in Sections 3.3 and 3.5.

Let $U(\theta, t)$ denote a fundamental solution of $\dot{u}(t) = Df(\gamma(\theta, t))u(t)$. Then Theorem 3.1.2 is valid for each parameter θ. Let $u_j(\theta, t)$ be the jth column of $U(\theta, t)$ and define $u_i^{\perp}(\theta, t)$ by $\langle u_i^{\perp}(\theta, t), u_j(\theta, t) \rangle = \delta_{ij}$. Now we can repeat the above procedure. Instead of (4.1.8), we make on each interval $[t_j, t_{j+1}]$, $j \in I_p$ the changes of variables

$$x(t + \bar{t}_j) = \gamma_j(\theta_j, t - \alpha_j) + \varepsilon z_j(t)$$

$$+ \frac{1}{2\tau_j} \left(\gamma_{j+1}(\theta_{j+1}, -\tau_{j+1} - \alpha_{j+1}) - \gamma_j(\theta_j, \tau_j - \alpha_j) \right)(t + \tau_j), \quad z_j \in Z_{\tau_j},$$

where $\gamma_j(\theta, t) = \gamma(\theta, t)$ for $e_j = 1$ and $\gamma_j(\theta, t) = 0$ for $e_j = 0$. In this way for $e_j = 1$, we arrive at a mapping $\tilde{M}_j : \mathbb{R}^d \to 2^{\mathbb{R}^d} \setminus \{\emptyset\}$ given by

$$
\begin{aligned}
\tilde{M}_j(\alpha_j, \theta_j) = \ & M(\bar{t}_j + \alpha_j, \theta_j), \quad M = (M_1, \cdots, M_d), \\
M_l(\alpha, \theta) = \ & \left\{ \int_{-\infty}^{\infty} \langle h(s), u_l^{\perp}(\theta, s) \rangle \, ds \ : \ h \in L^2_{\mathrm{loc}}(\mathbb{R}, \mathbb{R}^n), \right. \\
& \left. h(t) \in g(\gamma(\theta, t), 0, t + \alpha) \quad \text{a.e. on} \quad \mathbb{R} \right\}.
\end{aligned}
\tag{4.1.12}
$$

Following the proof of Theorem 4.1.4, we get the next result.

Theorem 4.1.6. *Let* (i), (ii), (iv) *hold. If there is a non-empty open bounded set* $\mathcal{B} \subset \mathbb{R}^{d-1}$ *along with a sequence of intervals* $\{(a_j, b_j)\}_{j \in \mathbb{Z}}$ *such that*

(i) $a_j \to +\infty$ *as* $j \to +\infty$, *and* $b_j \to -\infty$ *as* $j \to -\infty$

(ii) $\sup_{j \in \mathbb{Z}} (b_j - a_j) < \infty$

(iii) $\inf_{j \in \mathbb{Z}} \operatorname{dist}\left(0, M\left(\partial((a_j, b_j) \times \mathcal{B})\right)\right) > 0$

(iv) $\deg(M, (a_j, b_j) \times \mathcal{B}, 0) \neq 0$

where M *is given by* (4.1.12). *Then there is a constant* $K > 0$ *and for any sufficiently small* $\varepsilon \neq 0$ *there are increasing sequences* $\{t_j\}_{j \in \mathbb{Z}}$, $t_{j+1} - t_j \geq 2/|\varepsilon|$, $\forall j \in \mathbb{Z}$, *and* $\{k(j)\}_{j \in \mathbb{Z}}$, $k(j) \in \mathbb{Z}$, *such that for any infinite sequence* $E = \{e_j\}_{j \in \mathbb{Z}} \in \mathcal{E}$, *the differential inclusion* (4.1.2) *possesses a solution* $x_{E, \varepsilon}$ *satisfying*

$$
\sup_{t_j \leq t \leq t_{j+1}} \left| x_{E, \varepsilon}(t) - \gamma(\theta_j, t - \bar{\alpha}_{j, E, \varepsilon}) \right| \leq K|\varepsilon| \quad \text{when} \quad e_j = 1,
$$

$$
\sup_{t_j \leq t \leq t_{j+1}} \left| x_{E, \varepsilon}(t) \right| \leq K|\varepsilon| \quad \text{when} \quad e_j = 0,
$$

where $\bar{\alpha}_{j, E, \varepsilon} \in (a_{k(j)}, b_{k(j)}) \subset (t_j, t_{j+1})$ *and* $\theta_j \in \mathcal{B}$ *for any* $j \in \mathbb{Z}$ *with* $e_j = 1$. *The mapping* $E \to z_{E, \varepsilon}$ *is injective.*

4.1.5 Almost and Quasi Periodic Discontinuous O.D.Eqns

We apply in this subsection previous abstract results to concrete examples. For simplicity we first study the case when

$$
g(x, \varepsilon, t) = \mu_1 F(x) + \mu_2 h(t)
\tag{4.1.13}
$$

with $\mu_{1,2} \in \mathbb{R}$, $F : \mathbb{R}^n \to 2^{\mathbb{R}^n} \setminus \{\emptyset\}$ is upper-semicontinuous with compact and convex values, and $h \in C(\mathbb{R}, \mathbb{R}^n)$ is almost periodic, i.e. it fulfills the following definition [152].

Definition 4.1.7. $h \in C(\mathbb{R}, \mathbb{R}^n)$ is *almost periodic* if $\forall \zeta > 0 \ \exists L > 0$ such that $\forall a \in \mathbb{R} \ \exists \tau \in [a, a + L]$ such that $|h(x + \tau) - h(x)| \leq \zeta \ \forall x \in \mathbb{R}$.

Function (4.1.13) expresses usual forcing terms in mechanics with dry friction terms and almost periodic perturbations, for example a quasi-periodically forced Duffing's equation (see (3.1.36)):

$$\ddot{x} - x + 2x^3 + \varepsilon \mu_1 \operatorname{sgn} \dot{x} = \varepsilon \mu_2 \left(\cos t + \cos \sqrt{2} t \right) . \qquad (4.1.14)$$

For (4.1.13), formula (4.1.6) has the form

$$M_l(\alpha, \beta) = \mu_1 \left\{ \int_{-\infty}^{\infty} \langle h(s), u_l^{\perp}(s) \rangle \, ds \ : \ h \in L^2_{\text{loc}}(\mathbb{R}, \mathbb{R}^n), \right.$$
$$\left. h(t) \in F(\gamma(t)) \quad \text{a.e. on} \quad \mathbb{R} \right\} + \frac{1}{2} \sum_{i,r=1}^{d-1} \beta_i \beta_r b_{irl} + \mu_2 a_l(\alpha), \qquad (4.1.15)$$

where

$$b_{irl} = \int_{-\infty}^{\infty} \langle D^2 f(\gamma(t))(u_{d+i}(t), u_{d+r}(t)), u_l^{\perp}(t) \rangle \, dt$$
$$a_l(\alpha) = \int_{-\infty}^{\infty} \langle h(t + \alpha), u_l^{\perp}(t) \rangle \, dt ,$$

while (4.1.12) has the form

$$M_l(\alpha, \theta) = \mu_1 \left\{ \int_{-\infty}^{\infty} \langle h(s), u_l^{\perp}(\theta, s) \rangle \, ds \ : \ h \in L^2_{\text{loc}}(\mathbb{R}, \mathbb{R}^n), \right.$$
$$\left. h(t) \in F(\gamma(\theta, t)) \quad \text{a.e. on} \quad \mathbb{R} \right\} + \mu_2 a_l(\alpha, \theta), \qquad (4.1.16)$$

where

$$a_l(\alpha, \theta) = \int_{-\infty}^{\infty} \langle h(t + \alpha), u_l^{\perp}(\theta, t) \rangle \, dt .$$

Since all functions $u_l^{\perp}(t)$ and $u_l^{\perp}(\theta, s)$ are integrable over $\mathbb{R}$ with respect to s, it easy to show that functions $a(\alpha) = (a_1(\alpha), \cdots, a_d(\alpha))$ and $a(\alpha, \theta) = (a_1(\alpha, \theta), \cdots, a_d(\alpha, \theta))$ are also almost periodic in α, while $a(\alpha, \theta)$ uniformly on bounded sets of θ.

Theorem 4.1.8. *Let $d > 1$. If there is a non-empty open bounded set $\mathcal{B} \subset \mathbb{R}^{d-1}$ along with an interval $(a, b) \subset \mathbb{R}$ such that*

(i) $0 \notin M\left(\partial((a, b) \times \mathcal{B})\right)$

(ii) $\deg(M, (a, b) \times \mathcal{B}, 0) \neq 0$

where M is given by either (4.1.15) or (4.1.16). Then the statements of Theorems 4.1.4 and 4.1.6, respectively, are applicable to (4.1.2) with perturbation (4.1.13).

Proof. We consider the case (4.1.15), the case (4.1.16) is similar. We have to verify assumptions of Theorem 4.1.4. Since $a(\alpha)$ is almost periodic, by Definition 4.1.7, for any $\zeta > 0$ there is a sequence $\{c_k\}_{k \in \mathbb{Z}}$, $c_k \to \pm\infty$ as $k \to \pm\infty$ such that

$$|a(\alpha + c_k) - a(\alpha)| \leq \zeta, \quad \forall \alpha \in \mathbb{R}.$$

For any $j \in \mathbb{Z}$, we put

$$\mathcal{B}_j = \mathcal{B}, \quad a_j = c_j + a, \quad b_j = c_j + b.$$

If ζ is sufficiently small, then assumptions (i) and (ii) of this theorem clearly imply assumptions (i–iv) of Theorem 4.1.4, so the proof is finished. □

Similarly we have the next result.

Theorem 4.1.9. *Let* $d = 1$. *If there is an interval* $(a, b) \subset \mathbb{R}$ *such that* $M(a)M(b) \subset (-\infty, 0)$, *where* M *is given by*

$$M(\alpha) = \mu_1 \left\{ \int_{-\infty}^{\infty} \langle h(s), u_1^{\perp}(s) \rangle \, ds \, : \, h \in L_{loc}^2(\mathbb{R}, \mathbb{R}^n), \right.$$
$$\left. h(t) \in F(\gamma(t)) \quad a.e. \ on \quad \mathbb{R} \right\} + \mu_2 a_1(\alpha).$$

Then the statement of Theorem 4.1.5 is applicable for (4.1.2) *with perturbation* (4.1.13).

Now we intend to show that conditions of Theorems 4.1.8 and 4.1.9 imply the validity of these conditions also for any h^* in the hull $H(h)$ of h. We recall [152, 160] that the hull $H(h)$ of h is defined as a set

$$H(h) := \left\{ h^* \in C(\mathbb{R}, \mathbb{R}) \mid \exists \{\tau_n\}_{n \geq 1} \subset \mathbb{R} \text{ such that} \right.$$
$$\left. h(t + \tau_n) \to h^*(t) \text{ uniformly on } \mathbb{R} \text{ as } n \to \infty \right\}.$$

We note that h^* is also almost periodic. We again consider the case (4.1.15), so

$$a_l^h(\alpha) = \int_{-\infty}^{\infty} \langle h(t + \alpha), u_l^{\perp}(t) \rangle \, dt, \quad a_l^{h^*}(\alpha) = \int_{-\infty}^{\infty} \langle h^*(t + \alpha), u_l^{\perp}(t) \rangle \, dt$$

for $l = 1, 2, \cdots, d$. Theorem 3.1.2 implies that

$$|u_l^{\perp}(t)| \leq K_0 \, e^{-2M|t|} \quad \forall t \in \mathbb{R}.$$

For any $\delta > 0$, there is an $\tau \in \mathbb{R}$ such that

$$|h(t + \tau) - h^*(t)| \leq \delta \quad \forall t \in \mathbb{R}.$$

Hence

$$\int_{-\infty}^{\infty} \left| \langle h(t+\alpha) - h^*(t+\alpha-\tau), u_l^{\perp}(t) \rangle \right| \, dt \leq \delta K_0/M \,.$$

This gives

$$|a_l^h(\alpha) - a_l^{h^*}(\alpha-\tau)| \leq \delta K_0/M \,.$$

By taking $\delta > 0$ sufficiently small, the assumptions of Theorems 4.1.8 and 4.1.9 are satisfied for h^* in (4.1.13) instead of h with the set $\mathcal{B}$ and with an interval $(a - \tau, b - \tau)$. Consequently, Theorems 4.1.8 and 4.1.9 imply chaos of (4.1.2) with the perturbation (4.1.13) over any element of the hull $H(h)$. This result is a generalization of similar ones from [152, 160, 181, 195] to multivalued perturbations. We state this result in the next theorem.

Theorem 4.1.10. *Under conditions of Theorems 4.1.8 or 4.1.9, differential inclusion (4.1.2) is also chaotic with perturbation (4.1.13) when h is replaced in perturbation (4.1.13) by any $h^* \in H(h)$.*

Clearly the above results hold also for continuous $h(x,t)$ in perturbation (4.1.13) when $h(x,t)$ is uniformly almost periodic in t on any bounded set of variable x [152, p. 68].

Applying Theorem 4.1.9 to (4.1.14) and using computations to (3.1.36), we derive

$$M(\alpha) = -2\mu_1 + \mu_2 \left[\pi \operatorname{sech} \frac{\pi}{2} \sin \alpha + \sqrt{2}\pi \operatorname{sech} \frac{\sqrt{2}\pi}{2} \sin \sqrt{2}\alpha \right] \,.$$

Since $\sqrt{2}$ is irrational we have $\sup_{\alpha \in \mathbb{R}} \left| \pi \operatorname{sech} \frac{\pi}{2} \sin \alpha + \sqrt{2}\pi \operatorname{sech} \frac{\sqrt{2}\pi}{2} \sin \sqrt{2}\alpha \right| = \pi \operatorname{sech} \frac{\pi}{2} + \sqrt{2}\pi \operatorname{sech} \frac{\sqrt{2}\pi}{2}$. So for

$$\left[\pi \operatorname{sech} \frac{\pi}{2} + \sqrt{2}\pi \operatorname{sech} \frac{\sqrt{2}\pi}{2} \right] |\mu_2| > 2|\mu_1| \,, \tag{4.1.17}$$

the assumptions of Theorem 4.1.9 are satisfies and then (4.1.14) is chaotic. Moreover, we have [152, p. 70]

$$H(\cos t + \cos \sqrt{2}t) = \left\{ \cos(t + \xi_1) + \cos(\sqrt{2}t + \xi_2) \mid \xi_1, \xi_2 \in \mathbb{R}, \ \xi_{1,2} \equiv \operatorname{mod} 2\pi \right\} \,.$$

So (4.1.14) over its hull has the form

$$\ddot{x} - x + 2x^3 + \varepsilon\mu_1 \operatorname{sgn} \dot{x} = \varepsilon\mu_2 \left(\cos \Omega_1 + \cos \Omega_2 \right) \,,$$
$$\dot{\Omega}_1 = 1, \quad \dot{\Omega}_2 = \sqrt{2} \,, \tag{4.1.18}$$

and according to Theorem 4.1.10, system (4.1.18) is also chaotic when (4.1.17) holds.

Finally, we deal with weakly coupled oscillators with a symmetry given by (4.1.1) which is considered as a differential inclusion

$$\dot{x}_1 = x_2, \quad \dot{x}_2 - x_1 + 2x_1(x_1^2 + y_1^2) - \varepsilon\mu_1\cos\omega_1 t \in -\varepsilon\delta_1 \operatorname{Sgn} x_2,$$
$$\dot{y}_1 = y_2, \quad \dot{y}_2 - y_1 + 2y_1(x_1^2 + y_1^2) - \varepsilon\mu_2\cos\omega_2 t \in -\varepsilon\delta_2 \operatorname{Sgn} y_2,$$

where $\operatorname{Sgn} r$ is defined by (2.4.1). The unperturbed equation of (4.1.1)

$$\dot{x}_1 = x_2, \dot{x}_2 = x_1 - 2x_1(x_1^2 + y_1^2), \quad \dot{y}_1 = y_2, \dot{y}_2 = y_1 - 2y_1(x_1^2 + y_1^2) \quad (4.1.19)$$

has a rotational symmetry, i.e. it has a homoclinic manifold given by $\gamma(\theta, t) = \left(r(t)\cos\theta, \dot{r}(t)\cos\theta, r(t)\sin\theta, \dot{r}(t)\sin\theta\right)$ where $r(t) = \operatorname{sech} t$ (see example (3.5.22) and [106] for similar computations). Then

$$u_1^{\perp}(\theta, t) = \left(-\dot{r}(t)\sin\theta, r(t)\sin\theta, \dot{r}(t)\cos\theta, -r(t)\cos\theta\right),$$
$$u_2^{\perp}(\theta, t) = \left(-\ddot{r}(t)\cos\theta, \dot{r}(t)\cos\theta, -\ddot{r}(t)\sin\theta, \dot{r}(t)\sin\theta\right),$$

where $\theta \in [0, 2\pi)$. Now (4.1.16) has for $\theta \neq 0, \pi/2, \pi, 3\pi/2$ the form

$$M_1(\alpha, \theta) = \pi\mu_1 \operatorname{sech}\frac{\omega_1\pi}{2}\sin\theta\cos\omega_1\alpha - \pi\mu_2\operatorname{sech}\frac{\omega_2\pi}{2}\cos\theta\cos\omega_2\alpha,$$

$$M_2(\alpha, \theta) = \pi\mu_1\omega_1\operatorname{sech}\frac{\omega_1\pi}{2}\cos\theta\sin\omega_1\alpha \qquad (4.1.20)$$

$$+\pi\mu_2\omega_2\operatorname{sech}\frac{\omega_2\pi}{2}\sin\theta\sin\omega_2\alpha - 2\delta_1|\cos\theta| - 2\delta_2|\sin\theta|.$$

In order to apply Theorem 4.1.8, it is enough to find an isolated root of (4.1.20) with a nonzero Brouwer index. We consider that $\theta \in (0, \pi/2)$ and $\alpha \in \left(0, \frac{\pi}{2\omega_1}\right) = I_0$. Then equation $M(\alpha, \theta) = 0$ has the form

$$A_1(\alpha)\sin\theta + A_2(\alpha)\cos\theta = 0, \quad A_3(\alpha)\sin\theta + A_4(\alpha)\cos\theta = 0, \qquad (4.1.21)$$

where functions $A_i(\alpha)$, $i = 1, 2, 3, 4$ are given by the definition. Note that $A_1(\alpha) \neq 0$ and $A_2(\alpha) \neq 0$ on I_0. It is clear that (4.1.21) is equivalent to

$$B(\alpha) = A_1(\alpha)A_4(\alpha) - A_2(\alpha)A_3(\alpha) = 0$$
$$A_1(\alpha)\sin\theta + A_2(\alpha)\cos\theta = 0, \qquad (4.1.22)$$

where

$$B(\alpha) = \frac{\pi^2}{2}\mu_1^2\omega_1\left(\operatorname{sech}\frac{\omega_1\pi}{2}\right)^2\sin 2\omega_1\alpha + \frac{\pi^2}{2}\mu_2^2\omega_2\left(\operatorname{sech}\frac{\omega_2\pi}{2}\right)^2\sin 2\omega_2\alpha$$
$$-2\delta_1\mu_1\pi\operatorname{sech}\frac{\omega_1\pi}{2}\cos\omega_1\alpha - 2\delta_2\mu_2\pi\operatorname{sech}\frac{\omega_2\pi}{2}\cos\omega_2\alpha.$$

Since $B(0) < 0$, function $B(\alpha)$ is nonzero analytic and it could have only isolated zeroes with finite orders. Consequently, supposing

$$\frac{\pi^2}{2}\mu_1^2\omega_1\left(\operatorname{sech}\frac{\omega_1\pi}{2}\right)^2 + \frac{\pi^2}{2}\mu_2^2\omega_2\left(\operatorname{sech}\frac{\omega_2\pi}{2}\right)^2\sin\frac{\pi\omega_2}{2\omega_1} >$$
$$\sqrt{2}\delta_1\mu_1\pi\operatorname{sech}\frac{\omega_1\pi}{2} + 2\delta_2\mu_2\pi\operatorname{sech}\frac{\omega_2\pi}{2}\cos\frac{\pi\omega_2}{4\omega_1}, \qquad (4.1.23)$$

we see that $B\left(\frac{\pi}{4\omega_1}\right) > 0$ and $B(\alpha)$ changes the sign over I_0. Then there are $\alpha_0 \in I_0$, $\eta > 0$ such that $B(\alpha)$ has the only zero $\alpha = \alpha_0$ in the interval $[\alpha_0 - \eta, \alpha_0 + \eta] \subset I_0$ and $B(\alpha_0 - \eta)B(\alpha_0 + \eta) < 0$. We take

$$\theta(\alpha) = -\arctan(A_2(\alpha)/A_1(\alpha)), \quad \theta_0^- = \frac{\theta(\alpha_0)}{2}, \quad \theta_0^+ = \frac{2\theta_0^- + \pi}{4}.$$

Then on the set

$$\mathcal{I} = [\alpha_0 - \eta, \alpha_0 + \eta] \times [\theta_0^-, \theta_0^+] \subset I_0 \times (0, \pi/2),$$

mapping $M(\alpha, \theta)$ has the only zero point $\widetilde{\alpha}_0 := (\alpha_0, \theta(\alpha_0))$. To show the non-vanishing of the Brouwer index $I(\widetilde{\alpha}_0)$ of this isolated zero point of $M(\alpha, \theta)$ (cf. Section 2.2.3), we consider the equation

$$\begin{aligned}
A_1(\alpha)\sin\theta + A_2(\alpha)\cos\theta &= 0 \\
A_3(\alpha)\sin\theta + A_4(\alpha)\cos\theta &= c_1,
\end{aligned} \tag{4.1.24}$$

for $c_1 \neq 0$ small and $(\alpha, \theta) \in \mathcal{I}$. Equation (4.1.24) gives

$$c_1 = C(\alpha) := \frac{B(\alpha)}{\sqrt{A_1(\alpha)^2 + A_2(\alpha)^2}}. \tag{4.1.25}$$

Let $L(\alpha)$ be the linearization of the left hand side of (4.1.24) at point $(\alpha, \theta(\alpha))$. A boring computation gives

$$\det L(\alpha) = -C'(\alpha)\sqrt{A_1(\alpha)^2 + A_2(\alpha)^2}. \tag{4.1.26}$$

We note $A_1(\alpha)^2 + A_2(\alpha)^2 \neq 0$ on $[\alpha_0 - \eta, \alpha_0 + \eta]$. Let $\{(\alpha_i, \theta(\alpha_i))\}_{i=1}^{N}$ be the solutions of (4.1.24) in $\mathcal{I}$. Then for c_1 sufficiently small such that $C'(\alpha_i) \neq 0$, $i = 1, 2, \cdots, N$, by the definition of the Brouwer degree (cf. Section 2.2.3) and (4.1.26), we have

$$|I(\widetilde{\alpha}_0)| = \left|\sum_{i=1}^{N} \operatorname{sgn} \det L(\alpha_i)\right| = \left|\sum_{i=1}^{N} \operatorname{sgn} C'(\alpha_i)\right|$$

$$= \left|\deg\left(C(\alpha), (\alpha_0 - \eta, \alpha_0 + \eta), 0\right)\right| = 1,$$

since by (4.1.25), function $C(\alpha)$ has the only zero $\alpha = \alpha_0$ in $[\alpha_0 - \eta, \alpha_0 + \eta]$ and $C(\alpha_0 - \eta)C(\alpha_0 + \eta) < 0$. Hence we get from Theorem 4.1.8 the following result.

Theorem 4.1.11. *If* (4.1.23) *holds then* (4.1.1) *is chaotic for any* $\varepsilon > 0$ *small.*

Condition (4.1.23) means that the forcing terms must be sufficiently large in (4.1.1) with respect to damping terms, to get chaos in (4.1.1), and the rate is given by this condition.

We note that if the ratio ω_2/ω_1 is irrational, then the hull of the function $h(t) = (\mu_1\cos\omega_1 t, \mu_2\cos\omega_2 t)$ is given by

$$H(h) = \left\{(\mu_1\cos(\omega_1 t + \xi_1), \mu_2\cos(\omega_2 t + \xi_2)) \mid \xi_1, \xi_2 \in \mathbb{R}, \ \xi_{1,2} \equiv \bmod 2\pi\right\}.$$

So (4.1.1) considered over $H(h)$ has the form

$$\ddot{x} = x - 2x(x^2 + y^2) + \varepsilon\mu_1 \cos\Omega_1 - \varepsilon\delta_1 \operatorname{sgn}\dot{x},$$
$$\ddot{y} = y - 2y(x^2 + y^2) + \varepsilon\mu_2 \cos\Omega_2 - \varepsilon\delta_2 \operatorname{sgn}\dot{y}, \qquad (4.1.27)$$
$$\dot{\Omega}_1 = \omega_1, \quad \dot{\Omega}_2 = \omega_2.$$

Theorem 4.1.10 implies chaos of (4.1.27) under condition (4.1.23). This result is a generalization of similar ones of [152, 195] to multivalued and quasiperiodic perturbations. Of course, other discontinuous perturbations such as in Remark 3.1.29 could be considered in (4.1.1), but we do not carry out those computations.

4.2 Chaos in Periodic Differential Inclusions

4.2.1 Regular Periodic Perturbations

Using results of the previous Section 4.1, in this section we show chaotic solutions to the problem (3.1.3) studied in Chapter 3. We suppose assumptions (i–iv) from Subsection 3.1.1. Note that now all multivalued Melnikov functions are periodic. So applying Theorem 4.1.4 like in the proof of Theorem 4.1.8, we immediately have the following generalizations of Theorems 3.1.6–3.1.9 (see [76]).

Theorem 4.2.1. *Let $d > 1$. If there is a non–empty open bounded set $\mathcal{B} \subset \mathbb{R}^d$ and $\mu_0 \in S^{k-1}$ such that $0 \notin M_{\mu_0}(\partial\mathcal{B})$ and $\deg(M_{\mu_0}, \mathcal{B}, 0) \neq 0$, where M_μ is defined by (3.1.21). Then there is a constant $K > 0$ and a wedge-shaped region in $\mathbb{R}^k$ for μ of the form*

$$\mathcal{R} = \Big\{s^2\tilde{\mu} : s > 0, \text{ respectively } \tilde{\mu}, \text{ is from an open small connected}$$
$$\text{neighborhood } U_1, \text{ respectively } U_2 \subset S^{k-1}, \text{ of } 0 \in \mathbb{R}, \text{ respectively of } \mu_0\Big\}$$

such that for any $\mu \in \mathcal{R}$ of the form $\mu = s^2\tilde{\mu}$, $0 < s \in U_1$, $\tilde{\mu} \in U_2$, for any sequence $E = \{e_j\}_{j\in\mathbb{Z}} \in \mathcal{E}$ and for any $m \in \mathbb{N}$, $m \geq [1/s]$, the differential inclusion (3.1.3) possesses a solution $x_{m,E}$ satisfying

$$\sup_{(2j-1)m \leq t \leq (2j+1)m} \big|x_{m,E}(t) - \gamma(t - 2mj - \alpha_{m,j,E})\big| \leq Ks \quad \text{when} \quad e_j = 1,$$

$$\sup_{(2j-1)m \leq t \leq (2j+1)m} \big|x_{m,E}(t)\big| \leq Ks^2 \quad \text{when} \quad e_j = 0,$$

$$(4.2.1)$$

where $\alpha_{m,j,E} \in \mathbb{R}$ and $|\alpha_{m,j,E}| \leq K$.

Theorem 4.2.2. *Let $d = 1$. If there are constants $a < b$ and $\mu_0 \in S^{k-1}$ such that $M_{\mu_0}(a)M_{\mu_0}(b) \subset (-\infty, 0)$, where M_μ is defined by (3.1.24). Then there is a constant $K > 0$ and a wedge-shaped region in $\mathbb{R}^k$ for μ of the form*

$$\mathcal{R} = \Big\{\pm s^2\tilde{\mu} : s > 0, \text{ respectively } \tilde{\mu}, \text{ is from an open small connected}$$
$$\text{neighborhood } U_1, \text{ respectively } U_2 \subset S^{k-1}, \text{ of } 0 \in \mathbb{R}, \text{ respectively of } \mu_0\Big\}$$

such that for any $\mu \in \mathcal{R}$ of the form $\mu = \pm s^2 \tilde{\mu}$, $0 < s \in U_1$, $\tilde{\mu} \in U_2$, for any sequence $E = \{e_j\}_{j \in \mathbb{Z}} \in \mathcal{E}$ and for any $m \in \mathbb{N}$, $m \geq [1/s]$, the differential inclusion (3.1.3) possesses a solution $x_{m,E}$ satisfying

$$\sup_{(2j-1)m \leq t \leq (2j+1)m} \left| x_{m,E}(t) - \gamma(t - 2mj - \alpha_{m,j,E}) \right| \leq K s^2 \quad \text{when} \quad e_j = 1,$$

$$\sup_{(2j-1)m \leq t \leq (2j+1)m} \left| x_{m,E}(t) \right| \leq K s^2 \quad \text{when} \quad e_j = 0,$$

$$(4.2.2)$$

where $\alpha_{m,j,E} \in (a,b)$.

Note that now $t_j = (2j - 1)m$ in the notations of Theorem 4.1.4, since now M_μ is 2-periodic in α. This comes directly following the proof of Theorem 4.1.8 (see [76] for more details).

Hence for any $\mu \in \mathcal{R}$, $m \in \mathbb{N}$ sufficiently large and $E \in \mathcal{E}$, (3.1.3) possesses a solution $z_{m,E}$ satisfying either (4.2.1) or (4.2.2). These estimates (4.2.1–4.2.2) give the injectivity of the mapping $E \to z_{m,E}$ for $s > 0$ small. Let $\sigma : \mathcal{E} \to \mathcal{E}$ be the Bernoulli shift defined in Section 2.5.2. Now the estimates (4.2.1–4.2.2) imply that $x_{m,\sigma(E)}(t)$ is orbitally close to $x_{m,E}(t+2m)$. Summarizing we obtain the following result.

Theorem 4.2.3. *Under the assumptions either of Theorem 4.2.1 or of Theorem 4.2.2, for any $\mu \in \mathcal{R}$ and $m \in \mathbb{N}$ sufficiently large, (3.1.3) possesses a family of solutions $\{z_{m,E}\}_{E \in \mathcal{E}}$ such that*

(i) $E \to x_{m,E}$ is injective

(ii) $x_{m,\sigma(E)}(t)$ is orbitally close to $x_{m,E}(t + 2m)$

This result extends the deterministic chaos mentioned in Sections 2.5.2 and 2.5.3 to discontinuous o.d.eqns as follows. For simplicity, we consider (3.1.34) which we recall here for the reader convenience

$$\ddot{x} + g(x) + \mu_1 \operatorname{sgn} \dot{x} = \mu_2 \psi(t), \qquad (4.2.3)$$

where g satisfies conditions of (3.1.35), $\psi \in C^1(\mathbb{R}, \mathbb{R})$ is 2-periodic and μ_1, μ_2 are small parameters. When $\mu_1 = 0$, then (4.2.3) is a regular system of the form

$$\ddot{x} + g(x) = \mu_2 \psi(t). \qquad (4.2.4)$$

According to computations of Subsection 3.1.6, the corresponding Melnikov function for the problem (4.2.3) is as follows

$$M_\mu(\alpha) = \mu_2 \int_{-\infty}^{\infty} \dot{\omega}(s)\psi(s + \alpha)\, ds - 2\omega(t_0)\mu_1,$$

where $\mu := (\mu_1, \mu_2)$. The Melnikov function for (4.2.4) is just

$$\widetilde{M}(\alpha) = \int_{-\infty}^{\infty} \dot{\omega}(s)\psi(s + \alpha)\, ds.$$

Under the existence of a simple zero α_0 of $\widetilde{M}$, i.e. $\widetilde{M}(\alpha_0) = 0$ and $\widetilde{M}'(\alpha_0) \neq 0$, the assumption $\widetilde{M}(a)\widetilde{M}(b) < 0$ of Theorem 4.2.3 is satisfied for $a = \alpha_0 - \zeta$ and $b = \alpha_0 + \zeta$ with $\zeta > 0$ small. But since the existence of a simple zero of $\widetilde{M}$ is stronger than the above assumption of Theorem 4.2.3, when $\mu_2 \neq 0$ is small, the property (ii) of this theorem for (4.2.4) has the form (cf. [157, 195])

$$x_{m,\sigma(E)}(t) = x_{m,E}(t + 2m). \tag{4.2.5}$$

Put

$$\Lambda := \{(x_{m,E}(0), \dot{x}_{m,E}(0)) \mid E \in \mathcal{E}\} \subset \mathbb{R}^2.$$

Let F_{μ_2} be the time map of the flow of the first order system of (4.2.4), i.e. $F_{\mu_2}(x_0, y_0) := (x(2), y(2))$, where $x(t)$ and $y(t)$ solve the Cauchy problem

$$\dot{x}(t) = y(t), \quad \dot{y}(t) = \mu_2 \psi(t) - g(x(t)), \quad x(0) = x_0, \quad y(0) = y_0.$$

Then from (4.2.5) we immediately derive

$$F_{\mu_2}^m((x_{m,E}(0), \dot{x}_{m,E}(0))) = (x_{m,\sigma(E)}(0), \dot{x}_{m,\sigma(E)}(0)).$$

Hence

$$F_{\mu_2}^m : \Lambda \to \Lambda$$

and

$$F_{\mu_2}^m \circ \mathcal{J} = \mathcal{J} \circ \sigma, \tag{4.2.6}$$

where $\mathcal{J} : \mathcal{E} \to \Lambda$ is defined as follows

$$\mathcal{J}(E) = (x_{m,E}(0), \dot{x}_{m,E}(0)).$$

Then (4.2.6) means that $F_{\mu_2}^m$ has the same dynamics on Λ as the Bernoulli shift $\mathcal{J}$ on $\mathcal{E}$. So by Theorem 2.5.3, the time map $F_{\mu_2}^m$ is chaotic on Λ. Moreover, it is possible to show a sensitive dependence on initial conditions of $F_{\mu_2}^m$ on Λ in the sense that there is an $\varepsilon_0 > 0$ such that for any $(x, y) \in \Lambda$ and any neighborhood U of (x, y), there exists $(u, z) \in U \cap \Lambda$ and an integer $q \geq 1$ such that

$$|F_{\mu_2}^{mq}(x, y) - F_{\mu_2}^{mq}(u, z)| > \varepsilon_0.$$

Of course, these results are known [157, 195], and they are mentioned also in Section 2.5.3. Hence (4.2.4) is chaotic and sensitive depends on initial conditions as well. The set Λ is Smale's horseshoe of $F_{\mu_2}^m$ and $F_{\mu_2}^m$ has horseshoe dynamics on Λ.

Summarizing, when $\mu_1 \neq 0$ is small then Theorem 4.2.3 extends the deterministic chaos of (4.2.4) to (4.2.3), when in place of (4.2.5) we get property (ii) of Theorem 4.2.3.

4.2.2   Singular Differential Inclusions

In Subsection 4.2.1, Theorems 4.2.1–4.2.2 are obtained by extending Theorems
3.1.6–3.1.9 of Subsection 3.1.3. Now Theorems 3.1.12–3.1.13 have the following
extensions (see [76]).

Theorem 4.2.4. *Let $d > 1$. If there is a non–empty open bounded set $\mathcal{B} \subset \mathbb{R}^d$
and $* \in \{-, +\}$ such that $0 \notin M_*(\partial\mathcal{B})$ and $\deg(M_*, \mathcal{B}, 0) \neq 0$, where M_* is given
by (3.1.28). Then there is a constant $K > 0$ such that for any sufficiently small
$\varepsilon \neq 0$, $\operatorname{sgn} \varepsilon = *1$ and any $E = \{e_j\}_{j \in \mathbb{Z}} \in \mathcal{E}$, the differential inclusion (3.1.25)
possesses a solution $x_{m,\varepsilon,E}$ for any $m \in \mathbb{N}$ satisfying*

$$\sup_{(2j-1)m \leq t \leq (2j+1)m} \left| x_{m,\varepsilon,E}(t) - \gamma\left(\frac{t - 2mj - \alpha_{m,j,\varepsilon,E}}{\varepsilon}\right) \right| \leq K\sqrt{|\varepsilon|} \quad \text{when } e_j = 1,$$

$$\sup_{(2j-1)m \leq t \leq (2j+1)m} \left| x_{m,\varepsilon,E}(t) \right| \leq K|\varepsilon| \quad \text{when } e_j = 0,$$

where $\alpha_{m,j,\varepsilon,E} \in \mathbb{R}$ and $|\alpha_{m,j,\varepsilon,E}| \leq K$.

Theorem 4.2.5. *Let $d = 1$. If there are constants $a < b$ such that $M(a)M(b) \subset
(-\infty, 0)$, where M is given by (3.1.29). Then there is a constant $K > 0$ such
that for any sufficiently small $\varepsilon \neq 0$ and any $E = \{e_j\}_{j \in \mathbb{Z}} \in \mathcal{E}$, the differential
inclusion (3.1.25) possesses a solution $x_{m,\varepsilon,E}$ for any $m \in \mathbb{N}$ satisfying*

$$\sup_{(2j-1)m \leq t \leq (2j+1)m} \left| x_{m,\varepsilon,E}(t) - \gamma\left(\frac{t - 2mj - \alpha_{m,j,\varepsilon,E}}{\varepsilon}\right) \right| \leq K|\varepsilon| \quad \text{when } e_j = 1,$$

$$\sup_{(2j-1)m \leq t \leq (2j+1)m} \left| x_{m,\varepsilon,E}(t) \right| \leq K|\varepsilon| \quad \text{when } e_j = 0,$$

where $\alpha_{m,j,\varepsilon,E} \in (a, b)$.

Finally, we consider the singularly perturbed differential inclusion (3.2.4) of
Section 3.2. By combining the arguments of Sections 3.2 and 4.1, we obtain the
following extensions of Theorems 3.2.6–3.2.11.

Theorem 4.2.6. *Let $d > 1$. If there is a non–empty open bounded set $\mathcal{B} \subset \mathbb{R}^d$
such that $0 \notin M(\partial\mathcal{B})$ and $\deg(M, \mathcal{B}, 0) \neq 0$, where M is given by (3.2.20).
Then there are constants $K > 0$ and $\varepsilon_0 > 0$ such that for any $0 < \varepsilon < \varepsilon_0$
and any $E = \{e_j\}_{j \in \mathbb{Z}} \in \mathcal{E}$, the differential inclusion (3.2.4) possesses a solution
$(x_{m,\varepsilon,E}, y_{m,\varepsilon,E})$ for any $m \in \mathbb{N}$, $m \geq [1/\sqrt{\varepsilon}]$ satisfying*

$$\sup_{t \in \mathbb{R}} |y_{m,\varepsilon,E}(t)| \leq K\varepsilon$$

$$\sup_{(2j-1)m \leq t \leq (2j+1)m} \left| x_{m,\varepsilon,E}(t) - \gamma(t - 2mj - \alpha_{m,j,\varepsilon,E}) \right| \leq K\sqrt{\varepsilon} \text{ for } e_j = 1,$$

$$\sup_{(2j-1)m \leq t \leq (2j+1)m} \left| x_{m,\varepsilon,E}(t) \right| \leq K\varepsilon \quad \text{for } e_j = 0,$$

where $\alpha_{m,j,\varepsilon,E} \in \mathbb{R}$ and $|\alpha_{m,j,\varepsilon,E}| \leq K$.

Theorem 4.2.7. *Let $d = 1$. If there are constants $a < b$ such that $M(a)M(b) \subset (-\infty, 0)$, where M is given by (3.2.25). Then there are constants $K > 0$ and $\varepsilon_0 > 0$ such that for any $0 < \varepsilon < \varepsilon_0$ and any $E = \{e_j\}_{j \in \mathbb{Z}} \in \mathcal{E}$, the differential inclusion (3.2.4) possesses a solution $(x_{m,\varepsilon,E}, y_{m,\varepsilon,E})$ for any $m \in \mathbb{N}$, $m \geq [1/\sqrt{\varepsilon}]$ satisfying*

$$\sup_{t \in \mathbb{R}} |y_{m,\varepsilon,E}(t)| \leq K\varepsilon$$

$$\sup_{(2j-1)m \leq t \leq (2j+1)m} \left| x_{m,\varepsilon,E}(t) - \gamma(t - 2mj - \alpha_{m,j,\varepsilon,E}) \right| \leq K\varepsilon \text{ for } e_j = 1,$$

$$\sup_{(2j-1)m \leq t \leq (2j+1)m} \left| x_{m,\varepsilon,E}(t) \right| \leq K\varepsilon \text{ when } e_j = 0,$$

where $\alpha_{m,j,\varepsilon,E} \in (a, b)$.

4.3 More About Homoclinic Bifurcations

In Sections 4.1 and 4.2, we study bifurcation of chaotic solutions accumulating on bounded solutions for ordinary differential equations with small multivalued perturbations. We assume that the o.d.eqns have homoclinic solutions to hyperbolic equilibria. More recently, we have considered bifurcation from nonsmooth homoclinics, i.e. we considered parameterized discontinuous o.d.eqns with homoclinics crossing discontinuity levels. There are the following two main possibilities.

4.3.1 Transversal Homoclinic Crossing Discontinuity

In [22] we investigate the following problem: Let $G(x)$ be a C^r−function on $\Omega \subset \mathbb{R}^n$, $r \geq 2$ and let $\Omega_\pm = \{x \in \Omega \mid G(x) \gtrless 0\}$, $\Omega_0 := \{x \in \Omega \mid G(x) = 0\}$. Let $f_\pm(x) \in C^r(\bar{\Omega}_\pm)$ and consider the equation

$$\dot{x} = f_\pm(x) + \varepsilon g(t, x, \varepsilon), \quad x \in \bar{\Omega}_\pm, \tag{4.3.1}$$

where $g \in C_b^r(\mathbb{R} \times \Omega \times \mathbb{R})$ and $\varepsilon \in \mathbb{R}$ is a small parameter. Assume (see Fig. 4.1)

- For $\varepsilon = 0$ (4.3.1) has the hyperbolic equilibrium $x = 0 \in \Omega_-$ and a continuous (not necessarily C^1) homoclinic orbit $\gamma(t)$ to $x = 0$ that consists of three solutions

$$\gamma(t) = \begin{cases} \gamma_-(t) & \text{if } t \leq -T \\ \gamma_0(t) & \text{if } -T \leq t \leq T \\ \gamma_+(t) & \text{if } t \geq T \end{cases}$$

where $\gamma_\pm(t) \in \Omega_-$ for $|t| > T$, $\gamma_0(t) \in \Omega_+$ for $|t| < T$ and

$$\gamma_-(-T) = \gamma_0(-T) \in \Omega_0, \quad \gamma_+(T) = \gamma_0(T) \in \Omega_0.$$

Moreover we also assume that

$$G'(\gamma(-T))f_\pm(\gamma(-T)) > 0, \quad \text{and} \quad G'(\gamma(T))f_\pm(\gamma(T)) < 0.$$

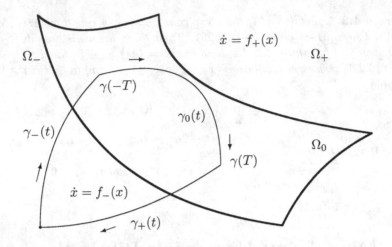

Figure 4.1: A transversal homoclinic cycle

We have derived a Melnikov bifurcation function to find a solution $x(t, \varepsilon)$ of (4.3.1) such that

$$\sup_{t \in \mathbb{R}} |x(t, \varepsilon) - \gamma(t - \alpha(\varepsilon))| \to 0, \text{ as } \varepsilon \to 0$$

for some $\alpha(\varepsilon) \in \mathbb{R}$. Note $\gamma(t)$ crosses transversally the discontinuity level Ω_0 in (4.3.1).

4.3.2   Homoclinic Sliding on Discontinuity

On the other hand, in [8], we study a case when a part of homoclinic orbit is sliding on a discontinuity level: Consider the planar discontinuous system

$$\begin{aligned}
\dot{z} &= f_+(z) + \varepsilon g(z, t, \varepsilon) \quad \text{for} \quad y > 1, \\
\dot{z} &= f_-(z) + \varepsilon g(z, t, \varepsilon) \quad \text{for} \quad y < 1,
\end{aligned} \tag{4.3.2}$$

where $z = (x, y) \in \mathbb{R}^2$, $f_\pm$, g are C^3-smooth and g is 1-periodic in t. While on $y = 1$ (cf. [134]), we consider the equation

$$\begin{aligned}
\dot{x} &= \frac{q_{+2}(x, 1, t, \varepsilon)}{q_{+2}(x, 1, t, \varepsilon) - q_{-2}(x, 1, t, \varepsilon)} q_{+1}(x, 1, t, \varepsilon) \\
&+ \frac{q_{-2}(x, 1, t, \varepsilon)}{q_{-2}(x, 1, t, \varepsilon) - q_{+2}(x, 1, t, \varepsilon)} q_{-1}(x, 1, t, \varepsilon),
\end{aligned} \tag{4.3.3}$$

where $q_\pm = (q_{\pm 1}, q_{\pm 2})$ and $q_\pm(z, t, \varepsilon) = f_\pm(z) + \varepsilon g(z, t, \varepsilon)$. We suppose the following conditions

(i) $f_-(0) = 0$ and $Df_-(0)$ has no eigenvalues on the imaginary axis.

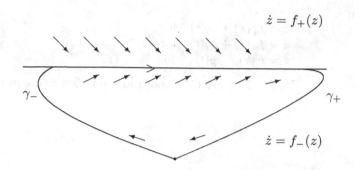

Figure 4.2: A planar sliding homoclinic cycle

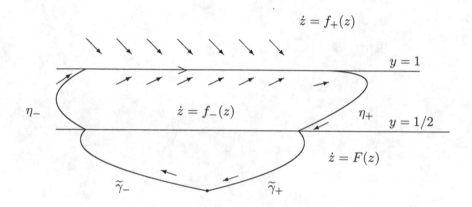

Figure 4.3: A planar homoclinic sliding crossing a line

(ii) There are two solutions $\gamma_-(s)$, $\gamma_+(s)$ of $\dot z = f_-(z)$, $y \le 1$ defined on $\mathbb{R}_- = (-\infty, 0]$, $\mathbb{R}_+ = [0, +\infty)$, respectively, such that $\lim_{s \to \pm\infty} \gamma_\pm(s) = 0$ and $\gamma_\pm(s) = (x_\pm(s), y_\pm(s))$ with $y_\pm(0) = 1$, $x_-(0) < x_+(0)$. Moreover, $f_\pm(z) = (f_{\pm 1}(z), f_{\pm 2}(z))$ with $f_{\pm 1}(x, 1) > 0$, $f_{+2}(x, 1) < 0$ for $x_-(0) \le x \le x_+(0)$. Furthermore, $f_{-2}(x, 1) > 0$ for $x_-(0) \le x < x_+(0)$, $f_{-2}(x_+(0), 1) = 0$ and $\partial_x f_{-2}(x_+(0), 1) < 0$.

Assumptions (i) and (ii) mean (see Fig. 4.2) that (4.3.2) for $\varepsilon = 0$ has a sliding homoclinic solution γ, created by $\gamma_\pm$, to a hyperbolic equilibrium 0. Conditions for the bifurcation of γ to bounded solutions on $\mathbb{R}$ of (4.3.2) under the perturbation $\varepsilon g(z, t, \varepsilon)$ are derived in [8]. Functional-analytical methods are used in [8, 22] based on the implicit function theorem.

Finally, we have also studied cases when homoclinic orbit $\gamma(s)$ transversally crosses another curves of discontinuity (see Fig. 4.3). For simplicity, we supposed that such a discontinuity in (4.3.2) occurs at the level $y = 1/2$, i.e. we considered the system

$$\dot{z} = f_+(z) + \varepsilon g(t) \quad \text{for} \quad y > 1,$$
$$\dot{z} = f_-(z) + \varepsilon g(t) \quad \text{for} \quad 1/2 < y < 1, \qquad\qquad (4.3.4)$$
$$\dot{z} = F(z) + \varepsilon g(t) \quad\;\; \text{for} \quad y < 1/2,$$

where $z = (x, y) \in \mathbb{R}^2$, $f_\pm$, F, g are C^3-smooth and g is 1-periodic in t.

Chapter 5

Topological Transversality

5.1 Topological Transversality and Chaos

5.1.1 Topologically Transversal Invariant Sets

We study bifurcation of chaotic solutions of discontinuous o.d.eqns in Chapter 4 using topological degree methods. Those results are extensions of a similar classical result, the Smale-Birkhoff homoclinic theorem for smooth o.d.eqns, based on the existence of Smale's horseshoe which is a consequence of a transversal intersection of stable and unstable manifolds of a hyperbolic fixed point of a diffeomorphism. When the smoothness of an o.d.eqn is dropped, then this classical approach fails. For this reason we use topological degree arguments. This is the aim of Chapter 4. Similar mathematical difficulties occur when a diffeomorphism possesses a hyperbolic fixed point, but the corresponding stable and unstable manifolds do not have a transversal intersection. So a natural question arises that which kind of intersection should have stable and unstable manifolds in order to have a chaotic behavior of a diffeomorphism near that intersection. The aim of this section is to give an answer on this question by extending the Smale-Birkhoff homoclinic theorem in this direction. We show that a topologically transversal intersection of stable and unstable manifolds guaranties chaotic behavior of the diffeomorphism.

In order to state our main results, we need the following notations and definitions. Let $\mathcal{M}$ be a C^1-smooth manifold without boundary. Consider a C^1-smooth diffeomorphism $f : \mathcal{M} \to \mathcal{M}$ possessing a hyperbolic fixed point p and let W_p^s, W_p^u be the global stable and unstable manifolds of p, respectively. Let $\widetilde{W}_p^s$, $\widetilde{W}_p^u$ be open subsets of W_p^s, W_p^u, respectively, which are submanifolds of $\mathcal{M}$, that is the immersed and induced topologies on $\widetilde{W}_p^s$ and $\widetilde{W}_p^u$, respectively, coincide. We assume that $\widetilde{W}_p^s \cap \widetilde{W}_p^u \setminus \{p\} \neq \emptyset$, i.e. there is a point q homoclinic to p. We also suppose the existence of a compact component $K \ni q$ of the set $\widetilde{W}_p^s \cap \widetilde{W}_p^u$, that is a compact subset $K \subset \widetilde{W}_p^s \cap \widetilde{W}_p^u \setminus \{p\}$ such that $q \in K$ and there exists an open connected precompact subset $U \subset \bar{U} \subset \mathcal{M} \setminus \{p\}$

M. Fečkan, *Topological Degree Approach to Bifurcation Problems*, 143–181.
© Springer Science + Business Media B.V., 2008

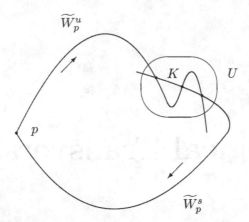

Figure 5.1: A transversal homoclinic set $K = \widetilde{W}_p^s \cap \widetilde{W}_p^u \cap U$

satisfying $U \cap \widetilde{W}_p^s \cap \widetilde{W}_p^u = K$ (see Fig. 5.1). Since K is compact there is an m_0 such that $f^{m_0}(K)$ is in a local chart U_p of p. By shrinking U, we can assume in addition that $\widetilde{W}_p^{s(u)} \cap \bar{U} = \overline{\widetilde{W}_p^{s(u)} \cap U}$ and as well as $f^{m_0}(\bar{U}) \subset U_p$, and consequently, U is orientable. Moreover, since U is precompact and open, $\widetilde{W}_p^{s(u)} \cap U$ are also submanifolds of $\mathcal{M}$ and there is an $N_0 > 0$ such that $\overline{\widetilde{W}_p^{s(u)} \cap U} \subset f^{\mp N_0}(W_{p,loc}^{s(u)}) \setminus \{p\}$. Hence $\widetilde{W}_p^{s(u)} \cap U$ are also orientable. Then we can define the local intersection number $\#\bigl(\widetilde{W}_p^s \cap U, \widetilde{W}_p^u \cap U\bigr)$ of the manifolds $\widetilde{W}_p^s \cap U$ and $\widetilde{W}_p^u \cap U$ in $\mathcal{M}$. The main purpose of this section is to prove the following result [18].

Theorem 5.1.1. *If* $\#\bigl(\widetilde{W}_p^s \cap U, \widetilde{W}_p^u \cap U\bigr) \neq 0$ *then there exists* $\omega_0 \in \mathbb{N}$ *such that for any* $\mathbb{N} \ni \omega \geq \omega_0$ *there is a set* $\Lambda_\omega \subset \mathcal{M}$ *and a mapping* $\pi_\omega : \Lambda_\omega \to \mathcal{E}$ *such that*

   (i) $f^{2\omega}(\Lambda_\omega) = \Lambda_\omega$

   (ii) π_ω *is continuous, one to one and onto*

   (iii) $\pi_\omega \circ f^{2\omega} = \sigma \circ \pi_\omega$, *where* $\sigma : \mathcal{E} \to \mathcal{E}$ *is the Bernoulli shift map*

Remark 5.1.2. Note that we do not know whether π_ω^{-1} is continuous. Thus we cannot say, in general, that Λ_ω is homeomorphic to $\mathcal{E}$. However, if q is a transversal homoclinic point, π_ω is a homeomorphism, since in the considerations that follow we can use the implicit function theorem instead of the Brouwer degree theory, getting the standard Smale horseshoe of the Smale-Birkhoff homoclinic theorem.

Remark 5.1.3. The diffeomorphism f of Theorem 5.1.1 has positive topological entropy. This follows from [49, Lemma 1.3].

Remark 5.1.4. Consider a C^1-smooth diffeomorphism $f : \mathcal{M} \to \mathcal{M}$ possessing two hyperbolic fixed points p_1 and p_2, $p_1 \neq p_2$. If $W_{p_1}^s$ and $W_{p_2}^u$, and $W_{p_2}^s$ and $W_{p_1}^u$, are topologically transversal, respectively, then we can prove a similar result for f like in Theorem 5.1.1.

Now we present a situation where topologically transversal intersections of stable and unstable manifolds naturally occur. Let $\mathcal{M}$ be a smooth symplectic surface with the symplectic area form ω. Let $f : \mathcal{M} \to \mathcal{M}$ be a smooth area-preserving diffeomorphism homotopic to identity and exactly symplectic, i.e. $f^*(\alpha) = \alpha + dS$ for some smooth function $S : \mathcal{M} \to \mathbb{R}$ and α is a differential 1-form such that $d\alpha = \omega$. Time-one-maps of 1-periodic Hamiltonian systems are such diffeomorphisms (see (5.1.2)). We note that any exactly symplectic map is also symplectic. If $\mathcal{M}$ is exactly symplectic, i.e. $\omega = d\alpha$, and simply connected then any symplectic map is also exactly symplectic. Following a proof of [196, Theorem 2.1], we have the next result.

Proposition 5.1.5. *Assume that f has two hyperbolic fixed points p_1, p_2, $p_1 \neq p_2$. Let us suppose that $W_{p_1}^s \cap W_{p_2}^u \neq \emptyset$ and $W_{p_2}^s \cap W_{p_1}^u \neq \emptyset$. If $W_{p_1}^s \neq W_{p_2}^u$ and $W_{p_2}^s \neq W_{p_1}^u$, and $S(p_1) = S(p_2)$, then $W_{p_1}^s$ and $W_{p_2}^u$, and $W_{p_2}^s$ and $W_{p_1}^u$, are topologically transversal, respectively. Hence Remark 5.1.4 gives a chaotic behavior of f.*

Example 5.1.6. Proposition 5.1.5 can be applied to the results of [41, Section 5, p. 703]. More precisely, let us consider the equation

$$\ddot{u} + W(u, t) = 0, \tag{5.1.1}$$

where $W : \mathbb{R} \times \mathbb{R} \to \mathbb{R}$ is C^1-smooth and 1-periodic in t. Suppose that (5.1.1) has two different hyperbolic periodic solutions u_1 and u_2. Let $\phi : \mathbb{R}^2 \times \mathbb{R} \to \mathbb{R}^2$ be the flow of $\dot{u} = v$, $\dot{v} = -W(u, t)$. Then $f(x, y) = \phi(x, y, 1)$ is exactly symplectic by taking

$$\omega = dx \wedge dy, \quad \alpha = -y\, dx, \quad S(x, y) = -\int_0^1 L(\phi(x, y, s))\, ds,$$

where L is the Lagrangian of (5.1.1) given by $L(\phi) = \frac{\dot{\phi}_2^2}{2} - G(\phi_1, t)$, $\phi = (\phi_1, \phi_2)$, $\partial G / \partial u = W$. Clearly the periodic solutions u_1, u_2 induce hyperbolic fixed points $(u_1(0), \dot{u}_1(0)) = w_1$ and $(u_2(0), \dot{u}_2(0)) = w_2$ of f. Then $S(w_1) = S(w_2)$ is a part of the condition (1) of [41, Definition 2.1]. Hence w_1 and w_2 are on the same action level for f in the terminology of [196, Theorem 8.1]. S is naturally related to the action functional over $H_1^1 = \{u \in H_{loc}^1(\mathbb{R}) : u(t+1) = u(t) \text{ a.e. in } \mathbb{R}\}$ defined as $u \mapsto \int_0^1 \left(\frac{\dot{u}(s)^2}{2} - G(u(s), s) \right) ds$ on [41, p. 679]. Hence the assumptions of Subsection 5 of [41] imply the validity of Theorem 5.1.1, which is stronger than the results of Subsection 5 of [41]. On the other hand, the main results of [41] deal with equations like (5.1.1) under assumptions on u_1 and u_2 weaker than in this subsection, namely u_1 and u_2 are not hyperbolic but they are the so-called consecutive minimizers, see [41, Definition 2.1]. By using variational

methods, chaotic bumping solutions are shown to exist in [41]. Finally we note
that for a C^1-smooth 1-periodic Hamiltonian system

$$\dot{x} = -\frac{\partial H}{\partial y}(x, y, t), \quad \dot{y} = \frac{\partial H}{\partial x}(x, y, t), \tag{5.1.2}$$

the time-one map is exactly symplectic with

$$\omega = dx \wedge dy, \quad \alpha = x\, dy, \quad S(x, y) = \int_0^1 \Big(\psi_1(x, y, t)\dot{\psi}_2(x, y, t) - H(\psi(x, y, t)) \Big)\, dt,$$

where $\psi = (\psi_1, \psi_2)$ is the flow of (5.1.2). The action functional for (5.1.2) over
H_1^1 is given by

$$(x, y) \mapsto \int_0^1 x(t)\dot{y}(t) - H(x(t), y(t), t)\, dt = \int_0^1 \frac{1}{2}\langle J\dot{u}(t), u(t) \rangle - H(u(t), t)\, dt,$$

where $J = \begin{pmatrix} 0 & 1 \\ -1 & 0 \end{pmatrix}$, $u = (x, y)$ and $\langle \cdot, \cdot \rangle$ is the usual inner product on $\mathbb{R}^2$.

Results similar to Theorem 5.1.1 have been proved by others authors. For
example, a semiconjugacy to the Bernoulli shift σ on $\mathcal{E}$ of some power of a
given map is proved in [154] provided an isolating neighborhood of the map
satisfies some conditions on the Conley indices of its subsets. On the other
hand, Lefschetz Fixed Point Theorem and Topological Principle of Wazewski is
applied in [180] to prove the existence of a compact invariant set for the Poincaré
map of a time-periodic vector field on which the same map is semiconjugated to
the Bernoulli shift σ on $\mathcal{E}$ and the counterimage (by the semiconjugacy) of any
periodic point of σ contains a periodic point of the Poincaré map. The notion
of periodic isolating segments is an essential tool for the proofs in [180]. Finally,
the same situation as in this section is studied in [49]. By using geometric and
homological methods, it is proved in [49] that, under the conditions of Theorem
5.1.1, there is an invariant set of some power of f on which the same power
of f is semiconjugated to the Bernoulli shift σ on $\mathcal{E}$. In all these papers by
semiconjugacy it is meant that the associated map between the invariant set and
the symbolic set (in this section it is the map π_ω) is shown to be continuous and
onto. Hence the semiconjugacy does not directly imply the existence of infinitely
many periodic orbits of a given map (apart from the result in [180]), but it
implies positive topological entropy of the map. Our approach instead, which
is based on an idea in [16], namely on the notion of exponential dichotomies of
difference equations, allow us to prove that π_ω is one to one, a result that was
not stated in [49,154,180]. Consequently, f has infinitely many periodic orbits as
well as quasiperiodic ones. Moreover we are able to identify the periodic points
of the map as solutions of a particular equation.

5.1.2   Difference Boundary Value Problems

To avoid the use of either the tangent vector bundle of $\mathcal{M}$ or local charts of $\mathcal{M}$,
we assume for simplicity in this subsection that $\mathcal{M} = \mathbb{R}^N$. This restriction is only

technical. Next, for any $\xi \in \widetilde{W}_p^s \cap \bar{U}$ and $\eta \in \widetilde{W}_p^u \cap \bar{U}$ we set $\xi_n = f^n(\xi)$, $n \in \mathbb{Z}_+$, $\eta_n = f^n(\eta)$, $n \in \mathbb{Z}_-$. Now we fix $\omega \in \mathbb{N}$ large and put

$$J_\omega = \{-\omega, -\omega+1, \ldots, \omega-1, \omega\}, \quad \tilde{J}_\omega = \{-\omega, -\omega+1, \ldots, \omega-1\},$$

$$J_\omega^- = \{-\omega, -\omega+1, \cdots, -1, 0\}, \quad I_\omega^- = \{-\omega, -\omega+1, \ldots, -1\},$$

$$J_\omega^+ = \{0, 1, \ldots, \omega-1, \omega\}, \quad I_\omega^+ = \{0, 1, \ldots, \omega-2, \omega-1\}.$$

In this subsection we study the nonlinear system

$$x_{n+1} = f(x_n) \tag{5.1.3}$$

near $\{\xi_n\}_{n \in J_\omega^+}$ and $\{\eta_n\}_{n \in J_\omega^-}$. As usually in the bifurcation theory, at first we consider linearizations of (5.1.3) along $\{\xi_n\}_{n \in \mathbb{Z}_+}$ and $\{\eta_n\}_{n \in \mathbb{Z}_-}$ given by

$$v_{n+1} = Df(\xi_n)v_n, \quad n \in \mathbb{Z}_+, \tag{5.1.4}$$

$$w_{n+1} = Df(\eta_n)w_n, \quad n \in \mathbb{Z}_-, \quad n \neq 0 \tag{5.1.5}$$

respectively. Since $\xi_n \to p$, $\eta_n \to p$ as $n \to \pm\infty$ and p is a hyperbolic fixed point of f, we have the following result.

Lemma 5.1.7. *Systems (5.1.4) and (5.1.5) have exponential dichotomies on $\mathbb{Z}_+$ and $\mathbb{Z}_-$, respectively, i.e. there are positive constants $L \geq 1$, $\delta \in (0,1)$ and orthogonal projections $P_\xi : \mathbb{R}^N \to T_\xi \widetilde{W}_p^s$, $Q_\eta : \mathbb{R}^N \to T_\eta \widetilde{W}_p^u$ such that the fundamental solutions $V_\xi(n)$ and $W_\eta(n)$ of (5.1.4) and (5.1.5) respectively, satisfy the following conditions:*

$$\|V_\xi(n)P_\xi V_\xi(m)^{-1}\| \leq L\delta^{n-m}, \quad m \leq n, \quad m, n \in \mathbb{Z}_+,$$
$$\|V_\xi(n)(\mathbb{I} - P_\xi)V_\xi(m)^{-1}\| \leq L\delta^{m-n}, \quad n \leq m, \quad m, n \in \mathbb{Z}_+, \tag{5.1.6}$$

$$\|W_\eta(n)(\mathbb{I} - Q_\eta)W_\eta(m)^{-1}\| \leq L\delta^{n-m}, \quad m \leq n, \quad m, n \in \mathbb{Z}_-,$$
$$\|W_\eta(n)Q_\eta W_\eta(m)^{-1}\| \leq L\delta^{m-n}, \quad n \leq m, \quad m, n \in \mathbb{Z}_-, \tag{5.1.7}$$

respectively, along with

$$V_\xi(n)P_\xi V_\xi(n)^{-1} \to P_p, \quad W_\eta(-n)Q_\eta W_\eta(-n)^{-1} \to \mathbb{I} - P_p \tag{5.1.8}$$

as $n \to \infty$ uniformly for ξ, η, where P_p is the exponential dichotomy projection of $\dot{z} = Df(p)z$. Moreover, L and δ can be chosen to be independent of $\xi \in \widetilde{W}_p^s \cap \bar{U}$ and $\eta \in \widetilde{W}_p^u \cap \bar{U}$. If $\mathcal{M}$ and f are C^r-smooth then P_ξ and Q_η are C^{r-1}-smooth in ξ, η respectively.

Proof. Let $\xi \in \widetilde{W}_p^s \cap \bar{U}$, then $f^n(\xi) \to p$ as $n \to +\infty$. From the roughness of exponential dichotomies in Lemma 2.5.1, it follows that there exists $r_\xi > 0$ such that when $|\tilde{\xi} - \xi| < r_\xi$, $\tilde{\xi} \in \widetilde{W}_p^s \cap \bar{U}$, $v_{n+1} = f'(\tilde{\xi}_n)v_n$ has an exponential dichotomy on $\mathbb{Z}_+$ with constants L_ξ and δ_ξ. Covering $\widetilde{W}_p^u \cap \bar{U}$ with a finite number of balls centered at ξ and of radius r_ξ the result follows as far as the dichotomy on $\mathbb{Z}_+$ is concerned. A similar argument applies for the dichotomy on $\mathbb{Z}_-$. Note that any projection having the same range as P_ξ, (resp. Q_η) satisfies condition (5.1.6) (resp. (5.1.7)). Thus, it is the additional requirement that P_ξ and Q_η are orthogonal that makes them unique. This uniqueness also implies that P_ξ and Q_η are continuous in ξ, η respectively. In fact let us prove this for P_ξ. Since $\widetilde{W}_p^s$ is C^1, we get that $T_\xi \widetilde{W}_p^s$ depends continuously on ξ and the same holds for its orthogonal complement $(T_\xi \widetilde{W}_p^s)^\perp$ in $\mathbb{R}^n$. So, if $\{v_1(\xi), \ldots, v_d(\xi)\}$ is a (local) orthonormal basis of $T_\xi \widetilde{W}_p^s$ that depends continuously on ξ in a neighborhood of some $\xi_0 \in \widetilde{W}_p^s$, we have $P_\xi v = \sum_{j=1}^d \langle v, v_j(\xi) \rangle v_j(\xi)$ and then P_ξ is continuous in ξ. Note that the uniqueness of P_ξ implies that $P_\xi v$ does not depend on the choice of the basis $\{v_1(\xi), \ldots, v_d(\xi)\}$. A similar argument holds for Q_η. Moreover, note that, when $\mathcal{M}$ and are C^r-smooth then P_ξ and Q_η are of class C^{r-1}. $\square$

Next we study nonhomogenous equations of (5.1.4) and (5.1.5).

Lemma 5.1.8. *There exist $\omega_0 \in \mathbb{N}$ and a constant $c > 0$ such that given any $\omega \in \mathbb{N}$, $\omega \geq \omega_0$, $(\xi, \eta) \in (\widetilde{W}_p^s \cap \bar{U}) \times (\widetilde{W}_p^u \cap \bar{U})$, and $b, h_n \in \mathbb{R}^N$, $n \in \tilde{J}_\omega$, $\phi \in RP_\xi$, $\psi \in RQ_\eta$, there exist unique solutions $\{v_n\}_{n \in J_\omega^+}$ and $\{w_n\}_{n \in J_\omega^-}$ of the linear systems*

$$v_{n+1} = Df(\xi_n)v_n + h_n, \quad n \in I_\omega^+,$$
$$w_{n+1} = Df(\eta_n)w_n + h_n, \quad n \in I_\omega^-, \tag{5.1.9}$$

respectively, together with the boundary value conditions

$$P_\xi v_0 = \phi, \quad Q_\eta w_0 = \psi, \quad v_\omega - w_{-\omega} = b. \tag{5.1.10}$$

Moreover such solutions are linear in (b, h, ϕ, ψ), $h = \{h_n\}_{n \in \tilde{J}_\omega}$ and satisfy

$$\max_{n \in J_\omega^+} |v_n|, \max_{n \in J_\omega^-} |w_n| \leq c \Big(\max_{n \in J_\omega^\pm} |h_n| + |b| + |\phi| + |\psi| \Big). \tag{5.1.11}$$

Proof. Uniqueness: When $h = 0$, $\phi = 0$, $\psi = 0$ and $b = 0$ then from (5.1.9) and (5.1.10) we get $v_n = V_\xi(n)v_0$, $w_n = W_\xi(n)w_0$, $P_\xi v_0 = 0$, $Q_\eta w_0 = 0$ and $V_\xi(\omega)v_0 = W_\xi(-\omega)w_0$. Then

$$\left[V_\xi(\omega)(\mathbb{I} - P_\xi)V_\xi(\omega)^{-1} \right] V_\xi(\omega)v_0 = \left[W_\eta(-\omega)(\mathbb{I} - Q_\eta)W_\eta(-\omega)^{-1} \right] W_\eta(-\omega)w_0$$

By (5.1.8) we have

$$\mathcal{R}\left[V_\xi(\omega)(\mathbb{I} - P_\xi)V_\xi(\omega)^{-1} \right] \cap \mathcal{R}\left[W_\eta(-\omega)(\mathbb{I} - Q_\eta)W_\eta(-\omega)^{-1} \right] = \{0\}$$

for ω large. Hence $V_\xi(\omega)v_0 = W_\eta(-\omega)w_0 = 0$, which give $v_n = w_n = 0$.

Existence: For $b, h_n \in \mathbb{R}^N, n \in \tilde{J}_\omega, \phi \in \mathcal{R}P_\xi, \psi \in \mathcal{R}Q_\eta, \phi_1 \in \mathcal{N}P_p, \psi_1 \in \mathcal{R}P_p$ we put

$$v_n = V_\xi(n)\phi + \sum_{k=0}^{n-1} V_\xi(n)P_\xi V_\xi(k+1)^{-1}h_k$$

$$-\sum_{k=n}^{\omega-1} V_\xi(n)(\mathbb{I} - P_\xi)V_\xi(k+1)^{-1}h_k + V_\xi(n)(\mathbb{I} - P_\xi)V_\xi(\omega)^{-1}\phi_1 \quad \text{for } n \in J_\omega^+$$

(5.1.12)

and

$$w_n = W_\eta(n)\phi + \sum_{k=-\omega}^{n-1} W_\eta(n)(\mathbb{I} - Q_\eta)W_\eta(k+1)^{-1}h_k$$

$$-\sum_{k=n}^{-1} W_\eta(n)Q_\eta W_\eta(k+1)^{-1}h_k + W_\eta(n)(\mathbb{I} - Q_\eta)W_\eta(-\omega)^{-1}\psi_1 \quad \text{for } n \in J_\omega^-$$

(5.1.13)

Clearly such v_n, w_n satisfy (5.1.9) and $P_\xi v_0 = \phi, Q_\eta w_0 = \psi$. To show $v_\omega - w_{-\omega} = b$, we solve

$$W_\eta(-\omega)(\mathbb{I} - Q_\eta)W_\eta(-\omega)^{-1}\psi_1 - V_\xi(\omega)(\mathbb{I} - P_\xi)V_\xi(\omega)^{-1}\phi_1 = V_\xi(\omega)\phi - b$$

$$-W_\eta(-\omega)\phi + \sum_{k=0}^{\omega-1} V_\xi(\omega)P_\xi V_\xi(k+1)^{-1}h_k + \sum_{k=-\omega}^{-1} W_\xi(-\omega)Q_\eta W_\eta(k+1)^{-1}h_k$$

(5.1.14)

By (5.1.8) we can uniquely solve ϕ_1, ψ_1 from (5.1.14) for ω large. This gives the desired solution of this Lemma. Estimates of (5.1.11) follow directly from (5.1.6), (5.1.7), (5.1.12), (5.1.13) and (5.1.14). The proof is finished. $\square$

Similarly we have the next result.

Lemma 5.1.9. *For any $(\xi, \eta) \in (\widetilde{W}_p^s \cap \bar{U}) \times (\widetilde{W}_p^u \cap \bar{U})$, $\phi \in \mathcal{R}P_\xi$, $\psi \in \mathcal{R}Q_\eta$, and for any bounded sequence $\{h_n\}_{n \in \mathbb{Z}}$, there exist unique solutions $\{v_n\}_{n \geq 0}$ and $\{w_n\}_{n \leq 0}$ of the linear systems*

$$v_{n+1} = Df(\xi_n)v_n + h_n, \quad n \geq 0,$$
$$w_{n+1} = Df(\eta_n)w_n + h_n, \quad n \leq -1,$$

respectively, together with the boundary value conditions

$$P_\xi v_0 = \phi, \quad Q_\eta w_0 = \psi.$$

Moreover such solutions are linear in $(h^\pm, \phi, \psi)$, $h^+ = \{h_n\}_{n \geq 0}$, $h^- = \{h_n\}_{n \leq 0}$, and there exists a constant $c > 0$, independent of $(h^\pm, \phi, \psi)$, such that:

$$\sup_{n \geq 0} |v_n| \leq c\left(\sup_{n \geq 0} |h_n| + |\phi|\right), \quad \sup_{n \leq 0} |w_n| \leq c\left(\sup_{n \leq 0} |h_n| + |\psi|\right)$$

Proof. The solutions are given by formulas

$$v_n = V_\xi(n)\phi + \sum_{k=0}^{n-1} V_\xi(n)P_\xi V_\xi(k+1)^{-1}h_k - \sum_{k=n}^{\infty} V_\xi(n)(\mathbb{I} - P_\xi)V_\xi(k+1)^{-1}h_k$$

for $n \geq 0$ and

$$w_n = W_\eta(n)\phi + \sum_{k=-\infty}^{n-1} W_\eta(n)(\mathbb{I}-Q_\eta)W_\eta(k+1)^{-1}h_k - \sum_{k=n}^{-1} W_\xi(n)Q_\eta W_\eta(k+1)^{-1}h_k$$

for $n \leq 0$. The proof is finished. $\square$

Now we are ready to study (5.1.3) near $\{\xi_n\}_{n \in J_\omega^+}$ and $\{\eta_n\}_{n \in J_\omega^-}$.

Theorem 5.1.10. *There exist $\omega_0 \in \mathbb{N}$ and a constant $c > 0$ such that, for any $\omega \in \mathbb{N}$, $\omega \geq \omega_0$, and $(\xi, \eta) \in (\widetilde{W}_p^s \cap \bar{U}) \times (\widetilde{W}_p^u \cap \bar{U})$, there exist unique $\{x_n^+(\omega, \xi, \eta)\}_{n \in J_\omega^+}$ and $\{x_n^-(\omega, \xi, \eta)\}_{n \in J_\omega^-}$ which satisfy (5.1.3) separately on I_ω^+ and I_ω^- such that*

$$P_\xi x_0^+(\omega, \xi, \eta) = P_\xi\xi, \quad Q_\eta x_0^-(\omega, \xi, \eta) = Q_\eta\eta, \quad x_\omega^+(\omega, \xi, \eta) = x_{-\omega}^-(\omega, \xi, \eta),$$

together with

$$\max_{n \in J_\omega^+} |x_n^+(\omega, \xi, \eta) - \xi_n| \leq c\delta^\omega, \quad \max_{n \in J_\omega^-} |x_n^-(\omega, \xi, \eta) - \eta_n| \leq c\delta^\omega.$$

Moreover, $x_n^{\pm}(\omega, \xi, \eta)$ are C^{r-1}-smooth with respect to ξ and η when $\mathcal{M}$ is a C^r-manifold and f is a C^r-diffeomorphism.

Proof. We apply the implicit function theorem to (5.1.3) near $\{\xi_n\}_{n \in J_\omega^+}$ and $\{\eta_n\}_{n \in J_\omega^-}$, respectively. By putting $x_n^+ = \xi_n + v_n$, $n \in J_\omega^+$ and $x_n^- = \eta_n + w_n$, $n \in J_\omega^-$, we get the system

$$\begin{aligned} v_{n+1} &= f(\xi_n + v_n) - f(\xi_n), \quad n \in I_\omega^+ \\ w_{n+1} &= f(\eta_n + w_n) - f(\eta_n), \quad n \in I_\omega^-. \end{aligned} \tag{5.1.15}$$

Since we are looking for solutions of (5.1.3) such that $x_\omega^+ = x_{-\omega}^-$, $P_\xi x_0^+ = P_\xi\xi$ and $Q_\eta x_0^- = Q_\eta\eta$ we add the boundary value conditions:

$$v_\omega - w_{-\omega} = \eta_{-\omega} - \xi_\omega = O(\delta^\omega), \quad P_\xi v_0 = 0, \quad Q_\eta w_0 = 0. \tag{5.1.16}$$

Let $v = (v_0, \ldots, v_\omega) \in \mathbb{R}^{N(\omega+1)}$, $w = (w_{-\omega}, \ldots, w_0) \in \mathbb{R}^{N(\omega+1)}$. To solve (5.1.15–5.1.16), we take the mapping $\Gamma_\omega : (\widetilde{W}_p^s \cap \bar{U}) \times \widetilde{W}_p^u \cap \bar{U}) \times \mathbb{R}^{2N(\omega+1)} \to \mathbb{R}^{2N(\omega+1)}$ defined by

$$\Gamma_\omega(\xi, \eta, v, w) = \begin{pmatrix} (v_{n+1} - f(\xi_n + v_n) + f(\xi_n))_{n \in I_\omega^+} \\ (w_{n+1} - f(\eta_n + w_n) + f(\eta_n))_{n \in I_\omega^-} \\ v_\omega - w_{-\omega} - (\eta_{-\omega} - \xi_\omega) \\ P_\xi v_0 \\ Q_\eta w_0 \end{pmatrix},$$

where $\begin{pmatrix} P_\xi v_0 \\ Q_\eta w_0 \end{pmatrix} \in \mathbb{R}^N = \mathcal{R}P_\xi \times \mathcal{R}Q_\eta$. Since P_ξ and Q_η are C^{r-1}-smooth, for any fixed $\omega \geq \omega_0$, Γ_ω is C^{r-1}-smooth in (ξ, η, v, w) as well. We take on $\mathbb{R}^{2N(\omega+1)}$ the maximum norm $\max_i\{|v_i|, |w_i|\}$. We have $\Gamma_\omega(\xi, \eta, 0, 0) = O(\delta^\omega)$ uniformly with respect to (ξ, η) and the linearized map $D_{(v,w)}\Gamma_\omega(\xi, \eta, 0, 0)$ has the form

$$D_{(v,w)}\Gamma_\omega(\xi, \eta, 0, 0)\begin{pmatrix} v \\ w \end{pmatrix} = \begin{pmatrix} (v_{n+1} - Df(\xi_n)v_n)_{n \in I_\omega^+} \\ (w_{n+1} - Df(\eta_n)w_n)_{n \in I_\omega^-} \\ v_\omega - w_{-\omega} \\ P_\xi v_0 \\ Q_\eta w_0. \end{pmatrix}$$

Lemma 5.1.8 implies that the map $D_{(v,w)}\Gamma_\omega(\xi, \eta, 0, 0)$ is invertible and that its inverse is bounded uniformly with respect to (ξ, η). Hence from the implicit function theorem we get that $c > 0$ and $\omega_0 \gg 1$ exist such that for $\omega \geq \omega_0$, the equation $\Gamma_\omega(\xi, \eta, v, w) = 0$ can be solved uniquely for (v, w) in a neighborhood of $(0, 0)$ in terms of (ξ, η, ω). Moreover $\max_i\{|v_i|, |w_i|\} < c\delta^\omega$, and the solution is C^{r-1}-smooth in (ξ, η), for any fixed $\omega \geq \omega_0$. The proof is finished. $\qquad\square$

5.1.3 Chaotic Orbits

In this section we prove Theorem 5.1.1.

Proof. We need the following technical arguments. Let $V \subset \mathcal{M}$ be an open subset such that $K \subset V \subset \bar{V} \subset U$ and ω_0 be as in Theorem 5.1.10. We also assume that ω_0 is large enough that $c\delta^{\omega_0}$ is less than the distance of V from ∂U and for any $\xi \in \widetilde{W}_p^s \cap \bar{V}$, $\eta \in \widetilde{W}_p^u \cap \bar{V}$ and $n \geq \omega_0$ we have $|\xi_n - p|, |\eta_n - p| \leq C\delta^n$ where C can be chosen independent of ξ, η because of the compactness of $\widetilde{W}_p^s \cap \bar{V}$ and $\widetilde{W}_p^u \cap \bar{V}$. Of course, here we assume that ξ_n, η_n, p are in the local chart U_p of p, for any $n \geq \omega_0$ so that we can consider their differences. Note that the solutions $\{x_n^\pm(\omega, \xi, \eta)\}_{n \in J^\pm}$ are defined for $(\xi, \eta) \in \widetilde{W}_p^s \cap \bar{V} \times \widetilde{W}_p^u \cap \bar{V}$, and $\#(\widetilde{W}_p^s \cap V, \widetilde{W}_p^u \cap V) = \#(\widetilde{W}_p^s \cap U, \widetilde{W}_p^u \cap U)$ because $K \subset V$ implies $K \cap \partial V = \emptyset$.

We split the proof into the following steps.

1. Step. *We show that f has enough periodic orbits oscillating between p and K:*

We recall that $f^{m_0}(\bar{U}) \subset U_p$ for some m_0, and then we can assume that U is embedded in $\mathbb{R}^N$, i.e. $U \hookrightarrow \mathbb{R}^N$. Let h, k be non negative integers. For any finite sequence $E = \{e_j\}_{j=-h}^k$, $e_j \in \{0, 1\}$ such that $e_0 = 1$, we set

$$\{j_1, j_2, \ldots, j_{i_E}\} = \left\{j \mid e_j = 1,\right\}$$

where $-h \leq j_1 < j_2 < \cdots < j_{i_E} \leq k$. Note that, being $e_0 = 1$, we have $j_1 \leq 0 \leq j_{i_E}$. Then we set:

$$j_0 = j_{i_E} - h - k - 1, \quad j_{i_E+1} = h + k + 1 + j_1. \tag{5.1.17}$$

Note that $j_0 \leq -h - 1 < j_1$ and $j_{i_E+1} \geq k + 1 > j_{i_E}$. Moreover:

$$j_{i_E+1} - j_{i_E} = j_1 - j_0. \tag{5.1.18}$$

Next, for $\omega \in \mathbb{N}$ fixed and large (that is greater than ω_0), we define:

$$F^E : \left((\widetilde{W}_p^s \cap \bar{V}) \times (\widetilde{W}_p^u \cap \bar{V}) \right)^{i_E} \to \mathbb{R}^{Ni_E}, \quad F^E = (F_1^E, F_2^E, \ldots, F_{i_E}^E)$$

where

$$F_r^E(\xi^1, \eta^1, \ldots, \xi^{i_E}, \eta^{i_E}) = x_0^-\left((j_r - j_{r-1})\omega, \xi^r, \eta^r\right) - x_0^+\left((j_{r+1} - j_r)\omega, \xi^{r+1}, \eta^{r+1}\right),$$

and

$$\xi^{i_E+1} = \xi^1, \quad \eta^{i_E+1} = \eta^1 \tag{5.1.19}$$

and $x_0^\pm((j_r - j_{r-1})\omega, \xi^r, \eta^r)$ are derived as in Theorem 5.1.10. We note that $x_0^+((j_r - j_{r-1})\omega, \xi^r, \eta^r)$ is at a distance from $\xi^r \in \widetilde{W}_p^s \cap \bar{V} \subset U \hookrightarrow \mathbb{R}^N$ less than $c\delta^\omega$ and that the same holds for $x_0^-((j_r - j_{r-1})\omega, \xi^r, \eta^r)$ and $\eta^r \in \widetilde{W}_p^u \cap \bar{V} \subset U \hookrightarrow \mathbb{R}^N$. Consequently, $x_0^\pm((j_r - j_{r-1})\omega, \xi^r, \eta^r) \in U$ and we can consider the above differences in the definition of F^E.

Let us now give a brief motivation for such a definition. Assume that the equation $F_r^E(\xi^1, \eta^1, \ldots, \xi^{i_E}, \eta^{i_E}) = 0$ has a solution $(\xi^1, \eta^1, \ldots, \xi^{i_E}, \eta^{i_E})$. Then, starting from:

$$x_0^-((j_r - j_{r-1})\omega, \xi^r, \eta^r) = x_0^+((j_{r+1} - j_r)\omega, \xi^{r+1}, \eta^{r+1})$$

and using

$$x_{(j_{r+1}-j_r)\omega}^+((j_{r+1} - j_r)\omega, \xi^{r+1}, \eta^{r+1}) = x_{-(j_{r+1}-j_r)\omega}^-((j_{r+1} - j_r)\omega, \xi^{r+1}, \eta^{r+1})$$

we obtain:

$$f^{2(j_{r+1}-j_r)\omega}x_0^-((j_r - j_{r-1})\omega, \xi^r, \eta^r) = f^{2(j_{r+1}-j_r)\omega}x_0^+((j_{r+1} - j_r)\omega, \xi^{r+1}, \eta^{r+1}) =$$
$$f^{(j_{r+1}-j_r)\omega}x_{(j_{r+1}-j_r)\omega}^+((j_{r+1} - j_r)\omega, \xi^{r+1}, \eta^{r+1}) =$$
$$f^{(j_{r+1}-j_r)\omega}x_{-(j_{r+1}-j_r)\omega}^-((j_{r+1} - j_r)\omega, \xi^{r+1}, \eta^{r+1}) = x_0^-((j_{r+1} - j_r)\omega, \xi^{r+1}, \eta^{r+1})$$

and then, using the induction:

$$f^{2(j_s-j_r)\omega}x_0^-((j_r - j_{r-1})\omega, \xi^r, \eta^r) = x_0^-((j_s - j_{s-1})\omega, \xi^s, \eta^s) \tag{5.1.20}$$

for any $0 \leq r \leq s \leq j_{i_E+1}$. Now, from $e_0 = 1$ we see that $\bar{\iota} \in \{1, \ldots, i_E\}$ exists such that $j_{\bar{\iota}} = 0$. Then we define:

$$x_0(\omega, E) = x_0^-(-j_{\bar{\iota}-1}\omega, \xi^{\bar{\iota}}, \eta^{\bar{\iota}}) = x_0^-((j_{\bar{\iota}} - j_{\bar{\iota}-1})\omega, \xi^{\bar{\iota}}, \eta^{\bar{\iota}})$$
$$= x_0^+((j_{\bar{\iota}+1} - j_{\bar{\iota}})\omega, \xi^{\bar{\iota}+1}, \eta^{\bar{\iota}+1}) = x_0^+(j_{\bar{\iota}+1}\omega, \xi^{\bar{\iota}+1}, \eta^{\bar{\iota}+1}) \tag{5.1.21}$$

and note that from (5.1.17), (5.1.18) and (5.1.20) we obtain:

$$f^{2(h+k+1)\omega}x_0(\omega, E) = f^{2(j_{i_E+1}-j_1)\omega}x_0(\omega, E)$$
$$= f^{2(j_{\bar{\iota}}-j_1)\omega}[f^{2(j_{i_E+1}-j_{\bar{\iota}})\omega}x_0^-((j_{\bar{\iota}} - j_{\bar{\iota}-1})\omega, \xi^{\bar{\iota}}, \eta^{\bar{\iota}})]$$
$$= f^{2(j_{\bar{\iota}}-j_1)\omega}x_0^-((j_1 - j_0)\omega, \xi^1, \eta^1)$$
$$= x_0^-((j_{\bar{\iota}} - j_{\bar{\iota}-1})\omega, \xi^{\bar{\iota}}, \eta^{\bar{\iota}}) = x_0(\omega, E)$$

that is $x_0(\omega, E)$ is a $2(h + k + 1)\omega$-periodic point of the map $x_{n+1} = f(x_n)$. Next, for any $r \in \{1, \ldots, j_{i_E}\}$, we have, using (5.1.20):

$$f^{2j_r\omega}x_0(\omega, E) = f^{2(j_r - j_{\bar\imath})\omega}x_0^-((j_{\bar\imath} - j_{\bar\imath - 1})\omega, \xi^{\bar\imath}, \eta^{\bar\imath})$$
$$= x_0^-((j_r - j_{r-1})\omega, \xi^r, \eta^r) = x_0^+((j_{r+1} - j_r)\omega, \xi^{r+1}, \eta^{r+1})$$

and then Theorem 5.1.10 implies that

$$\|f^{2j_r\omega}x_0(\omega, E) - \eta^r\| \leq c\delta^{(j_r - j_{r-1})\omega} \leq c\delta^\omega$$
$$\|f^{2j_r\omega}x_0(\omega, E) - \xi^{r+1}\| \leq c\delta^{(j_r - j_{r-1})\omega} \leq c\delta^\omega$$

that is $f^{2j_r\omega}x_0(\omega, E)$ belongs to a (small when $\omega > \omega_0$ is sufficiently large) neighborhood of K for any $r = 1, \ldots, i_E$. Moreover, for any $j \in \mathbb{N}$ such that $0 < j < j_{r+1} - j_r$, we have:

$$f^{2(j_r + j)\omega}x_0(\omega, E) = f^{2j\omega}x_0^+((j_{r+1} - j_r)\omega, \xi^{r+1}, \eta^{r+1})$$
$$= x_{j\omega}^+((j_{r+1} - j_r)\omega, \xi^{r+1}, \eta^{r+1})$$

and then, again from Theorem 5.1.10,

$$\|f^{2(j_r + j)\omega}x_0(\omega, E) - p\| \leq \|f^{2(j_r + j)\omega}x_0(\omega, E) - \xi_{j\omega}^{r+1}\| + \|\xi_{j\omega}^{r+1} - p\| \leq 2c\delta^\omega.$$

Thus the map F^E is constructed so that if $F^E(\xi^1, \eta^1, \ldots, \xi^{i_E}, \eta^{i_E}) = 0$, then the diffeomorphism f has a periodic orbit attracting and repelling several times by the hyperbolic fixed point p. More precisely, if the initial point of this periodic orbit is given by (5.1.21), the point $f^{2j\omega}x_0(\omega, E)$ is near the set K if $e_j = 1$ and it is near the fixed point p if $e_j = 0$. Using this it is easy to see that starting from different E we get different periodic orbits. To solve $F_E = 0$, we take the simple homotopy

$$H^E : \left((\widetilde{W}_p^s \cap \bar{V}) \times (\widetilde{W}_p^u \cap \bar{V})\right)^{i_E} \times [0, 1] \to \mathbb{R}^{Ni_E}, \quad H^E = (H_1^E, H_2^E, \ldots, H_{i_E}^E)$$

given by

$$H_r^E(\xi^1, \eta^1, \ldots, \xi^{i_E}, \eta^{i_E}, \lambda) = \lambda F_r^E(\xi^1, \eta^1, \ldots, \xi^{i_E}, \eta^{i_E}) + (1 - \lambda)(\eta^r - \xi^{r+1}).$$

for $0 \leq \lambda \leq 1$. Theorem 5.1.10 gives

$$\left|F_r^E(\xi^1, \eta^1, \ldots, \xi^{i_E}, \eta^{i_E}) - \eta^r + \xi^{r+1}\right| \leq 2c\delta^\omega,$$

where the constant c is the same as in Theorem 5.1.10. Hence we get

$$\left|H_r^E(\xi^1, \eta^1, \ldots, \xi^{i_E}, \eta^{i_E}, \lambda) - \eta^r + \xi^{r+1}\right| \leq 2c\delta^\omega.$$

Consequently $H^E(\cdot, \lambda) \neq 0$ on the boundary $\partial\left((\widetilde{W}_p^s \cap V) \times (\widetilde{W}_p^u \cap V)\right)^{i_E}$ for any $0 \leq \lambda \leq 1$. By Section 2.3.4, this gives for the Brouwer degree

$$\left|\deg\left(F^E, ((\widetilde{W}_p^s \cap V) \times (\widetilde{W}_p^u \cap V))^{i_E}, 0\right)\right| = \left|\#(\widetilde{W}_p^s \cap V, \widetilde{W}_p^u \cap V)^{i_E}\right| \neq 0.$$

Summarizing, we see that, under the assumptions of Theorem 5.1.1, the equation $F_E = 0$ is always solvable in the set $\left((\widetilde{W}_p^s \cap V) \times (\widetilde{W}_p^u \cap V) \right)^{i_E}$ for any sequence $E = \{e_j\}_{j=1}^k \in \{0, 1\}^k$, $e_1 = 1$ and any $k \in \mathbb{N}$ for a fixed large (i.e. greater than ω_0) $\omega \in \mathbb{N}$. Thus we have seen that the map f has enough periodic orbits oscillating between p and K.

2. Step. *We show more oscillatory orbits of f oscillating between the homoclinic set K and the hyperbolic fixed point p. This is done by constructing the set Λ_ω and the mapping π_ω in Theorem 5.1.1:*

Let $\sim$ be the equivalence relation on the set $\mathcal{E}$ defined as follows:

let $E, E' \in \mathcal{E}$. We say that $E \sim E'$ if $n_0 \in \mathbb{Z}$ exists such that $E = \sigma^{n_0}(E')$.

Then we choose a unique element for any equivalence class in $\mathcal{E}/\sim$ and form a metric subspace $\mathcal{E}_\sim$. Without loss of generality we can also assume that $\mathcal{E}_\sim \subset \mathcal{E}_1 := \{E = \{e_j\}_{j\in\mathbb{Z}} \in \mathcal{E} \ : \ e_j = 0 \text{ for any } j \in \mathbb{Z} \text{ or } e_0 = 1\}$. We obtain in this way a subspace $\mathcal{E}_\sim \subset \mathcal{E}_1$ such that if $E_1, E_2 \in \mathcal{E}_\sim$, then either $E_1 = E_2$ or $E_1 \neq \sigma^n(E_2)$ for any $n \in \mathbb{Z}$. Now we define a map $E \mapsto \mathcal{O}_E$ from $\mathcal{E}_1$ in the space of orbits of f as follows. If $e_j = 0$, $\forall j \in \mathbb{Z}$ then we put $\mathcal{O}_E = \{p\}$ the fixed point orbit of f. On the other hand, if $e_0 = 1$, we have the following two possibilities: either E is periodic with the minimal period m, i.e. $\sigma^m(E) = E$ and $\sigma^k(E) \neq E$ for $1 \leq k < m$, or E is nonperiodic, that is there is no $m \in \mathbb{N}$ such that $\sigma^m(E) = E$. In the first case we apply the above procedure to the finite sequence $\{e_j\}_{j=0}^{m-1}$ (m being the minimal period of E). We obtain then a $2m\omega$-periodic orbit $\mathcal{O}_E$ such that $f^{2j\omega}(x_0)$ is either near the set K or the point p according to $e_j = 1$ or $e_j = 0$, respectively. In the second case we consider, for any $m \in \mathbb{N}$, the finite sequence $E_m = \{e_j^m\}_{j=-m}^m := \{e_j\}_{j=-m}^m$, $m \in \mathbb{N}$, to obtain a periodic orbit $\mathcal{O}_{E_m}$ of $x_{n+1} = f(x_n)$ with the same oscillation property between K and p as above. We set $\mathcal{O}_{E_m} = \{x_n^m\}_{n\in\mathbb{Z}}$. Then take a convergent subsequence $x_0^{m_i}$ of x_0^m and let x_0 be its limit as $i \to \infty$. Note that, $\mathcal{O}_{E_m}$ being an orbit of $x_{n+1} = f(x_n)$, we have $x_j^m = f^j(x_0^m)$ for any $j \in \mathbb{Z}$. Thus $x_j^{m_i}$ converges to $f^j(x_0)$. Hence we set: $\mathcal{O}_E = \{f^j(x_0)\}_{j\in\mathbb{Z}}$. Note that $\mathcal{O}_E$ is an orbit of the map f such that $f^{2j\omega}(x_0)$ is either near the set K or the point p according to $e_j = 1$ or $e_j = 0$, respectively. In fact for any given $j \in \mathbb{Z}$ there exists $m_0 \in \mathbb{N}$ such that $e_j^m = e_j$ for any $m \geq m_0$. Thus the conclusion follows because it is satisfied by $f^{2j\omega}(x_0^{m_i})$ for any i sufficiently large. Observe also that if E is not periodic (that is $\sigma^n(E) \neq E$ for any $n \in \mathbb{Z}$) then $\mathcal{O}_E$ is also a non periodic orbit of f because of the stated oscillation properties. Moreover, if $\mathcal{O}_E = \mathcal{O}_{E'}$ then $E = E'$, that is the map $E \mapsto \mathcal{O}_E$ is one to one. Finally, for $\mathcal{O}_E = \{f^i(x_0)\}_{i\in\mathbb{Z}}$ we set

$$f^{2j\omega}(\mathcal{O}_E) = \{f^{2j\omega+i}(x_0)\}_{i\in\mathbb{Z}}. \tag{5.1.22}$$

At this point we would like that the following holds: $\mathcal{O}_{\sigma^n(E)} = f^{2\omega n}(\mathcal{O}_E)$ when E and $\sigma^n(E)$ belong to $\mathcal{E}_1$. However this is not generally true even if it is true that $\mathcal{O}_{\sigma^n(E)}$ and $f^{2\omega n}(\mathcal{O}_E)$ have the same oscillating properties between K and p. The point is that in order to define the orbit $\mathcal{O}_E$ we actually use the axiom of choice to choose a convergent subsequence $x_0^{m_i}$ of x_0^m. Thus, in general

$\mathcal{O}_{\sigma(E)} \neq f^{2\omega}(\mathcal{O}_E)$, because we can perhaps choose convergent subsequences of x_0^m and x_1^m such that their limits do not satisfy the equality $x_1 = f(x_0)$ (of course, when the sequence x_0^m is itself convergent this does not happen). For this reason, in order to extend the map $E \mapsto \mathcal{O}_E$ to $\mathcal{E}$ we have to pass through $\mathcal{E}_\sim$.

Let $E = \{e_n\}_{n \in \mathbb{Z}} \in \mathcal{E}$ be a doubly infinite sequence of 0 and 1. If $e_j = 0$ for any $j \in \mathbb{Z}$ we set $J_\omega(E) = \{p\}$, the fixed point orbit of f. If $j \in \mathbb{Z}$ exists such that $e_j = 1$, a unique $E' \in \mathcal{E}_\sim$ exists such that $E = \sigma^{n_0}(E')$ for some $n_0 \in \mathbb{Z}$. Such a n_0 is unique when E is nonperiodic and is defined up to a multiple of the least period, when E is periodic. Then we set

$$J_\omega(E) = f^{2\omega n_0}(\mathcal{O}_{E'}). \tag{5.1.23}$$

This definition does not depend on n_0. We only have to prove this in the case where E is periodic with least period, say, m. We have:

$$f^{2\omega(km+n_0)}(\mathcal{O}_{E'}) = f^{2\omega n_0}(f^{2\omega km}(\mathcal{O}_{E'})) = f^{2\omega n_0}(\mathcal{O}_{E'})$$

for any $k \in \mathbb{Z}$, since $\mathcal{O}_{E'}$ is $2\omega m$-periodic. Thus the definition (5.1.23) is independent of n_0. Moreover, if $E = \sigma^{n_0}(E')$, $E' \in \mathcal{E}_\sim$, then $\sigma(E) = \sigma^{n_0+1}(E')$ and:

$$J_\omega(\sigma(E)) = f^{2\omega(n_0+1)}(\mathcal{O}_{E'}) = f^{2\omega}(J_\omega(E))$$

that is

$$J_\omega \circ \sigma = f^{2\omega} \circ J_\omega. \tag{5.1.24}$$

Now we prove that J_ω is one to one. Because of the oscillating property, it follows immediately that $J_\omega(E) \neq \{p\}$ when E is not the identically zero sequence. Now, let $E_1, E_2 \in \mathcal{E}$ be two, non identically zero, sequences such that $J_\omega(E_1) = J_\omega(E_2)$. Write $E_1 = \sigma^{n_1}(E_1')$ and $E_2 = \sigma^{n_2}(E_2')$, with $E_1' = \{e_n'^{(1)}\}$, $E_2' = \{e_n'^{(2)}\} \in \mathcal{E}_\sim$. Then $J_\omega(E_1) = J_\omega(E_2)$ implies $\mathcal{O}_{E_1'} = f^{2\omega(n_2-n_1)}(\mathcal{O}_{E_2'})$. From this equation and the oscillating property we see that $e_{(n_2-n_1)}'^{(2)} = 1$, that is $\sigma^{n_2-n_1}(E_2') \in \mathcal{E}_1$. Moreover, as we have already observed, $f^{2\omega(n_2-n_1)}(\mathcal{O}_{E_2'})$ has the same oscillating properties between K and p as $\mathcal{O}_{\sigma^{(n_2-n_1)}(E_2')}$. Thus E_1' and $\sigma^{(n_2-n_1)}(E_2')$ are two elements of $\mathcal{E}_1$ such that $\mathcal{O}_{E_1'}$ and $\mathcal{O}_{\sigma^{(n_2-n_1)}(E_2')}$ have the same oscillating properties between K an p. But this means that $E_1' = \sigma^{(n_2-n_1)}(E_2')$ from which we get immediately $E_1 = E_2$. So J_ω is one to one and satisfies (5.1.24).

Now we consider the map $\mathcal{P} : J_\omega(\mathcal{E}) \to \mathbb{R}^N$ given by $\mathcal{P}(J_\omega(E)) = x_0$, where $J_\omega(E) = \{x_j\}_{j \in \mathbb{Z}}$. We set $\Lambda_\omega = \mathcal{P}(J_\omega(\mathcal{E}))$, we define $\mathcal{Q} : \Lambda_\omega \to J_\omega(\mathcal{E})$ as $\mathcal{Q}(x_0) = \{f^j(x_0)\}_{j \in \mathbb{Z}}$. Finally, we define $\pi_\omega : \Lambda_\omega \to \mathcal{E}$ as $\pi_\omega(x_0) = J_\omega^{-1}(\mathcal{Q}(x_0))$.

3. Step. *Verification of properties of π_ω and Λ_ω in Theorem 5.1.1:*

(i) π_ω is one to one. This easily follows from the fact that different initial points give different orbits (that is $\mathcal{Q}$ is one to one).

(ii) π_ω is continuous. To show this, let x_0^0, $\{x_0^i\}_{i\in\mathbb{N}} \subset \Lambda_\omega$ and $x_0^i \to x_0^0$ as $i \to$
∞. Then $f^j(x_0^i) \to f^j(x_0^0)$ as $i \to \infty$ for any $j \in \mathbb{Z}$. Hence for any $N_0 \in \mathbb{N}$,
and $|j| \le N_0$, the points $f^{2j\omega}(x_0^i)$ of the orbit $\mathcal{Q}(x_0^i) = J_\omega(E^i) \in J_\omega(\mathcal{E})$
and $f^{2j\omega}(x_0^0)$ of the orbit $\mathcal{Q}(x_0^0) = J_\omega(E^0) \in J_\omega(\mathcal{E})$ have, for i large, the
same kind of oscillation between K and p. Consequently, the sequences E^i
and E^0, for i large, have the same elements in the first j, $|j| \le N_0$ places.
This implies that $E^i \to E^0$ as $i \to \infty$.

(iii) $\sigma(\pi_\omega(x_0)) = \pi_\omega(f^{2\omega}(x_0))$. In fact, we know that $J_\omega(\sigma(E)) = f^{2\omega}(J_\omega(E))$
for any $E \in \mathcal{E}$. Thus if $E = \pi_\omega(x_0)$ we have $J_\omega(E) = \mathcal{Q}(x_0) = \{f^j(x_0)\}_{j\in\mathbb{Z}}$,
then

$$J_\omega(\sigma(\pi_\omega(x_0)))=J_\omega(\sigma(E))=f^{2\omega}(J_\omega(E))=\left\{f^j(f^{2\omega}(x_0))\right\}_{j\in\mathbb{Z}}=\mathcal{Q}(f^{2\omega}(x_0)).$$

Thus $\sigma(\pi_\omega(x_0)) = \pi_\omega(f^{2\omega}(x_0))$ for any $x_0 \in \Lambda_\omega$.

Summarizing, π_ω is continuous, one to one and $\pi_\omega \circ f^{2\omega} = \sigma \circ \pi_\omega$. By the
construction it is also clear that π_ω is onto. The proof of Theorem 5.1.1 is
completed. $\square$

5.1.4   Periodic Points and Extensions on Invariant Compact Subsets

In this subsection we study periodic orbits of f on Λ_ω more closely. First we
prove the following result.

Proposition 5.1.11. *Periodic points of f are dense in the set Λ_ω from Theorem
5.1.1.*

Proof. Let $x_0 \in \Lambda_\omega$. Then there is an $E \in \mathcal{E}$ such that $\mathcal{P}(J_\omega(E)) = x_0$ where
$J_\omega(E) = \{x_j\}_{j\in\mathbb{Z}}$. Let $E = \{e_j\}_{j\in\mathbb{Z}}$. If E is periodic then x_0 is a periodic point
of f. Let E be non-periodic. There are unique $E' \in \mathcal{E}_\sim$ and $n_0 \in \mathbb{Z}$ such that
$E = \sigma^{n_0}(E')$. Now E' is also non-periodic. We have $J_\omega(E') = f^{-2n_0\omega}(J_\omega(E))$.
The point $x_0' = \mathcal{P}(J_\omega(E'))$ can be approximated by the proof of Theorem 5.1.1
with periodic points of f from Λ_ω. Of course the same hold for the point $x_0 =
f^{2n_0\omega}(x_0')$. This finishes the proof of Proposition 5.1.11. $\square$

We also get that the only isolated points of the set Λ_ω could be periodic
points of f and f depends sensitively on the set Λ_ω' of all non-isolated points of
Λ_ω, that is there is a constant $d > 0$ such that in any neighborhood of $x_0 \in \Lambda_\omega'$
there are $x_0' \in \Lambda_\omega$ and $n_0' \in \mathbb{N}$ such that the distance between $f^{n_0'}(x_0)$ and
$f^{n_0'}(x_0')$ is greater than d. We do not know whether the periodic points of f in
Λ_ω are non-isolated or not.

On the other-hand, let either $\Upsilon_\omega = \Lambda_\omega$ or $\Upsilon_\omega = \Lambda_\omega'$. We extend the map π_ω
on the closure $\overline{\Upsilon}_\omega$ of Υ_ω, and we denote this extension again by π_ω. So $\overline{\Upsilon}_\omega$ is
compact but we do not know whether the unique continuous extension of π_ω is
one-to-one or not. The extension is made as follows: For any $x_0 \in \overline{\Upsilon}_\omega \setminus \Upsilon_\omega$, we
take a sequence $\{x_j\}_{j\in\mathbb{N}} \subset \Upsilon_\omega$ such that $x_j \to x_0$. Hence $f^{2\omega k}(x_j) \to f^{2\omega k}(x_0)$.

Consequently, for any $N \in \mathbb{N}$, the orbits $\{f^{2\omega k}(x_j)\}_{k=-N}^{k=N}$ and $\{f^{2\omega k}(x_0)\}_{k=-N}^{k=N}$ have the same oscillating properties between set K and point p for j large. This implies the existence of the limit $\lim_{j\to\infty} \pi_\omega(x_j) := \pi_\omega(x_0)$, which is independent of $\{x_j\}_{j\in\mathbb{N}}$. The continuity of π_ω follows as in the proof of Theorem 5.1.1. Clearly the extension π_ω is onto $\mathcal{E}$ for the case $\Upsilon_\omega = \Lambda_\omega$. If $\Upsilon_\omega = \Lambda'_\omega$ then $\pi_\omega(\Upsilon_\omega)$ is dense in $\mathcal{E}$ and $\pi_\omega(\overline{\Upsilon}_\omega)$ is compact in $\mathcal{E}$. This implies $\pi_\omega(\overline{\Upsilon}_\omega) = \mathcal{E}$ also for this case. The property $\pi_\omega \circ f^{2\omega} = \sigma \circ \pi_\omega$ follows from the limit procedure $x_j \to x_0$. Of course, $\overline{\Upsilon}_\omega$ is invariant for $f^{2\omega}$. For the case $\Upsilon_\omega = \Lambda_\omega$, we again have infinitely many periodic points of f which are dense in $\overline{\Upsilon}_\omega$. For the case $\Upsilon_\omega = \Lambda'_\omega$, we have that any point $x_0 \in \overline{\Lambda'_\omega}$ is an accumulating point of periodic points of f with periods tending to infinity. Iterations of those periodic points oscillate differently between the set K and the point p. Consequently, the map f is sensitive on $\overline{\Lambda'_\omega}$ in the following sense: there is a constant $d > 0$ such that in any neighborhood of $x_0 \in \overline{\Lambda'_\omega}$ there are x'_0 and $n'_0 \in \mathbb{N}$ such that the distance between $f^{n'_0}(x_0)$ and $f^{n'_0}(x'_0)$ is greater than d.

5.1.5 Perturbed Topological Transversality

Checking the topological transversality of stable and unstable manifold, is not an easy task. This is the reason why in this subsection we study the case where W_p^s and W_p^u intersect on a homoclinic manifold and consider a C^2-smooth perturbation of f. Then we have the following result.

Theorem 5.1.12. *Let $f(x,\varepsilon)$ be a C^2 map in its arguments, and assume there exist open, connected, bounded subsets $\Omega \subset \overline{\Omega} \subset U_\Omega \subset \mathbb{R}^\mu$ and C^2-smooth mappings $x_n(\alpha)$, $\alpha \in U_\Omega$, such that the following hold:*

(i) *$x_{n+1}(\alpha) = f(x_n(\alpha), 0)$, $n \in \mathbb{Z}$, $\lim_{n\to\pm\infty} x_n(\alpha) = p$ uniformly with respect to $\alpha \in U_\Omega$ for a hyperbolic fixed point p of the mapping $f(x, 0)$.*

(ii) *$\left\{ \frac{\partial x_n}{\partial \alpha_i}(\alpha), i = 1, 2, \ldots, \mu \right\}$ are linearly independent and they form a basis for the space of bounded solutions of the equation $v_{n+1} = f_x(x_n(\alpha), 0)v_n$ on $\mathbb{Z}$ for any $\alpha \in U_\Omega$. Moreover, the mapping $x_0 : U_\Omega \to \mathbb{R}^N$ is one to one.*

Assume, moreover, that the Melnikov function (see below (5.1.34)) associated to the perturbation $f(x,\varepsilon)$ satisfies the following conditions:

(H1) $M(\alpha) \neq 0$ on $\partial\Omega$

(H2) $\deg(M, \Omega, 0) \neq 0$

Then there exists $\varepsilon_0 > 0$ such that for $0 < |\varepsilon| \leq \varepsilon_0$, it is nonzero the local intersection number of the stable and unstable manifolds of the hyperbolic fixed point $p(\varepsilon)$ of the map $x_{n+1} = f(x_n, \varepsilon)$ which is located near the fixed point p of the map $x_{n+1} = f(x_n, 0)$.

Conditions (i) and (ii) mean that $f(x,0)$ has a non-degenerate homoclinic manifold given by $\{x_0(\alpha) \mid \alpha \in U_\Omega\}$. Note non-degenerate periodic and homoclinic manifolds of o.d.eqns are investigated in Sections 3.3 and 4.1. When a map satisfies the conditions of Theorem 5.1.12 we obtain, thanks to Theorem 5.1.1, a kind of chaotic behavior of the perturbed diffeomorphism $f(x,\varepsilon)$, when $\varepsilon \neq 0$.

Proof. We note that $x_0(\alpha) \in W_p^s \cap W_p^u$, for any $\alpha \in U_\Omega$. In the next constructions of this subsection, the set Ω is fixed but the neighborhood U_Ω of $\overline{\Omega}$ could be shrunk by keeping its connectedness. Let $U \subset \mathbb{R}^N$ be an open and bounded subset such that

$$\mathcal{H}_\Omega = \{x_0(\alpha) \mid \alpha \in U_\Omega\} = U \cap \widetilde{W}_p^s \cap \widetilde{W}_p^u, \qquad (5.1.25)$$

where again $\widetilde{W}_p^s$ and $\widetilde{W}_p^u$ are open subsets of W_p^s and W_p^u, respectively, which are submanifolds of $\mathbb{R}^N$.

Let P_ξ and Q_η be the projections of Subsection 5.1.2 for the open subset U, which are now C^1-smooth in ξ and η, respectively. Arguing as in Subsection 5.1.2, we get the following result.

Theorem 5.1.13. *There exist $\varepsilon_0 > 0$ and $\rho > 0$ such that for any $|\varepsilon| < \varepsilon_0$ and $\xi \in \widetilde{W}_p^s \cap U$, $\eta \in \widetilde{W}_p^u \cap U$ the equations*

$$x_{n+1}^+(\varepsilon,\xi) = f(x_n^+(\varepsilon,\xi),\varepsilon), \quad P_\xi x_0^+(\varepsilon,\xi) = P_\xi \xi \qquad (5.1.26)$$

for $n \geq 0$, and

$$x_{n+1}^-(\varepsilon,\eta) = f(x_n^-(\varepsilon,\eta),\varepsilon), \quad Q_\eta x_0^-(\varepsilon,\eta) = Q_\eta \eta \qquad (5.1.27)$$

for $n \leq -1$, have unique solutions $\{x_n^+(\varepsilon,\xi)\}_{n\geq 0}$ and $\{x_n^-(\varepsilon,\eta)\}_{n\leq 0}$ respectively, such that

$$\sup_{n\geq 0} |x_n^+(\varepsilon,\xi) - \xi_n| \leq \rho, \quad \sup_{n\leq 0} |x_n^-(\varepsilon,\eta) - \eta_n| \leq \rho \qquad (5.1.28)$$

Moreover $\{x_n^+(\varepsilon,\xi)\}_{n\geq 0}$ and $\{x_n^-(\varepsilon,\eta)\}_{n\leq 0}$ are C^1-smooth in their arguments and

$$\lim_{\varepsilon\to 0} \sup_{n\geq 0} |x_n^+(\varepsilon,\xi) - \xi_n| = 0, \quad \lim_{\varepsilon\to 0} \sup_{n\leq 0} |x_n^-(\varepsilon,\eta) - \eta_n| = 0 \qquad (5.1.29)$$

Proof. We give the proof for $n \geq 0$ the case $n \leq 0$ being handled similarly. Let $\xi \in \widetilde{W}_p^s \cap U$, and $x_n = \xi_n + v_n$. Then $\{v_n\}_{n\geq 0}$ satisfies the system:

$$v_{n+1} - f'(\xi_n)v_n = \{f(\xi_n + v_n,\varepsilon) - f(\xi_n) - f'(\xi_n)v_n\}, \quad P_\xi v_0 = 0. \qquad (5.1.30)$$

We are looking for solutions of (5.1.30) such that $\sup_{n\geq 0} |v_n| \to 0$ as $\varepsilon \to 0$. Let $\rho > 0$ be fixed. From Lemma 5.1.9 it follows that the map

$$\Gamma_\infty(v) = \begin{pmatrix} \{v_{n+1} - f'(\xi_n)v_n\}_{n\geq 0} \\ P_\xi v_0 \end{pmatrix}$$

has a bounded inverse. So, for any $\{v_n\}_{n\geq 0}$ such that $\sup_{n\geq 0} |v_n| < \rho$ we define $\{\hat{v}_n\}_{n\geq 0}$ as the unique solution of

$$\Gamma_\infty(\{\hat{v}_n\}_{n\geq 0}) = \left(\begin{array}{c} \{f(\xi_n + v_n, \varepsilon) - f(\xi_n) - f'(\xi_n)v_n\}_{n\geq 0} \\ 0 \end{array} \right).$$

From Lemma 5.1.9 it follows that

$$\sup_{n\geq 0} |\hat{v}_n| \leq c \sup_{n\geq 0} |f(\xi_n + v_n, \varepsilon) - f(\xi_n) - f'(\xi_n)v_n| \leq c\{\Delta(\rho) \sup_{n\geq 0} |v_n| + O(\varepsilon)\}$$

where $\Delta(\rho) \to 0$ as $\rho \to 0$. Thus it is easy to see that the map $\{v_n\}_{n\geq 0} \mapsto \{\hat{v}_n\}_{n\geq 0}$ is a contraction on the ball $\{\{v_n\}_{n\geq 0} : \sup_{n\geq 0} |v_n| < \rho\}$ provided ρ and ε_0 are sufficiently small. As a consequence there exists a unique fixed point $\{v_n(\varepsilon, \xi)\}_{n\geq 0}$ that gives rise to the solution $x_n(\varepsilon, \xi) = \xi_n + v_n(\varepsilon, \xi)$. From the smoothness of the map $\{v_n\}_{n\geq 0} \mapsto \{f(\xi_n + v_n, \varepsilon) - f(\xi_n) - f'(\xi_n)v_n\}_{n\geq 0}$, we obtain that $x_n(\varepsilon, \xi)$ is smooth and that (5.1.28), (5.1.29) hold. The proof is finished. $\square$

Now we consider the function $H : \widetilde{W}^s_p \cap U \times \widetilde{W}^u_p \cap U \times (-\varepsilon_0, \varepsilon_0) \to \mathbb{R}^N$ given by

$$H(\xi, \eta, \varepsilon) = x_0^+(\varepsilon, \xi) - x_0^-(\varepsilon, \eta). \tag{5.1.31}$$

Note that, because of the hyperbolicity of p, the map $x_{n+1} = f(x_n, \varepsilon)$ has, for small $|\varepsilon|$, a unique hyperbolic fixed point $p(\varepsilon)$ such that $p(\varepsilon) \to p$ as $\varepsilon \to 0$. Such a fixed point is C^2-smooth in ε and the solutions of $H(\xi, \eta, \varepsilon) = 0$ give rise to orbits $\{x_n(\varepsilon)\}_{n\in\mathbb{Z}}$ of the map $x_{n+1} = f(x_n, \varepsilon)$ that are homoclinic to $p(\varepsilon)$. Moreover, if $U_1 \subset \bar{U}_1 \subset U$ is an open, connected subset of U, the functions $x_0^+(\varepsilon, \xi)$, $\xi \in \widetilde{W}^s_p \cap U_1$ and $x_0^-(\varepsilon, \eta)$, $\eta \in \widetilde{W}^u_p \cap U_1$, describe open subsets of the stable and unstable manifolds $W^s_{p(\varepsilon)}$ and $W^u_{p(\varepsilon)}$ of $p(\varepsilon)$ that are also immersed submanifolds in $\mathbb{R}^n$. So, denoting with $\widetilde{W}^s_{p(\varepsilon)}$ and $\widetilde{W}^u_{p(\varepsilon)}$ these submanifolds of $\mathbb{R}^N$, by Section 2.3.4, the intersection number $\#\left(\widetilde{W}^s_{p(\varepsilon)} \cap U_1, \widetilde{W}^u_{p(\varepsilon)} \cap U_1\right)$ is found by computing the Brouwer degree $\deg(H(\xi, \eta, \varepsilon), (\widetilde{W}^s_p \cap U_1) \times (\widetilde{W}^u_p \cap U_1), 0)$.

Thus, let $d_s = \dim \widetilde{W}^s_p$, and $d_u = \dim \widetilde{W}^u_p$. From the hyperbolicity of p we get $d_s + d_u = N$, hence we can write $\mathbb{R}^N = W_\mu \oplus W_s \oplus W_u \oplus V_\mu$ where $\dim W_\mu = \dim V_\mu = \mu$, $V_\mu^\perp = W_\mu \oplus W_s \oplus W_u$, $\dim W_s = d_s - \mu$, $\dim W_u = d_u - \mu$, and U_Ω is an open subset of W_μ. Then, replacing U and U_Ω with smaller, open, connected and bounded subsets of $\mathbb{R}^n$ and $\mathbb{R}^\mu$ respectively, so that (5.1.25) and $\bar{\Omega} \subset U_\Omega$ are still satisfied, we can find open and convex subsets $O^s \subset W_s$, $O^u \subset W_u$, $O^* \subset V_\mu$, containing 0, and a C^1-diffeomorphism $\Phi : U_\Omega \oplus O^s \oplus O^u \oplus O^* \to U \subset \mathbb{R}^N$ such that the following holds:

$$\Phi(\alpha) = x_0(\alpha), \text{ for any } \alpha \in U_\Omega$$
$$\Phi(U_\Omega \oplus O^s) = \widetilde{W}^s_p \cap U, \quad \Phi(U_\Omega \oplus O^u) = \widetilde{W}^u_p \cap U \tag{5.1.32}$$

Let $\tilde{\xi}, \tilde{\eta}$ be the coordinates on $W_\mu \oplus W_s$ and $W_\mu \oplus W_u$ respectively. Then possibly shrinking U_Ω, O^s and O^u we consider, the function

$$\widetilde{H} : \overline{(U_\Omega \oplus O^s)} \times \overline{(U_\Omega \oplus O^u)} \to \mathbb{R}^N$$

given by:
$$\widetilde{H}(\widetilde{\xi},\widetilde{\eta},\varepsilon) := \Phi^{-1}(x_0^+(\varepsilon,\Phi(\widetilde{\xi}))) - \Phi^{-1}(x_0^-(\varepsilon,\Phi(\widetilde{\eta})))\,.$$

Obviously, $\widetilde{H}(\widetilde{\xi},\widetilde{\eta},\varepsilon) = 0$ if and only if $H(\xi,\eta,\varepsilon) = 0$, and then

$$\deg(H(\xi,\eta,\varepsilon),(\widetilde{W}_p^s\cap U)\times(\widetilde{W}_p^u\cap U),0)=\pm\deg(\widetilde{H}(\widetilde{\xi},\widetilde{\eta},\varepsilon),(U_\Omega\oplus O^s)\times(U_\Omega\oplus O^u),0).$$

Now, Theorem 5.1.13 implies that

$$\widetilde{H}(\widetilde{\xi},\widetilde{\eta},0) = \widetilde{\xi} - \widetilde{\eta}$$

from which we get $\widetilde{H}(\alpha,\alpha,0) = 0$ and

$$\widetilde{H}(\widetilde{\xi},\widetilde{\eta},\varepsilon) = \widetilde{\xi} - \widetilde{\eta}$$
$$+\varepsilon\left\{[\Phi'(\widetilde{\xi})]^{-1}\frac{\partial x_0^+}{\partial\varepsilon}(0,\Phi(\widetilde{\xi})) - [\Phi'(\widetilde{\eta})]^{-1}\frac{\partial x_0^-}{\partial\varepsilon}(0,\Phi(\widetilde{\eta}))\right\} + r(\widetilde{\xi},\widetilde{\eta},\varepsilon)\,,$$

where $\|r(\widetilde{\xi},\widetilde{\eta},\varepsilon)\| = o(\varepsilon)$ uniformly in $(\widetilde{\xi},\widetilde{\eta}) \in \overline{(U_\Omega \oplus O^s) \times (U_\Omega \oplus O^u)}$.

Let $L : (W_\mu \oplus W_s) \times (W_\mu \oplus W_u) \to \mathbb{R}^N$ be the linear map defined as $L(\widetilde{\xi},\widetilde{\eta}) = \widetilde{\xi} - \widetilde{\eta}$. We have $L(\widetilde{\xi},\widetilde{\eta}) = 0$ if and only if $\widetilde{\xi} = \widetilde{\eta} \in W_\mu$ and $\mathcal{R}L = W_\mu\oplus W_s\oplus W_u$, so we can write $\mathbb{R}^N = \mathcal{R}L\oplus V_\mu$. Next, let $W_\mu^\perp$ be a fixed subspace of $(W_\mu \oplus W_s) \times (W_\mu \oplus W_u)$ transversal to $\mathcal{N}L = \{(\widetilde{\xi},\widetilde{\xi}) : \widetilde{\xi} \in W_\mu\}$. Then, there exists an open convex set $O_1 \subset W_\mu^\perp$ such that $0 \in O_1$ and for any $(\hat{\xi},\hat{\eta}) \in \overline{O}_1$ and $\alpha \in \overline{\Omega}$ the point $(\widetilde{\xi},\widetilde{\eta}) = (\alpha + \hat{\xi}, \alpha + \hat{\eta})$ belongs to $(U_\Omega \oplus O^s) \times (U_\Omega \oplus O^u)$. We define a map $\hat{H} : \overline{O}_1 \times \overline{\Omega} \to \mathbb{R}^N$ as

$$\hat{H}(\hat{\xi},\hat{\eta},\alpha,\varepsilon) := \widetilde{H}(\alpha + \hat{\xi}, \alpha + \hat{\eta}, \varepsilon)\,.$$

Let $\mathcal{Q} : \mathbb{R}^N \to \mathbb{R}^N$ be the projection that corresponds to the splitting $\mathbb{R}^N = \mathcal{R}L \oplus V_\mu$ that is such that $\mathcal{N}Q = V_\mu$ and $\mathcal{R}Q = \mathcal{R}L$, and set $\hat{r} = \hat{r}(\hat{\xi},\hat{\eta},\alpha,\varepsilon) = \varepsilon^{-1}r(\alpha + \hat{\xi}, \alpha + \hat{\eta}, \varepsilon) = o(1)$. We write $\hat{H}(\hat{\xi},\hat{\eta},\alpha,\varepsilon)$ as

$$\hat{H}(\hat{\xi},\hat{\eta},\alpha,\varepsilon) = \hat{\xi} - \hat{\eta} + \varepsilon\hat{H}_1(\hat{\xi},\hat{\eta},\alpha,\varepsilon) + \varepsilon\hat{H}_2(\hat{\xi},\hat{\eta},\alpha,\varepsilon) \qquad (5.1.33)$$

where

$$\hat{H}_1(\hat{\xi},\hat{\eta},\alpha,\varepsilon) = \mathcal{Q}\left\{[\Phi'(\alpha+\hat{\xi})]^{-1}\frac{\partial x_0^+}{\partial\varepsilon}(0,\Phi(\alpha+\hat{\xi}))\right.$$
$$\left. -[\Phi'(\alpha+\hat{\eta})]^{-1}\frac{\partial x_0^-}{\partial\varepsilon}(0,\Phi(\alpha+\hat{\eta})) + \hat{r}\right\}$$

and

$$\hat{H}_2(\hat{\xi},\hat{\eta},\alpha,\varepsilon) = (\mathbb{I} - \mathcal{Q})\left\{[\Phi'(\alpha+\hat{\xi})]^{-1}\frac{\partial x_0^+}{\partial\varepsilon}(0,\Phi(\alpha+\hat{\xi}))\right.$$
$$\left. -[\Phi'(\alpha+\hat{\eta})]^{-1}\frac{\partial x_0^-}{\partial\varepsilon}(0,\Phi(\alpha+\hat{\eta})) + \hat{r}\right\}.$$

Note that $\hat{H}_1(\hat{\xi}, \hat{\eta}, \alpha, \varepsilon) \in \mathcal{RQ}$ and $\hat{H}_2(\hat{\xi}, \hat{\eta}, \alpha, \varepsilon) \in \mathcal{NQ}$. Thus $\hat{H}(\hat{\xi}, \hat{\eta}, \alpha, \varepsilon) = 0$ if and only if $\hat{\xi} - \hat{\eta} + \varepsilon \hat{H}_1 = 0$ and $\hat{H}_2 = 0$. Next we introduce the Melnikov function $M : \bar{\Omega} \to \mathcal{NQ}$:

$$M(\alpha) = (\mathbb{I} - \mathcal{Q})\Phi'(\alpha)^{-1} \left[\frac{\partial x_0^+}{\partial \varepsilon}(0, x_0(\alpha)) - \frac{\partial x_0^-}{\partial \varepsilon}(0, x_0(\alpha)) \right], \qquad (5.1.34)$$

whose components with respect to a fixed orthonormal basis $\{e_1, \ldots, e_\mu\}$ of V_μ are:

$$M_j(\alpha) = e_j^* \Phi'(\alpha)^{-1} \left[\frac{\partial x_0^+}{\partial \varepsilon}(0, x_0(\alpha)) - \frac{\partial x_0^-}{\partial \varepsilon}(0, x_0(\alpha)) \right]$$

$$= \left[(\Phi'(\alpha)^{-1})^* e_j \right]^* \left[\frac{\partial x_0^+}{\partial \varepsilon}(0, x_0(\alpha)) - \frac{\partial x_0^-}{\partial \varepsilon}(0, x_0(\alpha)) \right]$$

$$= \psi_j(\alpha)^* \left[\frac{\partial x_0^+}{\partial \varepsilon}(0, x_0(\alpha)) - \frac{\partial x_0^-}{\partial \varepsilon}(0, x_0(\alpha)) \right]$$

where $\psi_j(\alpha)$ are defined by the equality. Note that for any $v \in T_{x_0(\alpha)} \widetilde{W}_p^s$ we have:

$$\psi_j(\alpha)^* v = e_j^* \Phi'(\alpha)^{-1} v = 0$$

because $V_\mu^\perp = W_\mu \oplus W_s \oplus W_u$ and $\Phi'(\alpha)(W_\mu \oplus W_s) = T_{x_0(\alpha)} \widetilde{W}_p^s$. Similarly

$$\psi_j(\alpha)^* w = 0$$

for any $w \in T_{x_0(\alpha)} \widetilde{W}_p^u$. Thus the vectors $\psi_j(\alpha)$ are exactly the initial conditions to assign to the adjoint of the variational system

$$v_{n+1} = f'(x_n(\alpha))v_n$$

to obtain solutions that are bounded on $\mathbb{Z}$. Thus $M(\alpha)$ is the usual Melnikov function associated to the system $x_{n+1} = f(x_n, \varepsilon)$ (see the end of this subsection, or [16, 76, 88]).

From the smoothness of the functions $x_0^+(\varepsilon, \xi)$, $x_0^-(\varepsilon, \eta)$, $r(\xi, \eta, \varepsilon)$, $x_0(\alpha)$ and possibly changing O_1, we see that

$$\frac{\partial x_0^+}{\partial \varepsilon}(0, \hat{\xi} + x_0(\alpha)) - \frac{\partial x_0^-}{\partial \varepsilon}(0, \hat{\eta} + x_0(\alpha)) + \hat{r}(\hat{\xi}, \hat{\eta}, \alpha, \varepsilon)$$

is bounded on $\bar{O}_1 \times \bar{\Omega} \times [-\varepsilon_0, \varepsilon_0]$. Then we plug $\hat{H}(\hat{\xi}, \hat{\eta}, \alpha, \varepsilon)$ in the homotopy $\hat{H}(\hat{\xi}, \hat{\eta}, \alpha, \varepsilon, t)$, $0 \le t \le 1$ given by:

$$\hat{H}(\hat{\xi}, \hat{\eta}, \varepsilon, \alpha, t) = \hat{\xi} - \hat{\eta} + \varepsilon t \hat{H}_1(\hat{\xi}, \hat{\eta}, \varepsilon, \alpha) + k_\varepsilon(t) \hat{H}_2(t\hat{\xi}, t\hat{\eta}, \varepsilon, \alpha) \qquad (5.1.35)$$

where $k_\varepsilon(t) = \varepsilon t + 1 - t$ for $\varepsilon \ge 0$ and $k_\varepsilon(t) = \varepsilon t - 1 + t$ for $\varepsilon < 0$. Note that $|k_\varepsilon(t)| \ge |\varepsilon|$ and then $k_\varepsilon(t) \ne 0$ for $\varepsilon \ne 0$.

Lemma 5.1.14. *Assume (H1) holds. Then, if the neighborhood O_1 is chosen sufficiently small there is an $\varepsilon_0 > 0$ such that $\hat{H}(\hat{\xi}, \hat{\eta}, \alpha, \varepsilon, t) \ne 0$ for any $0 \le t \le 1$, $0 < |\varepsilon| < \varepsilon_0$ and $(\hat{\xi}, \hat{\eta}, \alpha) \in \partial(O_1 \times \Omega)$.*

Proof. We have already seen that $\hat{H}(\hat{\xi}, \hat{\eta}, \alpha, \varepsilon, t) = 0$ if and only if $\hat{\xi} - \hat{\eta} + \varepsilon t \hat{H}_1(\hat{\xi}, \hat{\eta}, \varepsilon, \alpha) = 0$ and $k_\varepsilon(t) \hat{H}_2(t\hat{\xi}, t\hat{\eta}, \varepsilon, \alpha) = 0$. Now, if $(\hat{\xi}, \hat{\eta}, \alpha) \in \partial(O_1 \times \Omega)$ then either $(\hat{\xi}, \hat{\eta}) \in \partial O_1$ or $\alpha \in \partial\Omega$.

If $(\hat{\xi}, \hat{\eta}) \in \partial O_1$, we have $\hat{\xi} \neq \hat{\eta}$ and then $\hat{\xi} - \hat{\eta} + \varepsilon t \hat{H}_1(\hat{\xi}, \hat{\eta}, \varepsilon, \alpha) \neq 0$ for ε_0 sufficiently small, because of the boundedness of

$$\frac{\partial x_0^-}{\partial \varepsilon}(0, \hat{\xi} + x_0(\alpha)) - \frac{\partial x_0^+}{\partial \varepsilon}(0, \hat{\eta} + x_0(\alpha)) + \hat{r}(\hat{\xi}, \hat{\eta}, \alpha, \varepsilon)$$

on $\overline{O}_1 \times \overline{\Omega} \times [-\varepsilon_0, \varepsilon_0]$. If $\alpha \in \partial\Omega$ then $M(\alpha) \neq 0$. Since $|k_\varepsilon(t)| \geq |\varepsilon|$, we get

$$k_\varepsilon(t) \hat{H}_2(t\hat{\xi}, t\hat{\eta}, \varepsilon, \alpha) \neq 0$$

provided O_1 and $|\varepsilon| \neq 0$ are sufficiently small. So again $H(\hat{\xi}, \hat{\eta}, \alpha, \varepsilon, t) \neq 0$ and the proof is finished. $\qquad\square$

Lemma 5.1.14 gives the next result.

Theorem 5.1.15. *Let O_1 be as in Lemma 5.1.14. Assume (H1), (H2). Then it follows that $\deg(\hat{H}(\hat{\xi}, \hat{\eta}, \alpha, \varepsilon), O_1 \times \Omega, 0) \neq 0$, for any $\varepsilon \neq 0$ sufficiently small.*

Proof. Lemma 5.1.14 implies

$$\deg(\hat{H}(\hat{\xi}, \hat{\eta}, \alpha, \varepsilon, 1), O_1 \times \Omega, 0) = \deg(\hat{H}(\hat{\xi}, \hat{\eta}, \alpha, \varepsilon, 0), O_1 \times \Omega, 0).$$

Now:

$$\hat{H}(\hat{\xi}, \hat{\eta}, \varepsilon, \alpha, 0) = \begin{pmatrix} L(\hat{\xi}, \hat{\eta}) \\ \operatorname{sgn} \varepsilon M(\alpha) \end{pmatrix}$$

and $L : W_\mu^\perp \to \mathcal{R}L$ is invertible. Thus

$$\deg(\hat{H}(\hat{\xi}, \hat{\eta}, \alpha, \varepsilon, 0), O_1 \times \Omega, 0) = \pm \deg(M, \Omega, 0) \neq 0.$$

The proof is finished. $\qquad\square$

Possibly shrinking U_Ω, O^s and O^u and using similar arguments like in the proof of Lemma 5.1.14 along with assumption (H1), we get $\widetilde{H}(\widetilde{\xi}, \widetilde{\eta}, \varepsilon) \neq 0$ for any $\varepsilon \neq 0$ sufficiently small and $(\widetilde{\xi}, \widetilde{\eta}) \in \overline{(U_\Omega \oplus O^s) \times (U_\Omega \oplus O^u)} \setminus \{(\alpha+\hat{\xi}, \alpha+\hat{\eta}) \mid \alpha \in \Omega, (\hat{\xi}, \hat{\eta}) \in O_1\}$. Then we have, because of Section 2.3.4, and the connectedness of U_Ω:

$$\left| \# \left(\widetilde{W}_{p(\varepsilon)}^s \cap U, \widetilde{W}_{p(\varepsilon)}^u \cap U \right) \right| = \left| \deg(H(\xi, \eta, \varepsilon), (\widetilde{W}_p^s \cap U) \times (\widetilde{W}_p^u \cap U), 0) \right|$$

$$= \left| \deg(\widetilde{H}(\widetilde{\xi}, \widetilde{\eta}, \varepsilon), (U_\Omega \oplus O^s) \times (U_\Omega \oplus O^u), 0) \right|$$

$$= \left| \deg(\hat{H}(\hat{\xi}, \hat{\eta}, \alpha, \varepsilon), O_1 \times \Omega, 0) \right| \neq 0$$

$$(5.1.36)$$

By (5.1.36) and Theorem 5.1.15 the proof of Theorem 5.1.12 is completed. $\qquad\square$

When f and $x_n(\alpha), n \in \mathbb{Z}$ are all C^3-smooth and $M(\alpha)$ has a simple root α_0, i.e. $M(\alpha_0) = 0$ and $DM(\alpha_0)$ is invertible, then assumptions (H1) and (H2) of Theorem 5.1.12 are satisfied when Ω is a small neighborhood of α_0. So we have a chaos of $f(x, \varepsilon)$ for $\varepsilon \neq 0$ small. But now we obtain much more: a transversal homoclinic orbit of $f(x, \varepsilon)$ for $\varepsilon \neq 0$ small with the corresponding Smale's horseshoe. Indeed, for finding a homoclinic orbit of $f(x, \varepsilon)$ to $p(\varepsilon)$, we have to solve $H(\xi, \eta, \varepsilon) = 0$ (see (5.1.31)) for $\varepsilon \neq 0$ small, which is equivalent to $\hat{H}(\hat{\xi}, \hat{\eta}, \alpha, \varepsilon) = 0$ (see (5.1.33)) and it is decomposed to

$$\hat{\xi} - \hat{\eta} + \varepsilon \hat{H}_1(\hat{\xi}, \hat{\eta}, \alpha, \varepsilon) = 0 \qquad (5.1.37)$$

and

$$\hat{H}_2(\hat{\xi}, \hat{\eta}, \alpha, \varepsilon) = 0. \qquad (5.1.38)$$

(5.1.37) can be solved by means of the implicit function theorem to get its C^1-smooth solutions $\hat{\xi}(\alpha, \varepsilon)$, $\hat{\eta}(\alpha, \varepsilon)$. Note that, because of uniqueness, we have: $\hat{\xi}(\alpha, 0) = \hat{\eta}(\alpha, 0) = 0$. Plugging these solutions into (5.1.38), we obtain the *bifurcation function* $B : \Omega \times (-\varepsilon_0, \varepsilon_0) \to \mathbb{R}^\mu, (\alpha, \varepsilon) \mapsto B(\alpha, \varepsilon)$, whose components $B_j(\alpha, \varepsilon)$ are:

$$B_j(\alpha, \varepsilon) = \psi_j(\alpha)^* \left[\frac{\partial x_0^-}{\partial \varepsilon}(0, \Phi(\alpha + \hat{\xi}(\alpha, \varepsilon))) - \frac{\partial x_0^+}{\partial \varepsilon}(0, \Phi(\alpha + \hat{\eta}(\alpha, \varepsilon))) \right.$$
$$\left. + r(\hat{\xi}(\alpha, \varepsilon), \hat{\eta}(\alpha, \varepsilon), \alpha, \varepsilon) \right].$$

Now, it is not difficult to see that, for $\varepsilon \to 0$, $B(\alpha, \varepsilon) \to M(\alpha)$ and $\frac{\partial}{\partial \alpha} B(\alpha, \varepsilon) \to DM(\alpha)$ uniformly on compact sets. So $B(\alpha, \varepsilon) = 0$ is uniquely solvable at $\alpha \sim \alpha_0$ for $\varepsilon \neq 0$ small. This gives a homoclinic orbit of $f(x, \varepsilon)$ for $\varepsilon \neq 0$. Its transversality can be proved like in [88, 107]. We conclude this part noting that the condition that $M(\alpha)$ has a simple zero at some α_0 is equivalent to the fact that the function

$$\widetilde{M}(\alpha) := \left(\psi_j(\alpha_0)^* \left[\frac{\partial x_0^+}{\partial \varepsilon}(0, x_0(\alpha)) - \frac{\partial x_0^-}{\partial \varepsilon}(0, x_0(\alpha)) \right] \right)_{j=1\ldots\mu}$$

has α_0 as a simple zero. In fact both $M(\alpha_0) = 0$ and $\widetilde{M}(\alpha_0) = 0$ mean that $\frac{\partial x_0^-}{\partial \varepsilon}(0, \Phi(\alpha_0)) = \frac{\partial x_0^+}{\partial \varepsilon}(0, \Phi(\alpha_0))$ and then the equality $M_\alpha(\alpha_0) = \widetilde{M}_\alpha(\alpha_0)$ easily follows from $\Phi(\alpha) = x_0(\alpha)$.

Theorem 5.1.12 is naturally applicable for the next problem. Consider the second order equation

$$\ddot{x} = g(x) + \varepsilon q(t),$$

where $x \in \mathbb{R}$, g, q are C^2-smooth and q is 2π-periodic. Suppose that the equation

$$\dot{x} = y, \quad \dot{y} = g(x)$$

has a homoclinic solution $(p(t), \dot{p}(t))$ to a hyperbolic fixed point. Then the Melnikov function has the form $M(\alpha) = \int_{-\infty}^{\infty} q(t + \alpha)\dot{p}(t)\, dt$, see computations for

(3.1.34). $M(\alpha)$ is 2π-periodic and $\int_0^{2\pi} M(\alpha)\, d\alpha = 0$. Hence if $M \neq 0$ then it changes the sign on $[0, 2\pi]$ and Theorem 5.1.12 can be applied. Concerning condition $M \neq 0$, we have the following result [19, Theorem 3.3].

Theorem 5.1.16. *Assume that $p(t) = \Phi(e^t)$, where $\Phi(u)$ is a rational function on $\mathbb{C}$ such that $\Phi(u) \to 0$ and $u\Phi'(u) \to 0$ as $u \to \infty$. Moreover, it has only the simple poles $w \neq 0$ and $\bar{w}$ (including the case that $\Phi(u)$ has only one simple pole $w = \bar{w}$). Then for any 2π-periodic nonconstant C^2-smooth function q, the associated Melnikov function $M(\alpha)$ is not identically zero.*

The next result [19, Theorem 4.1] allowing us to construct second order equations determined by prescribed homoclinic solutions.

Theorem 5.1.17. *Let $\Phi_0(u) = u^k G(u)$, $k \geq 1$, be a rational function such that $G(0) \neq 0$ and the following hold:*

(i) $\Phi_0(u) = \Phi_0(1/u)$ *(that is $G(1/u) = u^{2k} G(u)$)*

(ii) $\Phi_0(x) > 0$ *when x is real and $x > 0$*

(iii) $\Phi_0'(x) = 0$ *on $x > 0$ is equivalent to $x = 1$*

(iv) $\Phi_0''(1) \neq 0$

Then $\lim_{u \to \infty} u\Phi_0'(u) = 0$ and there exists a C^1-function $f(p)$ in a neighborhood of $[0, \Phi_0(1)]$ such that $p(t) = \Phi_0(e^t)$ is the solution of the equation $\ddot{p} = f(p)$. Moreover, if $G(u) = G_0(u^k)$ for some rational function $G_0(u)$, $G_0(0) \neq 0$, the function $f(p)$ is C^2 in a neighborhood of $[0, \Phi_0(1)]$.

Moreover we proved in [19] that the equation

$$\ddot{x} = 4x(2x^2 - 3x \coth(n\pi) + 1) + \varepsilon q(t) \tag{5.1.39}$$

has for even (odd) $n \in \mathbb{N}$ the Melnikov function vanishing identically on any 2π-periodic C^2-smooth functions $q(t)$ (or it is identically zero for infinitely many independent 2π-periodic C^2-smooth functions but not for all). The geometrical meaning of vanishing of the Melnikov function is that, in spite of the fact that the perturbation of (5.1.39) is of the order $O(\varepsilon)$, the distance between the stable and unstable manifolds of the perturbed equation, along a transverse direction, is of the order (at least) $O(\varepsilon^2)$. This means that in order to study the intersection of the stable and unstable manifolds, we have to look at the second order Melnikov function. This was also done in [19]. We refer the reader to more details about this subject to [19].

5.2   Topological Transversality and Reversibility

5.2.1   Period Blow-Up

In this section, we continue with the study of the relationship between topologically transversal intersections of certain sets for diffeomorphisms and the existence of oscillatory orbits. For illustration of the problem, let us consider a second order o.d.eqn of the form

$$\ddot{y} = g(y), \quad y \in \mathbb{R}^N. \tag{5.2.1}$$

For $N = 1$, a typical phase portrait of (5.2.1) consists from several families of periodic orbits, which are symmetric with respect to the y-axis, and those periodic orbits either terminate into equilibria or to heteroclinic/homoclinic cycles (see Fig. 1.2). In the last case, the minimal periods of periodic orbits tends to infinity as they accumulate on the heteroclinic/homoclinic cycle, i.e. we have a *period blow-up*. The symmetry of orbits of (5.2.1) follows from a simple observation that if $y(t)$ solves (5.2.1) then $y(-t)$ is also its solution. We say that (5.2.1) is *time reversible*. The time reversibility of (5.2.1) implies also the following antisymmetry. Rewriting (5.2.1) as a system

$$\dot{x} = h(x), \quad x := \begin{pmatrix} y \\ z \end{pmatrix}, \quad h(x) := \begin{pmatrix} z \\ g(y) \end{pmatrix} \tag{5.2.2}$$

and considering the involution $Rx := \begin{pmatrix} y \\ -z \end{pmatrix}$, we immediately see that

$$h(Rx) = -Rh(x) \quad \forall x \in \mathbb{R}^{2N}. \tag{5.2.3}$$

Then (5.2.3) implies that if $x(t)$ is a solution of (5.2.2) then so is $\tilde{x}(t) := Rx(-t)$. Or stated differently: Let $\varphi(x, t)$ be the time flow of (5.2.2) and set $f(x) := \varphi(x, 1)$, then (5.2.3) implies $Rf(x) = f^{-1}(Rx) \; \forall x \in \mathbb{R}^{2N}$. For $N > 1$ the phase portrait of (5.2.1) is very complicated but period blow-up phenomenon still may occur. This was first proved in [65–67] where it is called as a *blue sky catastrophe*. These results are generalized in [188, 189]. In this section, we proceed with the study of the period blow-up for diffeomorphisms.

5.2.2 Period Blow-Up for Reversible Diffeomorphisms

Let $R : \mathbb{R}^{2N} \to \mathbb{R}^{2N}$ be a *linear involution*, i.e $R^2 = \mathbb{I}$, such that $\dim \operatorname{Fix} R = N$, where $\operatorname{Fix} R = \{x \in \mathbb{R}^{2N} \mid Rx = x\}$. Any subset of $\mathbb{R}^{2N}$ invariant under the action of R is called *R-symmetric*. Consider a C^1-smooth diffeomorphism $f : \mathbb{R}^{2N} \to \mathbb{R}^{2N}$ which is *R-reversible*:

$$Rf(x) = f^{-1}(Rx), \quad \forall x \in \mathbb{R}^{2N},$$

and possessing a R-symmetric hyperbolic fixed point $p \in \operatorname{Fix} R$. Let W_p^s, W_p^u be the global stable and unstable manifolds of p, respectively. Let $\widetilde{W}_p^s$ be an open subset of W_p^s which is a submanifold of $\mathbb{R}^{2N}$ such that $\widetilde{W}_p^s \setminus \{p\} \cap \operatorname{Fix} R \neq \emptyset$. Since $RW_p^s = W_p^u$, we put $\widetilde{W}_p^u = R\widetilde{W}_p^s$. We also suppose the existence of a compact component $K \subset \widetilde{W}_p^s \setminus \{p\} \cap \operatorname{Fix} R$ and an open connected bounded subset $U \subset \bar{U} \subset \mathbb{R}^{2N} \setminus \{p\}$ satisfying $U \cap \widetilde{W}_p^s \cap \operatorname{Fix} R = K$ (see Fig. 5.2). By

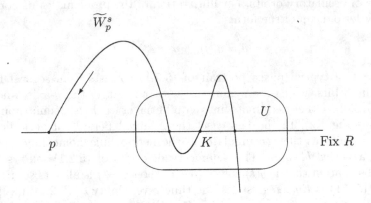

Figure 5.2: A transversal R-symmetric homoclinic set $K = \widetilde{W}_p^s \cap \mathrm{Fix}\,R \cap U$

shrinking U, we can assume that $\overline{\widetilde{W}_p^s \cap U} = \widetilde{W}_p^s \cap \bar{U}$. We note that $\widetilde{W}_p^s \cap U$ is an oriented submanifold of $\mathbb{R}^{2N}$. Then we can define the local intersection number $\#\left(\widetilde{W}_p^s \cap U, \mathrm{Fix}\,R \cap U\right)$ of the stable manifold W_p^s and the plain $\mathrm{Fix}\,R$ in $U \subset \mathbb{R}^{2N}$. Note $\dim \widetilde{W}_p^s = \dim \widetilde{W}_p^u = \dim \mathrm{Fix}\,R = N$. The main purpose of this section is to prove the following result [85].

Theorem 5.2.1. *If $\#\left(\widetilde{W}_p^s \cap U, \mathrm{Fix}\,R \cap U\right) \neq 0$ then there is an $\omega_0 \in \mathbb{N}$ such that for any $\mathbb{N} \ni \omega \geq \omega_0$, f possesses a 2ω-periodic orbit $\{x_n^\omega\}_{n \in \mathbb{Z}}$ such that $Rx_n^\omega = x_{-n}^\omega$, $n \in \mathbb{Z}$. Moreover, $x_0^\omega \in \mathrm{Fix}\,R$ is near to K, while $x_\omega^\omega \in \mathrm{Fix}\,R$ is near to p.*

Proof. Let $(\cdot, \cdot)$ be an inner product on $\mathbb{R}^{2N}$. Setting $\langle x, y \rangle := \frac{1}{2}((x,y) + (Rx, Ry))$ we have $\langle Rx, Ry \rangle = \langle x, y \rangle$, $x, y \in \mathbb{R}^{2N}$, and so $\|R\| = \|R^{-1}\| = 1$. Since $RK = K$, we can assume that $RU = U$.

For any $\xi \in \widetilde{W}_p^s \cap \bar{U}$ we set $\xi_n := f^n(\xi)$, $n \in \mathbb{Z}_+$ and then $\eta := R\xi$ with $\eta_n := f^n(\eta)$, $n \in \mathbb{Z}_-$. Clearly $\eta_{-n} = R\xi_n$, $n \in \mathbb{Z}_+$. Let $J_\omega^\pm$ and $I_\omega^\pm$ be the sets defined in Subsection 5.1.2. We study the nonlinear system

$$x_{n+1} = f(x_n) \tag{5.2.4}$$

near $\{\xi_n\}_{n \in J_\omega^+}$ and $\{\eta_n\}_{n \in J_\omega^-}$ following Subsection 5.1.2. According to Lemma 5.1.7, the linearization of (5.2.4) along $\{\xi_n\}_{n \in \mathbb{Z}_+}$

$$v_{n+1} = Df(\xi_n)v_n, \quad n \in \mathbb{Z}_+ \tag{5.2.5}$$

has an exponential dichotomy on $\mathbb{Z}_+$, i.e. there are positive constants L, $\delta \in (0,1)$ and the orthogonal projection $P_\xi : \mathbb{R}^{2N} \to T_\xi \widetilde{W}_p^s$ such that the fundamental solution $V_\xi(n)$ of (5.2.5) satisfies (5.1.6).

From the reversibility of f we immediately see that the linearization of (5.2.4) along $\{\eta_n\}_{n \in \mathbb{Z}_-}$

$$w_{n+1} = Df(\eta_n)w_n, \quad n \in \mathbb{Z}_-, \quad n \neq 0 \tag{5.2.6}$$

has the fundamental solution $W_\xi(n) = RV_\xi(-n)R^{-1}$, $n \in \mathbb{Z}_-$, and since $\|R\| = \|R^{-1}\| = 1$, the (5.2.6) has an exponential dichotomy on $\mathbb{Z}_-$ with the constants L, δ and the orthogonal projection $\mathbb{I} - Q_\eta$, where $Q_\eta = RP_\xi R^{-1}$, i.e. (5.1.7) holds. We note that the family $\{P_\xi \mid \xi \in \widetilde{W}_p^s \cap U\}$ is continuous on $\widetilde{W}_p^s \cap U$. Applying Theorem 5.1.10 to (5.2.4) with the above notation, we have that there exist $\omega_0 \in \mathbb{N}$ and a constant $c > 0$ such that, for any $\omega \in \mathbb{N}$, $\omega \geq \omega_0$, and $\xi \in \widetilde{W}_p^s \cap U$, there exist unique $\{x_n^+(\omega, \xi)\}_{n \in J_\omega^+}$ and $\{x_n^-(\omega, \xi)\}_{n \in J_\omega^-}$ such that

$$\begin{cases} x_{n+1}^\pm(\omega, \xi) = f(x_n^\pm(\omega, \xi)) \text{ separately on } I_\omega^\pm, \\[4pt] P_\xi x_0^+(\omega, \xi) = P_\xi \xi, \quad Q_{R\xi} x_0^-(\omega, \xi) = Q_{R\xi} R\xi, \quad x_\omega^+(\omega, \xi) = x_{-\omega}^-(\omega, \xi), \\[4pt] \max_{n \in J_\omega^+} |x_n^+(\omega, \xi) - \xi_n| \leq c\delta^\omega, \quad \max_{n \in J_\omega^-} |x_n^-(\omega, \xi) - \eta_n| \leq c\delta^\omega. \end{cases} \tag{5.2.7}$$

Moreover, $x_n^\pm(\omega, \xi)$ are continuous with respect to ξ. Since $Q_{R\xi} = RP_\xi R^{-1}$, $\eta_{-n} = R\xi_n$, $n \in \mathbb{Z}_+$, we see that the sequences given by

$$y_n^-(\omega, \xi) = Rx_{-n}^+(\omega, \xi), \quad n \in J_\omega^-; \qquad y_n^+(\omega, \xi) = Rx_{-n}^-(\omega, \xi), \quad n \in J_\omega^+$$

also satisfy (5.2.7). The uniqueness of such orbits implies that $Rx_n^\pm(\omega, \xi) = x_{-n}^\mp(\omega, \xi)$, $n \in J_\omega^\pm$.

In order to get a R-symmetric orbit of f, we have to solve the equation

$$(\mathbb{I} - R)x_0^+(\omega, \xi) = 0, \quad \xi \in \widetilde{W}_p^s \cap U. \tag{5.2.8}$$

Let V be an open subset such that $K \subset V \subset \bar{V} \subset U$. Note that the solution $x_0^+(\omega, \xi)$ is defined for $\xi \in \widetilde{W}_p^s \cap \bar{V}$ and

$$\#(\widetilde{W}_p^s \cap V, \operatorname{Fix} R \cap V) = \#(\widetilde{W}_p^s \cap U, \operatorname{Fix} R \cap U) \neq 0.$$

To solve (5.2.8), we put $F_\omega(\xi) := (\mathbb{I} - R)x_0^+(\omega, \xi)$ with $F_\omega : \widetilde{W}_p^s \cap \bar{V} \to R_- := \mathcal{R}(\mathbb{I} - R)$ and take the homotopy $H_\omega : \widetilde{W}_p^s \cap \bar{V} \times [0, 1] \to R_-$ given by

$$H_\omega(\xi, \lambda) = \lambda F_\omega(\xi) + (1 - \lambda)(\mathbb{I} - R)\xi.$$

Note $H_\omega(\xi, 1) = F_\omega(\xi)$ and $H_\omega(\xi, 0) = (\mathbb{I} - R)\xi$. Next (5.2.7) gives

$$|H_\omega(\xi, \lambda) - (\mathbb{I} - R)\xi| = |\lambda(F_\omega(\xi) - (\mathbb{I} - R)\xi)| \leq c\delta^\omega.$$

Consequently, $H_\omega(\cdot, \lambda) \neq 0$ on the boundary $\partial(\widetilde{W}_p^s \cap V)$ for any $0 \leq \lambda \leq 1$ and ω large. Note $P := \frac{1}{2}(\mathbb{I} + R)$ is a projection onto $\operatorname{Fix} R$ and $\mathbb{I} - P = \frac{1}{2}(\mathbb{I} - R)$.

By Section 2.3.4, this gives for the Brouwer degree

$$\left| \deg\left(F_\omega, \widetilde{W}_p^s \cap V, 0 \right) \right| = \left| \deg\left((\mathbb{I} - R), \widetilde{W}_p^s \cap V, 0 \right) \right|$$

$$= \left| \deg\left(\frac{1}{2} \left(\mathbb{I} - R\right), \widetilde{W}_p^s \cap V, 0 \right) \right| = \left| \deg\left((\mathbb{I} - P), \widetilde{W}_p^s \cap V, 0 \right) \right| \qquad (5.2.9)$$

$$= \left| \# \left(\widetilde{W}_p^s \cap V, \operatorname{Fix} R \cap V \right) \right| \neq 0.$$

Summarizing, we see that $F_\omega(\xi) = 0$ has a solution $\xi \in \widetilde{W}_p^s \cap V$ for any $\omega \geq \omega_0$, where ω_0 is sufficiently large. This proves Theorem 5.2.1. $\qquad \square$

Roughly speaking, Theorem 5.2.1 asserts that a combination of a homoclinic structure of a diffeomorphism and its reversibility may give infinitely many periodic orbits of the diffeomorphism.

Remark 5.2.2. When q is a transversal intersection of W_p^s and $\operatorname{Fix} R$, then Theorem 5.2.1 was proved in [65–67, 188, 189]. Then clearly $\#\left(\widetilde{W}_p^s \cap U, \operatorname{Fix} R \cap U \right) \neq 0$ for a small open neighborhood U of q.

Remark 5.2.3. Let $N = 1$. If p is a hyperbolic fixed point of f and W_p^s (or W_p^u) meets $\operatorname{Fix} R$ then a local intersection number of W_p^s (or W_p^u) with $\operatorname{Fix} R$ is nonzero. Indeed, let $q \in W_p^s \cap \operatorname{Fix} R$ be the first intersection starting on W_p^s from p. Since $Rf(q) = f^{-1}(Rq) = f^{-1}(q)$, the points $f^{-1}(q), f(q) \in W_p^s$ lie on the opposite half-plains separated by $\operatorname{Fix} R$. Hence an open bounded connected part $\widetilde{W}_p^s$ of W_p^s such that $f^{-1}(q), f(q) \in \widetilde{W}_p^s$ topologically nontrivially crosses $\operatorname{Fix} R$. Similarly for W_p^u.

Remark 5.2.4. Any accumulation point of the set $\{x_0^\omega\}_{\omega \geq \omega_0} \subset \operatorname{Fix} R$ from Theorem 5.2.1 is a starting point of a R-symmetric homoclinic orbit of f to p.

Next, if p is a non-R-symmetric hyperbolic fixed point of f, then Rp is also a non-R-symmetric hyperbolic fixed point of f. If $q \in W_p^s \cap \operatorname{Fix} R$ then $q \in W_p^s \cap W_{Rp}^u$, so q lies on a R-symmetric heteroclinic orbit connecting p and Rp. Consequently, like for Theorem 5.2.1, we can prove the following result.

Theorem 5.2.5. *Suppose f has a non-R-symmetric hyperbolic fixed point p. If W_p^s and W_p^u meet $\operatorname{Fix} R$ locally topologically transversally, then f has an infinite number of R-symmetric periodic orbits with periods tending to infinity.*

Heteroclinic period blow-up for symmetric o.d.eqns is studied in [17].

5.2.3 Perturbed Period Blow-Up

Verification of $\#\left(\widetilde{W}_p^s \cap U, \operatorname{Fix} R \cap U \right) \neq 0$ is not an easy task in Theorem 5.2.1. For this reason in this part, we consider a C^2-smooth perturbation $f(x, \varepsilon)$ of f:

$$f(x, 0) = f(x) \text{ and } Rf(x, \varepsilon) = f^{-1}(Rx, \varepsilon) \ \forall x \in \mathbb{R}^{2N} \text{ and any } \varepsilon \text{ small}.$$

Our aim is to find reasonable conditions to $f(x, \varepsilon)$ so that Theorem 5.2.1 is applicable for any $\varepsilon \neq 0$ small. To this end, concerning the unperturbed diffeomorphism f we suppose

(H1) There is an embedded compact C^2-smooth submanifold $\mathcal{M} \subset \widetilde{W}_p^s \setminus \{p\} \cap$
Fix R of an open subset $\widetilde{W}_p^s$ of W_p^s which is a submanifold of $\mathbb{R}^{2N}$ and
such that $\dim \operatorname{Fix} R \cap T_\xi \widetilde{W}_p^s = \dim \mathcal{M}$ for any $\xi \in \mathcal{M}$. Furthermore, there
is an oriented open bounded neighborhood $\mathcal{O}$ of $\mathcal{M}$, $\mathcal{M} \subset \mathcal{O} \subset \widetilde{W}_p^s \setminus \{p\}$.
We suppose that $\mathcal{M}$ is orientable embedded into $\mathcal{O}$.

Note always $T_\xi \mathcal{M} \subset \operatorname{Fix} R \cap T_\xi \widetilde{W}_p^s \ \forall \xi \in \mathcal{M}$ and $\dim \mathcal{M} = \dim T_\xi \mathcal{M}$, hence
(H1) implies $T_\xi \mathcal{M} = \operatorname{Fix} R \cap T_\xi \widetilde{W}_p^s \ \forall \xi \in \mathcal{M}$. Thus $\mathcal{M}$ is a *non-degenerate
homoclinic manifold* of $f(x)$.

By the implicit function theorem, $f(x, \varepsilon)$ has a unique hyperbolic fixed point
p_ε near p for ε small, i.e. $f(p_\varepsilon, \varepsilon) = p_\varepsilon$ which implies $Rp_\varepsilon = Rf(p_\varepsilon, \varepsilon) = f^{-1}(Rp_\varepsilon, \varepsilon)$ and the uniqueness gives $Rp_\varepsilon = p_\varepsilon$.

Next, Theorem 5.1.13 gives a C^1-mapping $x_0^+(\varepsilon, \xi)$, $\xi \in \widetilde{W}_p^s \cap U$ which determines an open subset of the stable manifold $W_{p_\varepsilon}^s$ of $f(x, \varepsilon)$ to the hyperbolic
symmetric fixed point p_ε of $f(x, \varepsilon)$ near p. Setting

$$F(\xi, \varepsilon) := (\mathbb{I} - R)x_0^+(\varepsilon, \xi)$$

we see that $F(\xi, \varepsilon) = 0$ is precisely the equation of R-symmetric homoclinic
solutions to p_ε. By Section 2.3.4 (see also (5.2.9)), in order to compute the local
intersection number of $\operatorname{Fix} R$ and $W_{p_\varepsilon}^s$, we have to compute a Brouwer degree of
$F(\cdot, \varepsilon)$ for $\varepsilon \neq 0$ small. To this end, we take a tubular neighborhood $\mathcal{V}$ of $\mathcal{M}$ in
$\widetilde{W}_p^s$, i.e. any $\xi \in \mathcal{V}$ can be uniquely expressed as a pair $\xi = (\tau, v)$, where $\tau \in \mathcal{M}$
and $v \in T_\tau \widetilde{W}_p^s / T_\tau \mathcal{M} = T_\tau \widetilde{W}_p^s / (\operatorname{Fix} R \cap T_\tau \widetilde{W}_p^s) = N_\tau$ - the fiber of the normal
vector bundle of $\mathcal{M}$ in $\widetilde{W}_p^s$, and $|v| < \Delta$ for some $\Delta > 0$. Hence we identify $\mathcal{V}$
with an open neighborhood of the zero section of the normal vector bundle of $\mathcal{M}$
in $\widetilde{W}_p^s$. We note that the assumption (H1) implies the invertibility of the linear
mapping $D_v F(\tau, 0, 0) : N_\tau \to \mathcal{R} D_v F(\tau, 0, 0)$. Indeed, from $x_0^+(0, \xi) = \xi$ we have
$D_v F(\tau, 0, 0)v = (\mathbb{I} - R)v$, $v \in T_\tau \widetilde{W}_p^s$ and hence $\mathcal{N} D_v F(\tau, 0, 0) = \operatorname{Fix} R \cap T_\tau \widetilde{W}_p^s$.
Next, since $\mathcal{M}$ is orientable embedded into $\mathcal{O}$ and $\mathcal{O}$ is oriented, the tangent
vector bundle $T\mathcal{M}$ and the normal vector bundle $\cup_{\tau \in \mathcal{M}} N_\tau$ are both oriented.
Hence the vector bundle $\cup_{\tau \in \mathcal{M}} (\mathbb{I} - R)N_\tau = \cup_{\tau \in \mathcal{M}} \mathcal{R} D_v F(\tau, 0, 0)$ is also oriented,
because $D_v F(\tau, 0, 0) : N_\tau \to \mathcal{R} D_v F(\tau, 0, 0)$ is invertible. Taking the orthogonal
projection

$$S_\tau : \operatorname{Fix}(-R) \to \mathcal{R} D_v F(\tau, 0, 0) \subset \operatorname{Fix}(-R)$$

and the oriented vector bundle $\cup_{\tau \in \mathcal{M}} \operatorname{Fix}(-R) = \mathcal{M} \times \operatorname{Fix}(-R)$, the vector
bundle $\cup_{\tau \in \mathcal{M}} \mathcal{R}(\mathbb{I} - S_\tau) \operatorname{Fix}(-R)$ is oriented as well. Consequently any section of
this vector bundle has a Brouwer degree. Now we can prove the main result of
this subsection.

Theorem 5.2.6. *Assume* (H1) *and the following ones*

(H2) *There is an open connected subset* $\Omega \subset \mathcal{M}$ *such that* $B(\tau) \neq 0$, $\forall \tau \in \partial\Omega$, *where* $B(\tau) = (\mathbb{I} - S_\tau)D_\varepsilon F(\tau, 0, 0)$ *is a section of the oriented vector bundle* $\cup_{\tau \in \mathcal{M}} \mathcal{R}(\mathbb{I} - S_\tau)\mathrm{Fix}(-R)$.

(H3) $\deg(B(\tau), \Omega) \neq 0$.

Then there exists $\varepsilon_0 > 0$ *such that for* $0 < |\varepsilon| \leq \varepsilon_0$, *the local intersection number of* $\mathrm{Fix}\, R$ *and* $W_{p_\varepsilon}^s$ *is nonzero.*

Proof. We have to compute a Brouwer degree of $F(\cdot, \varepsilon)$ for $\varepsilon \neq 0$ small. From $F(\tau, 0, 0) = 0$, we get $F(\tau, v, \varepsilon) = D_v F(\tau, 0, 0)v + \varepsilon D_\varepsilon F(\tau, 0, 0) + o(|v|) + o(\varepsilon)$. We consider the homotopy

$$H(\tau, v, \varepsilon, \lambda) = S_\tau\big(\lambda F(\tau, v, \varepsilon) + (1 - \lambda)D_v F(\tau, 0, 0)v\big)$$
$$+ (\mathbb{I} - S_\tau)\big(\lambda F(\tau, v, \varepsilon) + (1 - \lambda)\varepsilon D_\varepsilon F(\tau, 0, 0)\big).$$

Note $H(\tau, v, \varepsilon, 1) = F(\tau, v, \varepsilon)$. According to (H2), there is an open connected bounded neighborhood $U_1 \subset \mathcal{M}$ of $\bar{\Omega}$ such that $B(\tau) \neq 0$, $\forall \tau \in U_1 \setminus \Omega$. Now we take an open subset $V_\varepsilon = \{(\tau, v) \in V \mid \tau \in U_1 \text{ and } |v| < |\varepsilon|r_1\}$ for a positive constant r_1 and $0 < |\varepsilon| < \triangle/r_1$. We show that

$$H(\tau, v, \varepsilon, \lambda) \neq 0 \quad \forall (\tau, v, \lambda) \in \partial V_\varepsilon \times [0, 1] \text{ and } \varepsilon \neq 0 \text{ small.} \tag{5.2.10}$$

Indeed, from

$$S_\tau\big(\lambda F(\tau, v, \varepsilon) + (1 - \lambda)D_v F(\tau, 0, 0)v\big) = S_\tau D_v F(\tau, 0, 0)v + o(|v|) + O(\varepsilon)$$

and $S_\tau D_v F(\tau, 0, 0) : N_\tau \to \mathcal{R}D_v F(\tau, 0, 0)$ is invertible, we get that

$$S_\tau\big(\lambda F(\tau, v, \varepsilon) + (1 - \lambda)D_v F(\tau, 0, 0)v\big) \neq 0, \quad \forall (\tau, v) \in V_\varepsilon, \quad |v| = r_1|\varepsilon|$$

for r_1 sufficiently large and fixed. Furthermore, from $(\mathbb{I} - S_\tau)D_v F(\tau, 0, 0) = 0$ we have

$$(\mathbb{I} - S_\tau)\big(\lambda F(\tau, v, \varepsilon) + (1 - \lambda)\varepsilon D_\varepsilon F(\tau, 0, 0)\big)$$
$$= \varepsilon(\mathbb{I} - S_\tau)D_\varepsilon F(\tau, 0, 0) + o(|v|) + o(\varepsilon) = \varepsilon B(\tau) + o(|v|) + o(\varepsilon) \neq 0$$

for $(\tau, v) \in V_\varepsilon$, $|v| \leq r_1|\varepsilon|$ and $\tau \in U_1 \setminus \Omega$. Summarizing, we see that (5.2.10) holds. Consequently $\deg(F, V_\varepsilon, 0) = \deg(H(\cdot, \varepsilon, 0), V_\varepsilon, 0)$. Note S_τ, $(\mathbb{I} - S_\tau)$ are complementary orthogonal projections and

$$H(\tau, v, \varepsilon, 0) = S_\tau D_v F(\tau, 0, 0)v + \varepsilon(\mathbb{I} - S_\tau)D_\varepsilon F(\tau, 0, 0).$$

Since the linear map $S_\tau D_v F(\tau, 0, 0) : N_\tau \to \mathcal{R}D_v F(\tau, 0, 0)$ is invertible and U_1 is connected, we get

$$\deg(H(\cdot, \varepsilon, 0), V_\varepsilon, 0) = \pm \deg(B(\tau), \Omega) \neq 0.$$

Consequently, we obtain $\deg(F, V_\varepsilon, 0) \neq 0$, and so $\#(\widetilde{W}_{p_\varepsilon}^s \cap V_\varepsilon, \mathrm{Fix}\, R \cap V_\varepsilon) \neq 0$. Theorem 5.2.6 is proved. $\square$

Remark 5.2.7. Theorems 5.2.1 and 5.2.6 imply an infinite number of R-symmetric periodic orbits of $f(x, \varepsilon)$ accumulating on R-symmetric homoclinic orbits of $f(x, \varepsilon)$ for any $\varepsilon \neq 0$ small.

Remark 5.2.8. If (H1) holds and the Euler characteristic

$$\chi\big(\cup_{\tau \in \mathcal{M}} \mathcal{R}(\mathbb{I} - S_\tau)\mathrm{Fix}(-R)\big)$$

is nonzero, then any R-reversible C^2-smooth perturbation $f(x, \varepsilon)$ has a blue sky catastrophe in the sense of Remark 5.2.7. Indeed, then (H3) is satisfied by Section 2.3.6.

Now we show that $B(\tau)$ is the Melnikov function for bifurcation of R-symmetric homoclinic orbits.

Theorem 5.2.9. *Assume (H1). If $f(x, \varepsilon)$ is C^3-smooth and there is a simple zero τ_0 of $B(\tau)$, i.e. $B(\tau_0) = 0$ and $DB(\tau_0)$ is nonsingular, then there is a unique R-symmetric homoclinic orbit of $f(x, \varepsilon)$ to p_ε for any $\varepsilon \neq 0$ small bifurcating from the R-symmetric homoclinic orbit of $f(x)$ to p which starts from $\tau_0 \in \mathcal{M}$.*

Proof. Since $f(x, \varepsilon)$ is C^3-smooth then F is C^2-smooth. To find a R-symmetric homoclinic orbit of $f(x, \varepsilon)$ to p_ε for any $\varepsilon \neq 0$ small, we need to solve $F(\tau, v, \varepsilon) = 0$. Next, we decompose it as $F(\tau, v, \varepsilon) = S_\tau F(\tau, v, \varepsilon) + (\mathbb{I} - S_\tau)F(\tau, v, \varepsilon)$. Since $\mathcal{M}$ is compact, $S_\tau F(\tau, 0, 0) = 0$ and $S_\tau D_v F(\tau, 0, 0) : N_\tau \to \mathcal{R} D_v F(\tau, 0, 0)$ is invertible, using the implicit function theorem, we can solve the equation $S_\tau F(\tau, v, \varepsilon) = 0$ in v near 0 for ε small and $\tau \in \mathcal{M}$ to get its C^2-smooth solution $v = v(\tau, \varepsilon) = O(\varepsilon)$. Then we consider the *bifurcation equation*

$$C(\tau, \varepsilon) := (\mathbb{I} - S_\tau)F(\tau, v(\tau, \varepsilon), \varepsilon) = 0. \qquad (5.2.11)$$

From $(\mathbb{I} - S_\tau)D_v F(\tau, 0, 0) = 0$ we see

$$C(\tau, \varepsilon) = (\mathbb{I} - S_\tau)D_v F(\tau, 0, 0)v(\tau, \varepsilon) + (\mathbb{I} - S_\tau)\varepsilon D_\varepsilon F(\tau, 0, 0)$$
$$+ (\mathbb{I} - S_\tau)o(|v(\tau, \varepsilon)|) + o(\varepsilon) = \varepsilon B(\tau) + o(\varepsilon).$$

Hence $C(\tau, \varepsilon)/\varepsilon = B(\tau) + o(1)$ in the C^1-topology on $\mathcal{M}$ as $\varepsilon \to 0$. Consequently, the existence of a simple zero τ_0 of $B(\tau)$ implies the solvability of $C(\tau, \varepsilon) = 0$ in τ near τ_0 for $\varepsilon \neq 0$ small. This determines the desired R-symmetric homoclinic orbit of $f(x, \varepsilon)$ to p_ε for $\varepsilon \neq 0$ small. $\qquad \square$

Finally we simplify the formula of $B(\tau)$ using the following lemma.

Lemma 5.2.10. *It holds $\mathcal{R}(\mathbb{I} - S_\tau) = \left(T_\tau \widetilde{W}_p^s + \mathrm{Fix}\, R\right)^\perp$.*

Proof. Note $S_\tau : \mathrm{Fix}(-R) \to (\mathbb{I} - R)T_\tau \widetilde{W}_p^s \subset \mathrm{Fix}(-R)$ is the orthogonal projection. So $\mathcal{R}(\mathbb{I} - S_\tau) = \mathrm{Fix}(-R) \cap \left((\mathbb{I} - R)T_\tau \widetilde{W}_p^s\right)^\perp$. Since for any $a \in \mathrm{Fix}(-R)$ and for any $w \in T_\tau \widetilde{W}_p^s$ we have

$$\langle a, (\mathbb{I} - R)w \rangle = \langle (\mathbb{I} - R)a, w \rangle = 2\langle a, w \rangle,$$

we see $\mathrm{Fix}\,(-R) \cap \left((\mathbb{I} - R)T_\tau\widetilde{W}_p^s\right)^\perp = \mathrm{Fix}\,(-R) \cap \left(T_\tau\widetilde{W}_p^s\right)^\perp$. Next note $\mathrm{Fix}(-R) = (\mathrm{Fix}\,R)^\perp$ with the corresponding orthogonal projections $\frac{1}{2}(\mathbb{I} - R)$: $\mathbb{R}^{2N} \to \mathrm{Fix}(-R)$ and $\frac{1}{2}(\mathbb{I} + R) : \mathbb{R}^{2N} \to \mathrm{Fix}\,R$. Consequently, we get

$$\mathcal{R}\,(\mathbb{I} - S_\tau) = (\mathrm{Fix}\,R)^\perp \cap \left(T_\tau\widetilde{W}_p^s\right)^\perp = \left(T_\tau\widetilde{W}_p^s + \mathrm{Fix}\,R\right)^\perp .$$

The lemma is proved.   □

Using $\dim \widetilde{W}_p^s = \dim \mathrm{Fix}\,R = N$ we derive

$$\dim \left(T_\tau\widetilde{W}_p^s + \mathrm{Fix}\,R\right)^\perp = 2N - \dim \left(T_\tau\widetilde{W}_p^s + \mathrm{Fix}\,R\right)$$
$$= 2N - \dim T_\tau\widetilde{W}_p^s - \dim \mathrm{Fix}\,R + \dim T_\tau\widetilde{W}_p^s \cap \mathrm{Fix}\,R \qquad (5.2.12)$$
$$= \dim T_\tau\widetilde{W}_p^s \cap \mathrm{Fix}\,R = \dim \mathcal{M}.$$

Consequently, if $a_i(\tau)$, $i = 1, 2, \ldots, \dim \mathcal{M}$ is a continuous vector basis over $\mathcal{M}$ such that $a_i(\tau) \perp (T_\tau\widetilde{W}_p^s + \mathrm{Fix}\,R)$ for any $\tau \in \mathcal{M}$. Then

$$B(\tau) = (a_1(\tau)^*, \ldots, a_{\dim \mathcal{M}}(\tau)^*)\, D(\tau)^{-1}\widetilde{B}(\tau)$$

where components of $\widetilde{B}(\tau)$ are given by

$$\widetilde{B}_i(\tau) = \langle a_i(\tau), (\mathbb{I} - R)D_\varepsilon x_0^+(0, \tau)\rangle = 2\langle a_i(\tau), D_\varepsilon x_0^+(0, \tau)\rangle \qquad (5.2.13)$$

and $D(\tau) = (\langle a_i(\tau), a_j(\tau)\rangle)_{i,j=1}^{\dim \mathcal{M}}$ is the Gram matrix. Clearly $|\deg(B(\tau), \Omega)| = \left|\deg(\widetilde{B}(\tau), \Omega)\right|$ when $0 \notin B(\partial\Omega)$ for a connected open subset Ω of $\mathcal{M}$. So in place of $B(\tau)$ we consider $\widetilde{B}(\tau)$ for computations. Note $P_\tau x_0^+(\varepsilon, \tau) = P_\tau\tau$ (see (5.1.26)) implies $P_\tau D_\varepsilon x_0^+(0, \tau) = 0$, i.e.

$$D_\varepsilon x_0^+(0, \tau) \in \left(T_\tau\widetilde{W}_p^s\right)^\perp . \qquad (5.2.14)$$

5.2.4 Perturbed Second Order O.D.Eqns

In this part, we consider a perturbation of (5.2.1) of the form

$$\ddot{z} = g(z) + \varepsilon h(z), \quad z \in \mathbb{R}^N, \qquad (5.2.15)$$

where $g, h \in C^3(\mathbb{R}^N, \mathbb{R}^N)$, $g(0) = h(0) = 0$. We use Theorem 5.2.6 to construct a system of two perturbed second order o.d.eqns with a topologically transversal, but non-C^1-transversal, intersection of the stable manifold and $\mathrm{Fix}\,R$. We rewrite (5.2.15) as

$$\dot{z}_1 = z_2, \quad \dot{z}_2 = g(z_1) + \varepsilon h(z_1). \qquad (5.2.16)$$

Let $\phi(t, z_1, z_2, \varepsilon)$ be the flow of (5.2.16). Then $f(x, \varepsilon) = \phi(1, x, \varepsilon)$ and $x = (z_1, z_2)$. Here $R(z_1, z_2) = (z_1, -z_2)$ and

$$\mathrm{Fix}\,R = \{(z_1, 0) \mid z_1 \in \mathbb{R}^N\}, \quad \mathrm{Fix}\,(-R) = \{(0, z_2) \mid z_2 \in \mathbb{R}^N\}.$$

The inner product $\langle \cdot, \cdot \rangle$ is given by $\langle (z_1^1, z_2^1), (z_1^2, z_2^2) \rangle = (z_1^1, z_1^2) + (z_2^1, z_2^2)$, where $(\cdot, \cdot)$ is the usual inner product on $\mathbb{R}^N$. We assume that $p = (0,0)$ is a hyperbolic equilibrium of (5.2.16).

In the sequel, we intend to derive formula (5.2.13) for (5.2.16). To this end, we first make some general computations. For any $\tau \in \text{Fix } R \cap \widetilde{W}_p^s$, $\phi(t, \tau, 0)$ is a homoclinic solution to p of (5.2.16) with $\varepsilon = 0$. Moreover, if

$$\phi(t, \tau, 0) = (z_1^\tau(t), z_2^\tau(t))$$

then $\dot{z}_1^\tau(0) = z_2^\tau(0) = 0$ and $\ddot{z}_1^\tau = g(z_1^\tau)$, so $z_1^\tau(t)$ is even and $z_2^\tau(t) = \dot{z}_1^\tau(t)$ is odd. The linearization of (5.2.16) for $\varepsilon = 0$ along $\phi(t, \tau, 0)$ has the form

$$\dot{v} = w, \quad \dot{w} = Dg(z_1^\tau(t))v. \tag{5.2.17}$$

We know from Section 2.5.4 that

$$T_\tau \widetilde{W}_p^{s(u)} = \Big\{ (v(0), w(0)) \mid v(t), w(t)$$

$$\text{are bounded solutions of (5.2.17) on } \mathbb{R}_{+(-)} \Big\},$$

respectively. Furthermore, since $RT_\tau \widetilde{W}_p^s = T_{R\tau} \widetilde{W}_p^u = T_\tau \widetilde{W}_p^u$ and for any $w \in T_\tau \widetilde{W}_p^s$, $a \in \text{Fix}(-R)$, it holds $\langle a, Rw \rangle = -\langle Ra, Rw \rangle = -\langle a, w \rangle$, we get

$$\Big(T_\tau \widetilde{W}_p^s + \text{Fix } R \Big)^\perp = \Big(T_\tau \widetilde{W}_p^s + T_\tau \widetilde{W}_p^u + \text{Fix } R \Big)^\perp.$$

So the condition $a \perp (T_\tau \widetilde{W}_p^s + \text{Fix } R)$ is equivalent to

$$a \in \text{Fix}(-R) \cap \Big(T_\tau \widetilde{W}_p^s + \text{Fix } R \Big)^\perp,$$

i.e. now $a = (0, a_2)$, and $a \in \Big(T_\tau \widetilde{W}_p^s + T_\tau \widetilde{W}_p^u \Big)^\perp$. Hence according to arguments from Section 2.5.4, $v_1(0) = 0$ and $w_1(0) = a_2$ for bounded solutions $v_1(t)$ and $w_1(t)$ on $\mathbb{R}$ of the adjoint system of (5.2.17) given by

$$\dot{v}_1 = -Dg(z_1^\tau(t))^* w_1, \quad \dot{w}_1 = -v_1,$$

i.e. w_1 is the even bounded solution on $\mathbb{R}$ of $\ddot{w}_1 = Dg(z_1^\tau(t))^* w_1$ with $w_1(0) = a_2$. We note if $g(z) = \text{grad } G(z)$ for some $G \in C^4(\mathbb{R}^N, \mathbb{R})$ then $Dg(z) = Dg(z)^*$. We also see that

$$\dim \Big(T_\tau \widetilde{W}_p^s + \text{Fix } R \Big)^\perp = \dim \Big\{ w_1(0) \in \mathbb{R}^N \mid w_1 \text{ is an even}$$

$$\text{bounded solution on } \mathbb{R} \text{ of } \ddot{w}_1 = Dg(z_1^\tau(t))^* w_1 \Big\}. \tag{5.2.18}$$

Furthermore, since $x_0^+(\varepsilon, \tau) \in W_p^s$, $\psi(t, \varepsilon) := \phi(t, x_0^+(\varepsilon, \tau), \varepsilon)$ is a homoclinic solution to p of (5.2.16) with the initial value condition $\psi(0, \varepsilon) = x_0^+(\varepsilon, \tau)$. Then $(v_\tau(t), w_\tau(t)) := D_\varepsilon \psi(t, 0)$ is a bounded solutions on $\mathbb{R}_+$ of the system

$$\dot{v}_\tau = w_\tau, \quad \dot{w}_\tau = Dg(z_1^\tau(t))v_\tau + h(z_1^\tau(t)) \tag{5.2.19}$$

with $(v_\tau(0), w_\tau(0)) = D_\varepsilon x_0^+(0, \tau)$. From (5.2.14) we know $(v_\tau(0), w_\tau(0)) \perp T_\tau \widetilde{W}_p^s$. Consequently, the corresponding component (5.2.13) of $\tilde{B}(\tau)$ to a is given by

$$2\langle a, D_\varepsilon x_0^+(0, \tau) \rangle = 2\langle (0, a_2), (v_\tau(0), w_\tau(0)) \rangle = 2(w_1(0), w_\tau(0)).$$

On the other hand, using (5.2.19) along with $\lim\limits_{t \to +\infty} v_1(t) = 0$ and $\lim\limits_{t \to +\infty} w_1(t) = 0$, we derive

$$(w_1(0), w_\tau(0)) = -\int_0^\infty \frac{d}{dt} [(w_1(t), w_\tau(t))]\, dt$$

$$= -\int_0^\infty [(\dot{w}_1(t), w_\tau(t)) + (w_1(t), \dot{w}_\tau(t))]\, dt$$

$$= -\int_0^\infty [-(v_1(t), \dot{v}_\tau(t)) + (w_1(t), Dg(z_1^\tau(t))v_\tau(t) + h(z_1^\tau(t)))]\, dt$$

$$= -\int_0^\infty [-(v_1(t), \dot{v}_\tau(t)) + (Dg(z_1^\tau(t))^* w_1(t), v_\tau(t)) + (w_1(t), h(z_1^\tau(t)))]\, dt$$

$$= -\int_0^\infty [-(v_1(t), \dot{v}_\tau(t)) - (\dot{v}_1(t), v_\tau(t)) + (w_1(t), h(z_1^\tau(t)))]\, dt$$

$$= \int_0^\infty \frac{d}{dt} [(v_1(t), v_\tau(t))]\, dt - \int_0^\infty (w_1(t), h(z_1^\tau(t)))\, dt$$

$$= -\int_0^\infty (h(z_1^\tau(t)), w_1(t))\, dt.$$

$$(5.2.20)$$

Summarizing we have the following result.

Theorem 5.2.11. *Suppose $g(z) = \operatorname{grad} G(z)$ for some $G \in C^4(\mathbb{R}^N, \mathbb{R})$. Let*

$$\mathcal{M} = \Big\{ (z^\tau(0), 0) \mid z^\tau(t) \in C^3\left(O \times \mathbb{R}, \mathbb{R}^N\right),\ \tau \in O,\ z^\tau(t)\ \text{is an even}$$

$$\text{bounded solution on } \mathbb{R} \text{ of } (5.2.1) \Big\}$$

for some open subset $O \subset \mathbb{R}^m$. We suppose $\tau \to z^\tau(0)$ is injective. Then $\mathcal{M}$ is nondegenerate if $D_{\tau_i} z^\tau(t)$, $i = 1, \ldots, m$ form a basis of all even bounded solutions on $\mathbb{R}$ of $\ddot{w} = Dg(z_\tau(t))w$ for any $\tau = (\tau_1, \ldots, \tau_m) \in O$. The Melnikov mapping $M(\tau) = (M_1(\tau), \ldots, M_m(\tau))$ is given by

$$M_i(\tau) = \int_0^\infty (h(z^\tau(t)), D_{\tau_i} z^\tau(t))\, dt. \qquad (5.2.21)$$

Proof. According to (5.2.12), (5.2.18) and assumptions of this theorem, we get $\dim \mathcal{M} = m = \dim \left(T_\tau \widetilde{W}_p^s + \operatorname{Fix} R\right)^\perp = \dim T_\tau \widetilde{W}_p^s \cap \operatorname{Fix} R$. So $\mathcal{M}$ is nondegenerate and $a_i(\tau) = (0, D_{\tau_i} z^\tau(0))$, $i = 1, 2, \ldots, m$. Next, (5.2.21) follows from (5.2.13) and (5.2.20) when the factor -2 is dropped. The proof is finished. $\square$

To be more concrete, let $k \geq 2$, $k \in \mathbb{N}$. We consider the system

$$\ddot{x} = x - 2x(x^2 + y^2), \quad \ddot{y} = y - 2y(x^2 + y^2) + \varepsilon x^{2k}, \quad x, y \in \mathbb{R}, \qquad (5.2.22)$$

i.e. in the form of (5.2.16)

$$\dot{x}_1 = x_2, \quad \dot{y}_1 = y_2, \quad \dot{x}_2 = x_1 - 2x_1(x_1^2 + y_1^2),$$
$$\dot{y}_2 = y_1 - 2y_1(x_1^2 + y_1^2) + \varepsilon x_1^{2k}. \qquad (5.2.23)$$

Note $G(x, y) = \frac{x^2 + y^2}{2} - \frac{(x^2 + y^2)^2}{2}$ and $p_\varepsilon = p = 0$ in the above notations. Next, system (5.2.23) has for $\varepsilon = 0$ a homoclinic manifold to 0 (see also (4.1.19))

$$\gamma(\theta, t) := (x_\theta(t), y_\theta(t), \dot{x}_\theta(t), \dot{y}_\theta(t))$$

with $x_\theta(t) = \sin \theta r(t)$, $y_\theta(t) = \cos \theta r(t)$ and $r(t) = \operatorname{sech} t$, which intersects Fix R in a circle

$$\mathcal{M} = \{\tau = (\sin \theta, \cos \theta, 0, 0) \mid \theta \in \mathbb{R}\} .$$

Next, for $\tau \in \mathcal{M}$ we have

$$T_\tau \widetilde{W}_p^s = \operatorname{span} \{D_\theta \gamma(\theta, 0), \dot{\gamma}(\theta, 0)\}$$
$$= \operatorname{span} \{(\cos \theta, -\sin \theta, 0, 0), (0, 0, -\sin \theta, -\cos \theta)\} .$$

Then

$$T_\tau \widetilde{W}_p^s \cap \operatorname{Fix} R = \operatorname{span} \{(\cos \theta, -\sin \theta, 0, 0)\} = T_\tau \mathcal{M} ,$$

so (H1) holds. On the other hand, in notation of Theorem 5.2.11 we have $z^\theta(t) = (x_\theta(t), y_\theta(t))$ and $w_1(t) = D_\theta z^\theta(t) = (y_\theta(t), -x_\theta(t))$. By (5.2.21), function $M(\theta)$ has now the form

$$M(\theta) = \int_0^\infty ((0, x_\theta(t)^{2k}), (y_\theta(t), -x_\theta(t))) \, dt = -\int_0^\infty x_\theta(t)^{2k+1} \, dt$$
$$= -\int_0^\infty \sin^{2k+1} \theta r^{2k+1}(t) \, dt = -\frac{3.5 \ldots (2k-1)}{2^{k+1} k!} \pi \sin^{2k+1} \theta .$$

The bifurcation equation (5.2.11) is now analytical, so it is $\widetilde{C}(\theta, \varepsilon) := C(\tau, \varepsilon)/\varepsilon$ with $\tau = (\sin \theta, \cos \theta, 0, 0)$. Hence $\theta = 0$ is an isolated solution of $\widetilde{C}(\theta, \varepsilon) = 0$ for any $\varepsilon \neq 0$ small. The Brouwer degree of $\widetilde{C}(\theta, 0) = B(\theta)$ at $\theta = 0$ is nonzero, so Theorem 5.2.6 implies the following result.

Theorem 5.2.12. *The point* $(0, 1, 0, 0)$ *is an isolated topologically transversal intersection of* $W_{p_\varepsilon}^s$ *and* Fix R *for* (5.2.23) *with* $\varepsilon \neq 0$ *small. But this point is not a* C^1-*transversal intersection.*

Proof. To prove the non-C^1-transversal intersection, we consider a C^3- perturbation of (5.2.22) given by

$$\ddot{x} = x - 2x(x^2 + y^2), \quad \ddot{y} = y - 2y(x^2 + y^2) + \varepsilon \phi_\delta(x), \qquad (5.2.24)$$

where $x, y \in \mathbb{R}$, $\delta > 0$ is fixed and

$$\phi_\delta(x) := \begin{cases} 0 & \text{if } |x| \le \delta \\ (x - \delta \operatorname{sgn} x)^{2k} & \text{if } |x| \ge \delta. \end{cases}$$

We see that (5.2.24) has even homoclinics $x_\theta(t)$, $y_\theta(t)$ for $|\sin\theta| < \delta$ and any ε. Hence $(0, 1, 0, 0)$ is not an isolated reversible homoclinic point for the C^3-perturbation (5.2.24) of the system (5.2.22). The proof is finished.   □

5.3   Chains of Reversible Oscillators

Methods of Section 5.2 can be directly applied to chains of weakly coupled oscillators given by

$$\dot{x}_n = V(x_n) + \varepsilon H(x_{n-s}, x_{n-s+1}, \dots, x_{n+r}), \tag{5.3.1}$$

where $n \in \mathbb{Z}$, $V \in C^3(\mathbb{R}^{2N}, \mathbb{R}^{2N})$, $H \in C^3(\mathbb{R}^{2N(r+s+1)}, \mathbb{R}^{2N})$, $\varepsilon \ne 0$ is a small parameter and $s, r \in \mathbb{N}$ are fixed. Such systems as (5.3.1) are considered as ordinary differential systems on lattices or as chains of coupled ordinary differential equations. They naturally occur in spatially discretized nonlinear systems (see also Section 6.1) and they play a crucial role in modeling of many phenomena in different fields, ranging from condensed matter and biophysics to mechanical engineering [7, 14, 113, 142, 174]. Typical examples are discrete Klein-Gordon equations

$$\ddot{x}_n - x_n + 2x_n^3 - \varepsilon(x_{n+1} - 2x_n + x_{n-1}) = 0, \tag{5.3.2}$$

$$\ddot{x}_n + x_n - x_n^3 - \varepsilon(x_{n+1} - 2x_n + x_{n-1}) = 0, \tag{5.3.3}$$

or the discrete sine-Gordon equation

$$\ddot{x}_n + \sin x_n - \varepsilon(x_{n+1} - 2x_n + x_{n-1}) = 0. \tag{5.3.4}$$

The above equations (5.3.2–5.3.4) are spatial discretizations of a p.d.eqn

$$u_{tt} - u_{xx} + h(u) = 0, \tag{5.3.5}$$

with the corresponding function h, since the spatial discretization of (5.3.5) gives

$$\ddot{u}_n - \frac{1}{\theta}(u_{n+1} - 2u_n + u_{n-1}) + h(u_n) = 0. \tag{5.3.6}$$

For $\theta \to 0$; $(\theta \to \infty)$ we have an *integrable; (anti-integrable) case*, respectively. Now $\varepsilon = 1/\theta \ll 1$ so we deal with the anti-integrable case. The integrable one is studied in Section 6.1. There are several papers [112, 142, 177] showing *breathers* of (5.3.1) initializing from periodic solutions of *anti-integrable (or anti-continuum) limit equation*

$$\dot{x} = V(x). \tag{5.3.7}$$

We recall that *breathers* are spatially localized time-periodic solutions, that is, time-periodic solutions whose amplitudes decay exponentially in the space. The purpose of this section is to show that if (5.3.7) has a homoclinic/heteroclinic period blow-up (or blue sky catastrophe), then also (5.3.1) will have similar phenomena. We end this section with extending period blow-up phenomenon to *traveling waves* of (5.3.1).

5.3.1 Homoclinic Period Blow-Up for Breathers

First we study the homoclinic period blow-up supposing that:

(a) System (5.3.1) is *reversible*, i.e. there is a linear involution $S : \mathbb{R}^{2N} \to \mathbb{R}^{2N}$, $S^2 = \mathbb{I}$ such that

$$V(Sx) = -SV(x), \quad H(Sx_1, \ldots, Sx_{r+s+1}) = -SH(x_1, \ldots, x_{r+s+1}).$$

(b) $\dim \mathrm{Fix}\, S = N$.

(c) $H(0, \ldots, 0) = 0$, $V(0) = 0$ and the spectrum of $DV(0)$ lies off the imaginary axis.

(d) There is a transversal S-reversible homoclinic orbit γ of (5.3.7) to 0, i.e. there is an $0 \neq \gamma : \mathbb{R} \to \mathbb{R}^{2N}$ which is a solution of (5.3.7), $\lim\limits_{|t| \to \infty} \gamma(t) = 0$, $S\gamma(t) = \gamma(-t)$ and $T_{\gamma(0)} W_0^s \cap \mathrm{Fix}\, S = \{0\}$.

For (5.3.2), (5.3.7) has the form $\dot{x} = y$, $\dot{y} = x - 2x^3$ which is the Duffing equation with $\gamma(t) = (r(t), \dot{r}(t))$, $r(t) = \mathrm{sech}\, t$ and $S(x, y) = (x, -y)$. So (5.3.2) satisfies assumptions (a–d).

According to Subsection 5.2.2, we take an inner product $\langle \cdot, \cdot \rangle$ on $\mathbb{R}^{2N}$ such that $\langle Sx, Sy \rangle = \langle x, y \rangle$. Hence $\|S\| = \|S^{-1}\| = 1$. Let us fix an $\eta > 1$ and consider the Banach space

$$X_\eta = \left\{ x = \{x_n\}_{n \in \mathbb{Z}} \mid x_n \in \mathbb{R}^{2N}, \ |x|_\eta := \sup_n |x_n| \eta^n < \infty \right\}.$$

Then (5.3.1) has on X_η the form

$$\dot{x} = v(x) + \varepsilon h(x) \tag{5.3.8}$$

for $v(x) = \{V(x_n)\}_{n \in \mathbb{Z}}$ and $h(x) = \{H(x_{n-s}, \ldots, x_{n+r})\}_{n \in \mathbb{Z}}$. It is easy to see from assumption (c) that $v, h \in C^3(X_\eta, X_\eta)$. Hence (5.3.1) is a smooth dynamical system on X_η. By extending involution S onto X_η as $Sx = \{Sx_n\}_{n \in \mathbb{Z}}$, we see that (5.3.8) is S-reversible. We note $|Sx|_\eta = |x|_\eta$.

We denote by $\mathcal{E}_0$ the set of doubly infinite sequences of 0 and 1 with finite numbers of entries 1. For any $q = \{q_n\}_{n \in \mathbb{Z}} \in \mathcal{E}_0$, we put $\gamma_q(t) = \{\gamma_q(t)_n\}_{n \in \mathbb{Z}} \in X_\eta$ as follows

$$\gamma_q(t)_n = \begin{cases} \gamma(t) & \text{for} \quad q_n = 1, \\ 0 & \text{for} \quad q_n = 0. \end{cases}$$

We see that $\gamma_q(t)$ is a homoclinic solution of

$$\dot{x} = v(x) \tag{5.3.9}$$

to $x = 0$. So we take $\mathcal{M} := \{\gamma_q(0) \mid q \in \mathcal{E}_0\}$. $\mathcal{M}$ is not compact, but it consists
from isolated points, i.e. it is a 0-dimensional manifold and thus $T_\tau \mathcal{M} = \{0\}$
for any $\tau \in \mathcal{M}$. The linearization of (5.3.9) at $x = 0$ is $\{\dot{u}_n = DV(0)u_n\}_{n\in\mathbb{Z}}$.
Due to hypothesis (c), (5.3.9) has the global stable and unstable manifolds
$\mathcal{W}_0^s = \{W_0^s\}_{n\in\mathbb{Z}}$ and $\mathcal{W}_0^u = \{W_0^u\}_{n\in\mathbb{Z}}$, respectively. Note $ST_0W_0^s = T_0W_0^u$ and
$T_0W_0^s \cap T_0W_0^u = \{0\}$. Hence $T_0W_0^s \cap \operatorname{Fix} S = \{0\}$. Consequently, by assumption
(d) it holds

$$T_\tau \mathcal{W}_0^s \cap \operatorname{Fix} S = \{0\} = T_\tau \mathcal{M}.$$

Hence hypothesis (H1) of Subsection 5.2.3 is satisfied. Now there is no a bifurcation function $B(\tau)$, since $\mathcal{M}$ is now a parametric space. We also have a
C^1-transversal intersection of $T_\tau \mathcal{W}_0^s$ and $\operatorname{Fix} S$. So for any $\tau \in \mathcal{M}$ we can apply
Theorem 5.2.1 and Remark 5.2.2 to obtain the following theorem (see also [82]).

Theorem 5.3.1. *Let assumptions (a–d) hold for (5.3.1), (5.3.7). Then there
is an $\varepsilon_0 > 0$ such that for any $q \in \mathcal{E}_0$ and for any $|\varepsilon| \le \varepsilon_0$, there is a unique
homoclinic solution $x_n^{\varepsilon,q}(t)$, $n \in \mathbb{Z}$ of (5.3.1) such that*

(i) *$x_n^{\varepsilon,q}(t) \to 0$ exponentially fast as $|t| \to \infty$ and uniformly for $|\varepsilon| \le \varepsilon_0$ and
$n \in \mathbb{Z}$.*

(ii) *$x_n^{\varepsilon,q}(t)$ are S-reversible, i.e. $S x_n^{\varepsilon,q}(t) = x_n^{\varepsilon,q}(-t)$ for all $n \in \mathbb{Z}$.*

(iii) *$\{x_n^{\varepsilon,q}(t)\}_{n\in\mathbb{Z}} \in X_\eta$ is near to $\gamma_q(t)$ in X_η and $\{x_n^{0,q}(t)\}_{n\in\mathbb{Z}} = \gamma_q(t)$.*

(iv) *The homoclinic loop $\{x_n^{\varepsilon,q}(t)\}_{n\in\mathbb{Z}}$ is accumulated by ω-periodic S-reversible
solutions $\{x_n^{\omega,\varepsilon,q}(t)\}_{n\in\mathbb{Z}}$ of (5.3.1) for any $\omega > 1/\varepsilon_0$.*

Theorem 5.3.1 ensures the existence of continuum many S-reversible homoclinic solutions to 0 of (5.3.1)) and each of them is accumulated by continuum
many periodic solutions with periods tending to infinity. Consequently, under
assumptions of Theorem 5.3.1, dynamics of (5.3.1) is very rich with infinitely
many narrow layers of breathers with arbitrarily large periods. Due to the hyperbolicity of the equilibrium $x = 0$ of (5.3.8) for ε small, clearly these homoclinic
solutions are not stable. We already know that Theorem 5.3.1 is applicable to
(5.3.2).

5.3.2   Heteroclinic Period Blow-Up for Non-breathers

Assumptions (a) and (b) hold for (5.3.3) and (5.3.4), but (c) and (d) must be
replaced by the following one:

(e) There is a *heteroclinic loop* of (5.3.7), i.e. there is a hyperbolic equilibrium
x_0 of (5.3.7) such that $Sx_0 \ne x_0$ and a S-reversible heteroclinic solution
$\gamma_1 : \mathbb{R} \to \mathbb{R}^{2N}$ of (5.3.7) from equilibrium x_0 to equilibrium Sx_0 such
that the unstable manifold $W_{x_0}^u$ of (5.3.7) at x_0 C^1-transversally crosses
$\operatorname{Fix} S$ at $\gamma_1(0)$ along with the existence of another S-reversible heteroclinic
solution $\gamma_2 : \mathbb{R} \to \mathbb{R}^{2N}$ of (5.3.7) from equilibrium Sx_0 to equilibrium x_0
such that the stable manifold $W_{x_0}^s$ of (5.3.7) at x_0 C^1-transversally crosses
$\operatorname{Fix} S$ at $\gamma_2(0)$.

Again $\dim W_{x_0}^s = \dim W_{x_0}^u = N$. Now we consider (5.3.8) on X_1, which is the usual $\ell_\infty(2N)$ and instead of $\mathcal{E}_0$, we take $\mathcal{E}$-the set of doubly infinite sequences of 0 and 1. For any $q = \{q_n\}_{n \in \mathbb{Z}} \in \mathcal{E}$ we put $x_{0,q} = \{x_{0,q,n}\}_{n \in \mathbb{Z}} \in \ell_\infty(2N)$ by

$$x_{0,q,n} = \begin{cases} x_0 & \text{for } q_n = 1, \\ Sx_0 & \text{for } q_n = 0 \end{cases}$$

and $\gamma_{j,q}(t) = \{\gamma_{j,q}(t)_n\}_{n \in \mathbb{Z}} \in \ell_\infty(2N)$, $j = 1, 2$ as follows

$$\gamma_{1,q}(t)_n = \begin{cases} \gamma_1(t) & \text{for } q_n = 1, \\ \gamma_2(t) & \text{for } q_n = 0 \end{cases}$$

and

$$\gamma_{2,q}(t)_n = \begin{cases} \gamma_2(t) & \text{for } q_n = 1, \\ \gamma_1(t) & \text{for } q_n = 0. \end{cases}$$

Like above, we can easily verify that equilibria $x_{0,q}$ ($Sx_{0,q}$) and $Sx_{0,q}$ ($x_{0,q}$) of (5.3.9) are connected by a S-reversible heteroclinic solution $\gamma_{1,q}$ ($\gamma_{2,q}$) with a C^1-transversal crossing of the unstable (stable) manifold $W_{x_{0,q}}^u$ ($W_{x_{0,q}}^s$) of (5.3.9) at $x_{0,q}$ with $\operatorname{Fix} S$, respectively. Consequently, Theorem 5.2.5 can be applied to get the next result.

Theorem 5.3.2. *Let assumptions (a), (b), (e) hold for (5.3.1) and (5.3.7). Then there is an $\varepsilon_0 > 0$ such that for any $q \in \mathcal{E}$ and for any $|\varepsilon| \leq \varepsilon_0$, there are unique heteroclinic solutions $x_n^{1,\varepsilon,q}(t)$ and $x_n^{2,\varepsilon,q}(t)$, $n \in \mathbb{Z}$ of (5.3.1) such that*

(A) *$\{x_n^{1,\varepsilon,q}(t)\}_{n \in \mathbb{Z}}$ connects $x_{\varepsilon,q}$ and $Sx_{\varepsilon,q}$ with exponential decay uniformly for $|\varepsilon| \leq \varepsilon_0$. $\{x_n^{2,\varepsilon,q}(t)\}_{n \in \mathbb{Z}}$ connects $Sx_{\varepsilon,q}$ and $x_{\varepsilon,q}$ with exponential decay uniformly for $|\varepsilon| \leq \varepsilon_0$. Here $x_{\varepsilon,q}$ is a unique equilibrium of (5.3.1) near $x_{0,q}$.*

(B) *$x_n^{i,\varepsilon,q}(t)$, $i = 1, 2$ are S-reversible, i.e. $Sx_n^{i,\varepsilon,q}(t) = x_n^{i,\varepsilon,q}(-t)$ for all $n \in \mathbb{Z}$ and $i = 1, 2$.*

(C) *$\{x_n^{i,\varepsilon,q}(t)\}_{n \in \mathbb{Z}} \in \ell_\infty(2N)$ is near to $\gamma_{i,q}(t)$ in $\ell_\infty(2N)$ and $\{x_n^{i,0,q}(t)\}_{n \in \mathbb{Z}} = \gamma_{i,q}(t)$, $i = 1, 2$.*

(D) *The heteroclinic loop in $\ell_\infty(2N)$ created by $\{x_n^{i,\varepsilon,q}(t)\}_{n \in \mathbb{Z}}$, $i = 1, 2$ is accumulated by ω-periodic S-reversible solutions $\{x_n^{\omega,\varepsilon,q}(t)\}_{n \in \mathbb{Z}}$ of (5.3.1) for any $\omega > 1/\varepsilon_0$.*

Again Theorem 5.3.2 ensures the existence of continuum many S-reversible heteroclinic loops of (5.3.1) and each of them is accumulated by continuum many periodic solutions with periods tending to infinity. Due to $x_0 \neq 0$, of course these periodic solutions are not breathers, since they are not spatially localized.

To apply Theorem 5.3.2, we take the involution $S(x, y) = (-x, y)$ with $x_0 = (-1, 0)$ and $\gamma_1(t) = (r_1(t), \dot{r}_1(t))$, $\gamma_2(t) = -\gamma_1(t)$, $r_1(t) = \tanh \frac{\sqrt{2}}{2} t$ for (5.3.3), and with $x_0 = (-\pi, 0)$ and $\gamma_1(t) = (r_2(t), \dot{r}_2(t))$, $\gamma_2(t) = -\gamma_1(t)$, $r_2(t) = 2 \arctan(\sinh t)$ for (5.3.4), respectively.

More general equations on lattices can be studied than (5.3.1). For instance, let us consider the topological discrete sine-Gordon equation [113]

$$\dot{x}_n = y_n$$

$$\dot{y}_n = \varepsilon \cos x_n (\sin x_{n+1} + \sin x_{n-1}) - \left(\varepsilon + \frac{1}{2}\right) \sin x_n (\cos x_{n+1} + \cos x_{n-1}),$$

$$(5.3.10)$$

where ε is small and $n \in \mathbb{Z}$. For $\varepsilon = 0$, we get

$$\dot{x}_n = y_n, \quad \dot{y}_n = -\frac{1}{2} \sin x_n (\cos x_{n+1} + \cos x_{n-1}). \qquad (5.3.11)$$

We consider (5.3.10) on $\ell_\infty(2)$ with the involution

$$S(\{(x_n, y_n)\}_{n \in \mathbb{Z}}) := \{(-x_n, y_n)\}_{n \in \mathbb{Z}}.$$

Then (5.3.10) is S-reversible. For any increasing sequence $q = \{n_i\}_i$, $n_i \in \mathbb{Z}$, we put $x_{0,q} = \{x_{0,q,n}\}_{n \in \mathbb{Z}} \in \ell_\infty(2)$ by

$$x_{0,q,n} = \begin{cases} (-\pi, 0) & \text{for} \quad n = 2n_i + 1, \\ (\pi, 0) & \text{for odd} \quad n \neq 2n_i + 1, \\ (0, 0) & \text{for even} \quad n \end{cases}$$

and $\gamma_{j,q}(t) = \{\gamma_{j,q}(t)_n\}_{n \in \mathbb{Z}} \in \ell_\infty(2)$, $j = 1, 2$ as follows

$$\gamma_{j,q}(t)_n = \begin{cases} (-1)^{j+1}(\phi(t), \dot{\phi}(t)) & \text{for} \quad n = 2n_i + 1, \\ (-1)^j(\phi(t), \dot{\phi}(t)) & \text{for odd} \quad n \neq 2n_i + 1, \\ (0, 0) & \text{for even} \quad n \end{cases}$$

for $\phi(t) = 2 \arctan(\sinh t)$. We can easily verify like above that both $W^s_{x_{0,q}}$ and $W^u_{x_{0,q}}$ C^1-transversally cross the set Fix S at $\gamma_{2,q}(0)$ and $\gamma_{1,q}(0)$, respectively. Summarizing, a statement similar to Theorem 5.3.2 holds also for (5.3.10) with ε sufficiently small. Those heteroclinic and periodic solutions of (5.3.10) are again spatially not localized.

In this section, we consider for simplicity only transversal intersections of stable and unstable manifolds with Fix S *of (5.3.7), but topologically transversal intersections could be dealt similarly.*

Finally, similar approach is used in [83] to show breathers for diatomic lattices modeling two one-dimensional interacting sublattices of harmonically coupled protons and heavy ions [161, 162] representing the Bernal-Flower filaments in ice or more complex biological macromolecules in membranes in which only the degrees of freedom that contribute predominantly to proton mobility have been conserved. In these systems, each proton lies between a pair of "oxygens". The following two coupled infinite chains of oscillators is considered

$$\ddot{u}_n = \frac{k_1}{m}(u_{n+1} - 2u_n + u_{n-1}) + \frac{4\xi_0}{md_0^2}u_n\left(1 - \frac{u_n^2}{d_0^2}\right) - 2\frac{\chi}{m}\rho_n u_n,$$

$$\ddot{\rho}_n = \frac{K_1}{M}(\rho_{n+1} - 2\rho_n + \rho_{n-1}) - \Omega_0^2 \rho_n - \frac{\chi}{M}(u_n^2 - d_0^2),$$

where u_n denotes the displacement of the nth proton with respect to the center of the oxygen pair, k_1 is the coupling between neighboring protons and m is the mass of protons, ξ_0 is the potential barrier, $2d_0$ is the distance between two minima of the double-well potential of protons, ρ_n is the displacement between two oxygens, M is the mass of oxygens, Ω_0 is the frequency of the optical mode, K_1 is the harmonic coupling between neighboring oxygens and χ measures the strength of the coupling. It is supposed that couplings are small $k_1, K_1, \chi \sim 0$.

5.3.3   Period Blow-Up for Traveling Waves

We finish this section with the study of traveling waves of (5.3.1) of the form $x_n(t) = v(\nu t - n)$ for which we get

$$\nu v'(z) = V(v(z)) + \varepsilon H\left(v(z+s), v(z+s-1), \ldots, v(z-r)\right), \qquad (5.3.12)$$

where $z = \nu t - n$ and $\nu \neq 0$. We are looking for *S-reversible solutions* of (5.3.12), i.e. $v(-z) = Sv(z)$, by applying results and methods of Section 5.2. Under either assumptions (a–d) or (a), (b), (e) of this section, we only need to verify (see (5.2.3)) that the perturbation of (5.3.12) is also *S-antireversible*:

$$v(-z) = Sv(z) \Rightarrow S\widetilde{v}(z) = -\widetilde{v}(-z) \qquad (5.3.13)$$

for $\widetilde{v}(z) := H\left(v(z+s), v(z+s-1), \ldots, v(z-r)\right)$. Since

$$S\widetilde{v}(z) = -H\left(v(-z-s), v(-z-s+1), \ldots, v(-z+r)\right),$$

we see that (5.3.13) is satisfied provided we suppose

(f) $r = s$ and H is symmetric, i.e. it holds

$$H(x_1, x_2, \ldots, x_{2s}, x_{2s+1}) = H(x_{2s+1}, x_{2s}, \ldots, x_2, x_1)$$

for any $x_1, x_2, \ldots, x_{2s+1} \in \mathbb{R}^{2N}$. We have the next result.

Theorem 5.3.3. *(a) Under assumptions (a–d) and (f) of this section, for any $\nu \neq 0$ there is an $\varepsilon_0 > 0$ that for all $|\varepsilon| < \varepsilon_0$, (5.3.1) has a traveling wave solution of a form $x_n(t) = v(\nu t - n)$ which is near to $\gamma\left(t - \frac{n}{\nu}\right)$ such that $v(z)$ is S-reversible and asymptotic to 0 as $z \to \pm\infty$, respectively. This traveling wave is accumulated by periodic traveling waves with minimal periods tending to infinity.*

(b) Under assumptions (a), (b), (e) and (f) of this section, for any $\nu \neq 0$ there is an $\varepsilon_0 > 0$ that for all $|\varepsilon| < \varepsilon_0$, (5.3.1) has traveling wave solutions (kinks) of forms $x_n(t) = v_{1,2}(\nu t - n)$ which are near to $\gamma_{1,2}\left(t - \frac{n}{\nu}\right)$ such that $v_{1,2}(z)$ are S-reversible and asymptotic to x_ε and Sx_ε as $z \to \pm\infty$, respectively. These kinks are accumulated by periodic traveling waves with minimal periods tending to infinity. Here x_ε is a unique solution of $V(x_\varepsilon) + \varepsilon H\left(x_\varepsilon, x_\varepsilon, \ldots, x_\varepsilon\right) = 0$ near x_0.

The constant ε_0 depends on ν, but it is uniform for ν from a bounded set.

Chapter 6

Traveling Waves on Lattices

6.1 Traveling Waves in Discretized P.D.Eqns

Most nonlinear lattice systems are non-integrable even if a p.d.eqn model in the continuum limit is integrable (see [7, 99, 175] and references therein). Prototype models for such nonlinear lattices are various discrete nonlinear Schrödinger and Klein-Gordon equations or systems. There is a particularly important class of solutions so called *discrete breathers* which are homoclinic in space and oscillatory in time (see Section 5.3) [99]. Other questions involve the existence and propagation of topological defects or *kinks* which mathematically are heteroclinic connections between a ground and an excited steady state [176, 182] (see Theorem 5.3.3). They have applications to problems such as dislocation and mass transport in solids, charge-density waves, commensurable-incommensurable phase transitions, Josephson transmission lines etc. Prototype models here are discrete sine-Gordon equations, also known as Frenkel-Kontorova models [2].

In this section, we consider a chain of coupled particles subjected to an external potential (see Fig. 6.1). A Hamiltonian $\mathcal{H}$ of such system can be written as:

$$\mathcal{H} = \sum_{n \in \mathbb{Z}} \left(\frac{1}{2} \dot{u}_n^2 + \frac{1}{2\varepsilon^2} (u_{n+1} - u_n)^2 - \mathcal{F}(u_n) \right), \qquad (6.1.1)$$

where u_n is the displacement of the n–th particle from its equilibrium position. This gives the discrete nonlinear Klein-Gordon equation:

$$\ddot{u}_n - \frac{1}{\varepsilon^2} (u_{n+1} - 2u_n + u_{n-1}) - h(u_n) = 0, \qquad (6.1.2)$$

where $h(u_n) = \mathcal{F}'(u_n), n \in \mathbb{Z}$. Equation (6.1.2) is also a spatial discretization of a p.d.eqn

$$u_{tt} - u_{xx} - h(u) = 0. \qquad (6.1.3)$$

M. Fečkan, *Topological Degree Approach to Bifurcation Problems*, 183–198.
© Springer Science + Business Media B.V., 2008

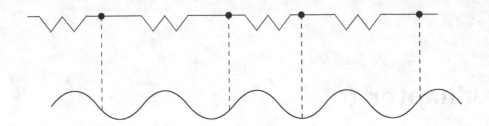

Figure 6.1: The model of discrete sine-Gordon equation

We get (6.1.2) from (6.1.3) putting

$$u_n(t) = u(\varepsilon n, t),$$

$$u_{xx}(\varepsilon n, t) \sim \frac{u(\varepsilon(n+1), t) - 2u(\varepsilon n, t) + u(\varepsilon(n-1), t)}{\varepsilon^2}. \qquad (6.1.4)$$

Since $\varepsilon \ll 1$ we study the *continuum/integrable* case. We suppose $h \in C^1$ along with

(A) $h(0) = 0$, $h'(0) = -a^2 < 0$ and there is a homoclinic solution ϕ of $\ddot{x} + h(x) = 0$ such that $\phi(t) = \phi(-t)$ and $\phi(t) \to 0$ as $t \to \pm\infty$.

Then (6.1.3) admits traveling wave solutions

$$u(x, t) = \phi\left(\frac{x - \nu t}{\sqrt{1 - \nu^2}}\right), \quad 0 < \nu < 1$$

We also consider for (6.1.2) traveling wave solutions

$$u_n(t) = V\left(n - \frac{\nu}{\varepsilon}t\right) \equiv V(z), \quad z = n - \frac{\nu}{\varepsilon}t, \quad 0 < \nu < 1.$$

Substituting this into (6.1.2) we obtain the following functional differential equation:

$$\nu^2 V''(z) - V(z+1) + 2V(z) - V(z-1) - \varepsilon^2 h(V(z)) = 0. \qquad (6.1.5)$$

The discrete sine-Gordon equation for $h(u) = -\sin u$ in (6.1.2) of the form

$$\ddot{u}_n = u_{n+1} - 2u_n + u_{n-1} - \varepsilon^2 \sin u_n \qquad (6.1.6)$$

has been numerically investigated in [71, 176]: As $\varepsilon \to 0$, we get the continuum sine-Gordon equation $u_{tt} - u_{xx} + \sin u = 0$ with the supporting moving kinks of the form

$$4 \arctan\left[\exp\left(\frac{x - \nu t}{\sqrt{1 - \nu^2}}\right)\right].$$

Thus it was natural in [71, 176] to seek numerically solutions of

$$\nu^2 U''(z) = U(z+1) - 2U(z) + U(z-1) - \varepsilon^2 \sin U(z), \qquad (6.1.7)$$

where $U(z) = U(n - \nu t) = u_n(t)$, with the boundary conditions $U(z) \to 0$ mod 2π as $z \to \pm\infty$. They did not find such solutions. Their closest result is that the numerical solution of (6.1.7) near

$$4 \arctan \left[\exp \left(\varepsilon \frac{x - \nu t}{\sqrt{1 - \nu^2}} \right) \right]$$

has tails of periodic waves of small amplitude. But according to the form of (6.2.5) below, that result is consistent with our analytical result, since the y-part of (6.2.5) is oscillatory with small amplitude. Recent numerical simulations of discrete lattices such as (6.1.6) are studied in [2] of the form

$$\ddot{u}_n = u_{n+1} - 2u_n + u_{n-1} - \vartheta \frac{(1 + 2\alpha) \sin u}{(1 + \alpha(1 - \cos u))^2}, \qquad (6.1.8)$$

where $\vartheta > 0$ measures the onsite potential strength and $\alpha \geq 0$ measures the degree of anharmonicity. For $\alpha = 0$ this is (6.1.6).

Finally, the *anticontinuum/anti-integrable* case of (6.1.2) is

$$\ddot{u}_n - \theta(u_{n+1} - 2u_n + u_{n-1}) - h(u_n) = 0, \qquad (6.1.9)$$

for $\theta \to 0$. Then Theorem 5.3.3 can be applied to get a traveling wave solution $u_n(t) = V(\nu t - n)$ of (6.1.9) near a $\tilde{\gamma} \left(t - \frac{n}{\nu} \right)$ where $\nu \neq 0$ and $\tilde{\gamma}(t)$ is a homoclinic/heteroclinic solution of $\ddot{u} - h(u) = 0$. Note $V(z)$ now satisfies

$$\nu^2 V''(z) - \theta(V(z+1) + 2V(z) - V(z-1)) - h(V(z)) = 0$$

and $u(x, t) = \tilde{\gamma} \left(t - \frac{n}{\nu} \right)$ is a degenerate traveling wave of (6.1.3). Applying these arguments to (6.1.6) as $\varepsilon \to \infty$, we get its traveling wave (a kink) near

$$2 \arctan \left[\sinh \left(\varepsilon t - \frac{n}{\nu} \right) \right]$$

accumulated by periodic traveling waves with minimal periods tending to infinity. Of course, the magnitude of ε depends on ν.

6.2 Center Manifold Reduction

Now we use the method of center manifolds of Section 2.6 in order to show existence of periodic solutions of (6.1.5) [94].

1. Step: *The idea is to rewrite (6.1.5) as an evolution equation on an appropriate functional Banach space.* To this end, we introduce the Banach spaces $\mathbb{H}$ and $\mathbb{D}$ for $U = \left(x, \xi, X(v) \right)$

$$\mathbb{H} = \mathbb{R}^2 \times C([-1, 1]),$$
$$\mathbb{D} = \{ U \in \mathbb{R}^2 \times C^1[-1, 1] \mid X(0) = x \}$$

with the usual maximum norms. Then (6.1.5) can be written as follows

$$U_t = LU + \frac{\varepsilon^2}{\nu^2} M(U), \quad U(t, v) = (x(t), \xi(t), X(t, v))^*, \qquad (6.2.1)$$

where

$$L = \begin{pmatrix} 0 & 1 & 0 \\ -\frac{2}{\nu^2} & 0 & \frac{1}{\nu^2}\delta^1 + \frac{1}{\nu^2}\delta^{-1} \\ 0 & 0 & \partial_v \end{pmatrix}$$

$$M(U) = (0, h(x), 0)^*, \quad \delta^{\pm}X(v) = X(\pm 1)$$

with $L \in \mathcal{L}(\mathbb{D}, \mathbb{H})$ and $M \in C^1(\mathbb{D}, \mathbb{D})$. We consider (6.2.1) on $\mathbb{D}$.

2. Step: *A linear analysis of* (6.2.1). The spectrum $\sigma(L)$ of L is given by the resolvent equation

$$(\lambda\mathbb{I} - L)U = F, \quad F \in \mathbb{H}, \lambda \in \mathbb{C}, U \in \mathbb{D}.$$

This is solvable if and only if $N(\lambda) = 0$ for $N(\lambda) = \lambda^2 + \frac{2}{\nu^2}(1 - \cosh\lambda)$. Clearly $\sigma(L)$ is invariant under $\lambda \to \bar{\lambda}$ and $\lambda \to -\lambda$. The central part $\sigma_0(L) = \sigma(L) \cap \imath\mathbb{R}$ is determined by the equation $N(\imath q) = 0$, i.e.

$$q^2 + \frac{2}{\nu^2}(\cos q - 1) = 0, \quad q \in \mathbb{R}. \tag{6.2.2}$$

The basic properties of $\sigma(L)$ are given in [120, Lemma 1]:

Lemma 6.2.1.   *(i) For each $\nu > 0$, there exists $p_0 > 0$ such that $\forall\lambda \in \sigma(L) \setminus \sigma_0(L)$, $|\Re\lambda| \geq p_0$.*

  (ii) If $\lambda = p + \imath q \in \sigma(L)$ then $|q| \leq 2\frac{\sqrt{e^2+4\nu^2}}{\nu e}\cosh(p/2)$.

  (iii) For $\nu > 1$, 0 is the only eigenvalue on the imaginary axis with the multiplicity 2. There are only two real eigenvalues $\pm\lambda$ tending to 0 as $\nu \to 1$. For $\nu \leq 1$, the eigenvalue 0 is the only real one.

  (iv) For $\nu = 1$, the eigenvalue 0 is quadruple with a 4×4 Jordan block.

  (v) There is a decreasing sequence ν_n, $n = 0, 1, 2 \ldots$ such that $\nu_0 = 1$ and $\nu_n \to 0$, and for $\nu = \nu_n$ with $n \geq 1$, there is a pair $\pm\imath q_n$ of double non-semi-simple imaginary eigenvalues in addition to the double non-semi-simple eigenvalue at 0, and $2n - 1$ pairs of simple imaginary eigenvalues $\pm\imath\widetilde{g}_j$ such that $0 < \widetilde{q}_j < q_n$.

In this section, we assume that $\nu_1 < \nu < 1$. Note $\nu = \nu_1$ is the first value from the left of 1 for which the equations

$$\lambda^2 + \frac{2}{\nu^2}(\cos\lambda - 1) = 0, \quad \lambda - \frac{1}{\nu^2}\sin\lambda = 0 \tag{6.2.3}$$

have a common nonzero solution $\lambda \neq 0$. Then by (v) of Lemma 6.2.1 we have $\sigma_0(L) = \{0, \pm\imath q\}$. After some computations we see that the corresponding 4th-dimensional central subspace $\mathbb{H}_c$ has a basis $(\xi_1, \xi_2, \xi_3, \xi_4)$ defined by

$$\xi_1 = (1, 0, 1), \quad \xi_2 = (0, 1, v), \quad \xi_3 = (1, 0, \cos qv), \quad \xi_4 = (0, q, \sin qv)$$

with $L\xi_1 = 0$, $L\xi_2 = \xi_1$, $L\xi_3 = -q\xi_4$, $L\xi_4 = q\xi_3$. So L on $\mathbb{H}_c$ has the form

$$L_c = L/\mathbb{H}_c = \begin{pmatrix} 0 & 1 & 0 & 0 \\ 0 & 0 & 0 & 0 \\ 0 & 0 & 0 & q \\ 0 & 0 & -q & 0 \end{pmatrix}.$$

The corresponding spectrum projections are derived as the residues of the inverse $(\lambda\mathbb{I}-L)^{-1}$ at $\lambda = 0$, $\pm\imath q$, respectively, of the resolvent operator. Performing these computations, the projection $P_c : \mathbb{H} \to \mathbb{H}_c$ is given by

$$P_c(U) = P_1(U)\xi_1 + P_2(U)\xi_2 + P_3(U)\xi_3 + P_4(U)\xi_4,$$

where

$$P_1(U) = \frac{\nu^2}{\nu^2 - 1}x - \frac{1}{\nu^2 - 1}\int_0^1 (1 - s)\big[X(s) + X(-s)\big]\,ds,$$

$$P_2(U) = \frac{\nu^2}{\nu^2 - 1}\xi + \frac{1}{\nu^2 - 1}\int_0^1 \big[X(-s) - X(s)\big]\,ds,$$

$$P_3(U) = \left(\nu^2 qx - \int_0^1 \sin q(1 - s)\big[X(s) + X(-s)\big]\,ds\right)\big/(q\nu^2 - \sin q),$$

$$P_4(U) = \left(\nu^2 \xi + \int_0^1 \cos q(1 - s)\big[X(-s) - X(s)\big]\,ds\right)\big/(q\nu^2 - \sin q).$$

So the condition (i) of a hypothesis (H) of Theorem 2.6.2 is satisfied. The last one (ii) is shown in [122]. Hence we can proceed to the next step.

3. Step: *The center manifold reduction method to* (6.2.1) *for simplifying it.* Since $M(U)$ is Lipschitz, for any bounded ball Ω of $\mathbb{H}_c$ centered at 0, we can apply the procedure of a center manifold method of Theorem 2.6.2 to get for ε small the reduced equation of (6.2.1) over Ω given by

$$\dot{u}_c = L_c u_c + \frac{\varepsilon^2}{\nu^2}P_c M\big(u_c + \varepsilon^2 \Phi_\varepsilon(u_c)\big) = L_c u_c + \frac{\varepsilon^2}{\nu^2}P_c(M(u_c)) + O(\varepsilon^4),$$

$$(6.2.4)$$

where $u_c = u_1\xi_1 + u_2\xi_2 + u_3\xi_3 + u_4\xi_4$ and Φ_ε is the graph map of the center manifold. So any solution of (6.2.4) in Ω determines a solution $U(t, v) = u_c(t) + \varepsilon^2 \Phi_\varepsilon(u_c(t))$ of (6.2.1). Using the above formulas for P_c, (6.2.4) has the form

$$\dot{u}_1 = u_2, \quad \dot{u}_2 = \frac{\varepsilon^2}{\nu^2 - 1}\tilde{h}(u_1, u_2, u_3, u_4, \varepsilon^2)$$

$$\dot{u}_3 = qu_4, \quad \dot{u}_4 = -qu_3 + \frac{\varepsilon^2}{q\nu^2 - \sin q}\tilde{h}(u_1, u_2, u_3, u_4, \varepsilon^2),$$

for a C^1-function $\tilde{h}$. Considering

$$x(t) = x_1(t) = u_1(t/\varepsilon), \quad x_2(t) = u_2(t/\varepsilon)/\varepsilon,$$
$$y(t) = y_1(t) = u_3(t/\varepsilon), \quad y_2(t) = u_4(t/\varepsilon),$$

(6.2.4) takes the form

$$\dot{x}_1 = x_2, \quad \dot{x}_2 = \frac{1}{\nu^2 - 1}\tilde{h}(x_1, \varepsilon x_2, y_1, y_2, \varepsilon^2)$$
$$\dot{y}_1 = \frac{q}{\varepsilon}y_2, \quad \dot{y}_2 = -\frac{q}{\varepsilon}y_1 + \frac{\varepsilon}{q\nu^2 - \sin q}\tilde{h}(x_1, \varepsilon x_2, y_1, y_2, \varepsilon^2),$$

which gives

$$\ddot{x} = \frac{1}{1 - \nu^2}f(x, \varepsilon\dot{x}, y, \varepsilon\dot{y}/q, \varepsilon),$$
$$\varepsilon^2\ddot{y} + q^2 y = \frac{\varepsilon^2 q}{\sin q - \nu^2 q}f(x, \varepsilon\dot{x}, y, \varepsilon\dot{y}/q, \varepsilon),$$

(6.2.5)

where $f(x_1, x_2, y_1, y_2, \varepsilon) = -h(x_1 + y_1) + O(\varepsilon^2)$. Consequently, for $\varepsilon = 0$ and $y = 0$, the limit equation of (6.2.5) has the form

$$(1 - \nu^2)\ddot{x} + h(x) = 0.$$

(6.2.6)

Note looking for a traveling wave solution $u(x,t) = w(\nu t - x)$ of (6.1.3), we get $(1 - \nu^2)\ddot{w} + h(w) = 0$. So (6.2.6) is precisely the *traveling wave equation* of (6.1.3). It is possessing a homoclinic solution $x(t) = \phi(t/\sqrt{1 - \nu^2})$. Summarizing we get the following result.

Proposition 6.2.2. *A dynamics of (6.2.1) can be reduced to (6.2.5) which is a singular perturbation of the traveling wave equation (6.2.6) of (6.1.3).*

4. Step: *Symmetries of the reduced equation.* (6.2.5) is still rather complicated to study it. For this reason we consider the symmetry

$$S(U) = (x, -\xi, X(-v))$$

on $\mathbb{H}$. Then (6.2.1) is reversible with respect to S, i.e. $S \circ L = -L \circ S$, $M \circ S = -S \circ M$. Moreover, we have $P_c \circ S = S \circ P_c$ and $S\xi_1 = \xi_1$, $S\xi_2 = -\xi_2$, $S\xi_3 = \xi_3$,

$S\xi_4 = -\xi_4$. Since S is unitary, by Section 2.6, the map Φ_ε can be chosen in such a way that $S \circ \Phi_\varepsilon = \Phi_\varepsilon \circ S_c$ for $S_c := S/\mathbb{H}_c$. Note $S_c : \mathbb{H}_c \to \mathbb{H}_c$. This implies

$$L_c S_c u_c + \frac{\varepsilon^2}{\nu^2} P_c M\big(S_c u_c + \varepsilon^2 \Phi_\varepsilon(S_c u_c)\big) = -S_c\bigg(L_c u_c + \frac{\varepsilon^2}{\nu^2} P_c M\big(u_c + \varepsilon^2 \Phi_\varepsilon(u_c)\big)\bigg).$$

Hence (6.2.4) is reversible with respect to S_c. Moreover, S_c has in the coordinates (x_1, x_2, y_1, y_2) on $\mathbb{H}_c$ the form $S_c(x_1, x_2, y_1, y_2) = (x_1, -x_2, y_1, -y_2)$.

6.3 A Class of Singularly Perturbed O.D.Eqns

Motivated by (6.2.5), we consider a system

$$\ddot{x} + h(x) = f(x, \dot{x}, y, \varepsilon\dot{y}, \varepsilon), \quad \varepsilon^2 \ddot{y} + y = \varepsilon^2 g(x, \dot{x}, y, \varepsilon\dot{y}, \varepsilon), \tag{6.3.1}$$

where $\varepsilon > 0$ is a small parameter, $h \in C^1$ satisfies (A) and with f, g such that

(B) $f, g \in C^1$, $f(x_1, x_2, 0, 0, 0) = 0$

(C) $f(x_1, x_2, y_1, y_2, \varepsilon)$, $g(x_1, x_2, y_1, y_2, \varepsilon)$ are even in the variables x_2 and y_2, i.e. $f(x_1, -x_2, y_1, -y_2, \varepsilon) = f(x_1, x_2, y_1, y_2, \varepsilon)$ and $g(x_1, -x_2, y_1, -y_2, \varepsilon) = g(x_1, x_2, y_1, y_2, \varepsilon)$

Note in [15, 70, 109, 133] there are examined the existence or nonexistence of homoclinic solutions of singular ordinary differential systems of the following type

$$\varepsilon^2 y^{(4)} + \ddot{y} - y + y^2 = 0 \tag{6.3.2}$$

which arises in the theory of water-waves in the presence of surface tension [4]. Setting $v = y$, $u = \ddot{y} - y + y^2$, (6.3.2) leads to

$$\ddot{v} = u + v - v^2, \quad \varepsilon^2 \ddot{u} + u = \varepsilon^2\big[2\dot{v}^2 - (1 - 2v)(u + v - v^2)\big], \tag{6.3.3}$$

which has a form of (6.3.1). Next, in [15] it is shown that bifurcation functions of homoclinic solutions of (6.3.1) under the above assumptions are exponentially small in addition that h, f, g are analytic. In [70, 109] it is established the nonexistence of certain homoclinic solutions of (6.3.2).

6.4 Bifurcation of Periodic Solutions

In this section, we study the existence of periodic solutions of (6.3.1) near $(\phi(t), 0)$. Substituting $y = 0, \varepsilon = 0$ into (6.3.1), we get the equation

$$\ddot{x} + h(x) = 0. \tag{6.4.1}$$

Equation (6.4.1) has a hyperbolic fixed point $(0,0)$ with the homoclinic solution $(\phi, \dot{\phi})$ which is accumulated by periodic solutions with periods tending to infinity. We show that in spite of the fact that generally the homoclinic solution of

(6.4.1) does not survive under the singular perturbation (6.3.1). The problem (6.3.1) has many layers of continuum periodic solutions near the solution $(\phi, 0)$: The smaller ε the more layers of continuum periodic solutions of (6.3.1) exist near $(\phi, 0)$ with very large periods. This is some kind of blue sky catastrophe bifurcation to (6.3.1) studied in Section 5.2. Results of [15, 70, 109, 133] are now not applicable since h is only C^1-smooth.

In order to find periodic solutions of (6.3.1) near $(\phi, 0)$, we make the change of variables

$$x(t) = \phi(t) + \varepsilon^{1/4} u(t), \quad y(t) = \sqrt{\varepsilon} v(t),$$

and we get

$$\varepsilon^2 \ddot{v} + v = \varepsilon^{3/2} g(\phi + \varepsilon^{1/4} u, \dot{\phi} + \varepsilon^{1/4} \dot{u}, \sqrt{\varepsilon} v, \varepsilon^{3/2} \dot{v}, \varepsilon)$$

$$\ddot{u} + h'(\phi) u = -\frac{1}{\varepsilon^{1/4}} \left\{ h(\phi + \varepsilon^{1/4} u) - h(\phi) - h'(\phi) \varepsilon^{1/4} u \right\} \qquad (6.4.2)$$

$$+ \frac{1}{\varepsilon^{1/4}} f(\phi + \varepsilon^{1/4} u, \dot{\phi} + \varepsilon^{1/4} \dot{u}, \sqrt{\varepsilon} v, \varepsilon^{3/2} \dot{v}, \varepsilon).$$

We are looking for solutions of (6.3.1) satisfying $\dot{x}(0) = \dot{x}(T) = 0$, $\dot{y}(0) = \dot{y}(T) = 0$. This gives

$$\dot{u}(0) = 0, \quad \dot{u}(T) = -\dot{\phi}(T)/\varepsilon^{1/4}, \quad \dot{v}(0) = 0, \quad \dot{v}(T) = 0. \qquad (6.4.3)$$

First we study linear parts of (6.4.2). We take the linearization of (6.4.1) along $\phi(t)$ and consider the variational equation

$$\ddot{u} + h'(\phi(t)) u = z(t), \quad 0 \leq t \leq T \qquad (6.4.4)$$

with the boundary value conditions

$$\dot{u}(0) = 0, \quad \dot{u}(T) = b. \qquad (6.4.5)$$

Since $h'(0) = -a^2 < 0$, $a > 0$, we have $\phi(t), \dot{\phi}(t) \sim e^{-at}$ as $t \to +\infty$, i.e. it holds that

$$\phi(t)/e^{-at} \to c_1 \neq 0 \quad \text{and} \quad \dot{\phi}(t)/e^{-at} \to c_2 \neq 0 \quad \text{as} \quad t \to +\infty.$$

The homogeneous equation (6.4.4) with $z = 0$ has solutions $w_i(t)$, $i = 1, 2$ such that:

- w_1 is odd, $w_1(0) = 0$, $\dot{w}_1(0) = 1$, $w_1(t), \dot{w}_1(t) \sim e^{-at}$ as $t \to +\infty$

- w_2 is even, $w_2(0) = -1$, $\dot{w}_2(0) = 0$, $w_2(t), \dot{w}_2(t) \sim e^{at}$ as $t \to +\infty$

The general solution of (6.4.4) has the form

$$u(t) = L_T(z, b) \equiv c_1 w_1(t) + c_2 w_2(t) + z_1(t),$$

$$z_1(t) = \int_0^t [w_2(t) w_1(s) - w_1(t) w_2(s)] z(s) \, ds.$$

The condition (6.4.5) gives $c_1 = 0$ and $c_2 = -\frac{\dot{z}_1(T)}{\dot{w}_2(T)} + \frac{b}{\dot{w}_2(T)}$. Hence, we get

$$u(t) = b\frac{w_2(t)}{\dot{w}_2(T)} - \int_t^T w_2(t)w_1(s)z(s)\,ds$$

$$+\frac{\dot{w}_1(T)}{\dot{w}_2(T)}\int_0^T w_2(t)w_2(s)z(s)\,ds - \int_0^t w_1(t)w_2(s)z(s)\,ds \qquad (6.4.6)$$

and

$$\dot{u}(t) = b\frac{\dot{w}_2(t)}{\dot{w}_2(T)} - \int_t^T \dot{w}_2(t)w_1(s)z(s)\,ds$$

$$+\frac{\dot{w}_1(T)}{\dot{w}_2(T)}\int_0^T \dot{w}_2(t)w_2(s)z(s)\,ds - \int_0^t \dot{w}_1(t)w_2(s)z(s)\,ds. \qquad (6.4.7)$$

By using the above asymptotic properties of w_1 and w_2, there is a constant $C_1 > 0$ such that for any $t, s \in [0, T]$ and $T > 0$ large, we get

$$|w_2(t)/\dot{w}_2(T)| \le C_1\,e^{a(t-T)}, \qquad |w_2(t)w_1(s)| \le C_1\,e^{a(t-s)},$$

$$\left|\frac{\dot{w}_1(T)}{\dot{w}_2(T)}w_2(t)w_2(s)\right| \le C_1\,e^{a(-2T+t+s)}, \qquad |w_1(t)w_2(s)| \le C_1\,e^{a(s-t)},$$

$$|\dot{w}_2(t)w_1(s)| \le C_1\,e^{a(t-s)}, \qquad |\dot{w}_1(t)w_2(s)| \le C_1\,e^{a(s-t)},$$

$$|\dot{w}_2(t)/\dot{w}_2(T)| \le C_1\,e^{a(t-T)}, \qquad \left|\frac{\dot{w}_1(T)}{\dot{w}_2(T)}\dot{w}_2(t)w_2(s)\right| \le C_1\,e^{a(-2T+t+s)}.$$

These estimates imply together with (6.4.6–6.4.7) the existence of a constant $c > 0$ such that

$$||u|| + ||\dot{u}|| \le c(|b| + ||z||), \qquad (6.4.8)$$

where $||x|| = \max_{[0,T]} |x(t)|$. Summarizing, we get the next result.

Lemma 6.4.1. *Problem* (6.4.4–6.4.5) *has a unique solution* $u = L_T(z, b)$ *satisfying* (6.4.8).

Now, we consider the problem

$$\varepsilon^2\ddot{v} + v = \varepsilon z(t), \quad \dot{v}(0) = \dot{v}(T) = 0, \quad 0 \le t \le T. \qquad (6.4.9)$$

We can immediately see that the solution of (6.4.9) is given by

$$v(t) = L_{\varepsilon,T}(z) \equiv \frac{1}{\sin(T/\varepsilon)}\int_0^T \cos\frac{T-s}{\varepsilon}z(s)\,ds\,\cos(t/\varepsilon) + \int_0^t \sin\frac{t-s}{\varepsilon}z(s)\,ds.$$

If T satisfies

$$\left|\frac{T}{\varepsilon} - 2k\pi \pm \frac{\pi}{2}\right| \le \pi/4, \quad k \in \mathbb{N} \qquad (6.4.10)$$

then $1 \geq |\sin(T/\varepsilon)| \geq \sqrt{2}/2$, and we obtain the estimate

$$||v|| + ||\varepsilon\dot{v}|| \leq 2T||z||(\sqrt{2}+1). \tag{6.4.11}$$

Summarizing, we get the next result.

Lemma 6.4.2. *If condition* (6.4.10) *holds then problem* (6.4.9) *has a unique solution* $v = L_{\varepsilon,T}(z)$ *satisfying* (6.4.11).

Note (6.4.10) is a nonresonance condition. Now we are ready to prove the following bifurcation result.

Theorem 6.4.3. *For any* $k_0 \in \mathbb{N}$ *there is an* $\varepsilon_0 > 0$ *such that for any* $0 < \varepsilon < \varepsilon_0$ *and* $T = \varepsilon\big(2k[1/\varepsilon^{3/2}]\pi + \tau\big)$ *with* $k \in \mathbb{N}$, $k \leq k_0$, $\tau \in [\pi/4, 3\pi/4] \cup [5\pi/4, 7\pi/4]$, *system* (6.3.1) *has a* $2T$-*periodic solution near* $(\phi(t), 0)$, $-T \leq t \leq T$. *Here* $[1/\varepsilon^{3/2}]$ *is the integer part of* $1/\varepsilon^{3/2}$. *Moreover,* $x_{T,\varepsilon}(t) - \phi(t) = O(\varepsilon^{1/4})$, $\dot{x}_{T,\varepsilon}(t) - \dot{\phi}(t) = O(\varepsilon^{1/4})$, $y_{T,\varepsilon}(t) = O(\sqrt{\varepsilon})$, $\varepsilon\dot{y}_{T,\varepsilon}(t) = O(\sqrt{\varepsilon})$ *uniformly for* $-T \leq t \leq T$.

Proof. First of all, we show the existence of a solution of (6.4.2–6.4.3) applying the Schauder fixed point theorem. We take the Banach space $X_\varepsilon = C^1([0,T], \mathbb{R})^2$ with the norm $|||(v,u)||| = ||u|| + ||\dot{u}|| + ||v|| + ||\varepsilon\dot{v}||$. Using Lemmas 6.4.1 and 6.4.2, we rewrite (6.4.2)-(6.4.3) in the form

$$v = L_{\varepsilon,T}\left(\sqrt{\varepsilon}g(\phi + \varepsilon^{1/4}u, \dot{\phi} + \varepsilon^{1/4}\dot{u}, \sqrt{\varepsilon}v, \varepsilon^{3/2}\dot{v}, \varepsilon)\right)$$

$$u = L_T\Big(-\frac{1}{\varepsilon^{1/4}}\big\{h(\phi + \varepsilon^{1/4}u) - h(\phi) - h'(\phi)\varepsilon^{1/4}u\big\} \tag{6.4.12}$$

$$+\frac{1}{\varepsilon^{1/4}}f(\phi + \varepsilon^{1/4}u, \dot{\phi} + \varepsilon^{1/4}\dot{u}, \sqrt{\varepsilon}v, \varepsilon^{3/2}\dot{v}, \varepsilon), -\dot{\phi}(T)/\varepsilon^{1/4}\Big)$$

as a fixed point problem in X_ε. Now we fix $k_0 \in \mathbb{N}$ and take $T = \varepsilon\big(2k[1/\varepsilon^{3/2}]\pi + \tau\big)$ with $k \in \mathbb{N}$, $k \leq k_0$ and $\tau \in [\pi/4, 3\pi/4] \cup [5\pi/4, 7\pi/4]$. Let $B_K = \big\{(v,u) \in X_\varepsilon \mid |||(v,u)||| \leq K\big\}$ be a ball in X_ε. Since $T \sim 1/\sqrt{\varepsilon}$ and $\dot{\phi}(T) \sim e^{-aT}$, we get $\dot{\phi}(T)/\varepsilon^{1/4} \sim e^{-a/\sqrt{\varepsilon}}/\varepsilon^{1/4} = O(\varepsilon)$. From the C^1-smoothness of f, g, h, there is a constant $M > 0$ such that $\forall K > 0$, $\exists\varepsilon_0 > 0$, $\forall\varepsilon \in (0, \varepsilon_0]$, $\forall(v,u) \in B_K$, it holds that

$$\left|g(\phi + \varepsilon^{1/4}u, \dot{\phi} + \varepsilon^{1/4}\dot{u}, \sqrt{\varepsilon}v, \varepsilon^{3/2}\dot{v}, \varepsilon)\right| \leq M,$$

$$\left|\frac{1}{\varepsilon^{1/4}}\big\{h(\phi + \varepsilon^{1/4}u) - h(\phi) - h'(\phi)\varepsilon^{1/4}u\big\}\right| \leq 1,$$

$$\left|\frac{1}{\varepsilon^{1/4}}f(\phi + \varepsilon^{1/4}u, \dot{\phi} + \varepsilon^{1/4}\dot{u}, \sqrt{\varepsilon}v, \varepsilon^{3/2}\dot{v}, \varepsilon)\right| \leq 1.$$

For any $(u,v) \in B_K$, $0 < \varepsilon \leq \varepsilon_0$, we put

$$u_1 = L_T\Big(-\frac{1}{\varepsilon^{1/4}}\big\{h(\phi + \varepsilon^{1/4}u) - h(\phi) - h'(\phi)\varepsilon^{1/4}u\big\}$$

$$+\frac{1}{\varepsilon^{1/4}}f(\phi + \varepsilon^{1/4}u, \dot{\phi} + \varepsilon^{1/4}\dot{u}, \sqrt{\varepsilon}v, \varepsilon^{3/2}\dot{v}, \varepsilon), -\dot{\phi}(T)/\varepsilon^{1/4}\Big),$$

$$v_1 = L_{\varepsilon,T}\left(\sqrt{\varepsilon}g(\phi + \varepsilon^{1/4}u, \dot{\phi} + \varepsilon^{1/4}\dot{u}, \sqrt{\varepsilon}v, \varepsilon^{3/2}\dot{v}, \varepsilon)\right).$$

Then estimate (6.4.8) implies

$$\|u_1\| + \|\dot{u}_1\| \le c(2 + O(\varepsilon)),$$

and estimates (6.4.10–6.4.11) imply

$$\|v_1\| + \|\varepsilon \dot{v}_1\| \le 2\sqrt{\varepsilon} T M (1 + \sqrt{2}) \le 2M(\sqrt{2} + 1)\left(2\pi k_0 + \frac{7\pi}{4}\varepsilon^{3/2}\right).$$

Consequently, we obtain

$$\|\|(v_1, u_1)\|\| \le 2c + 4M(\sqrt{2} + 1)\pi k_0 + O(\varepsilon).$$

Hence for K such that $2c + 4M(\sqrt{2} + 1)\pi k_0 < K$ and $\varepsilon_0 > 0$ sufficiently small, B_K is mapped to itself with the compact operator defined by the right-hand side of (6.4.12). We fix such a K and apply the Schauder fixed point theorem to get a solution of (6.4.12) in X_ε, i.e. there is a solution of (6.3.1) satisfying $\dot{x}(0) = \dot{x}(T) = 0$, $\dot{y}(0) = \dot{y}(T) = 0$. Since h, f, g are C^1, we get the uniqueness of the Cauchy problem for (6.3.1). Then the evenness of f, g in x_2, y_2 and the conditions $\dot{x}(0) = 0$, $\dot{y}(0) = 0$ imply that x, y are even functions. This implies

$$x(-T) = x(T), \quad \dot{x}(-T) = -\dot{x}(T) = 0,$$
$$y(-T) = y(T), \quad \dot{y}(-T) = -\dot{y}(T) = 0.$$

Consequently, the uniqueness of the Cauchy problem for (6.3.1) implies that x and y are $2T$-periodic. The proof is finished.   □

Remark 6.4.4. If $h \in C^2$ then, we can apply the implicit function theorem to (6.4.12) for getting a unique $2T$-periodic and even solution of (6.3.1) near $(\phi(t), 0)$ for $-T \le t \le T$.

6.5 Traveling Waves in Homoclinic Cases

From Section 6.2 we get that assumptions (A), (B), (C) of Sections 6.1 and 6.3 are satisfied for (6.2.5). Applying Theorem 6.4.3, (6.2.5) has a $2T$-periodic solution $(x_{T,\varepsilon}(t), y_{T,\varepsilon}(t))$ near $(\phi(t/\sqrt{1 - \nu^2}), 0)$, $-T \le t \le T$ for any T satisfying the assumption of Theorem 6.4.3. They have the form

$$u_c^{T,\varepsilon}(t) = x_{T,\varepsilon}(\varepsilon t)\xi_1 + \varepsilon \dot{x}_{T,\varepsilon}(\varepsilon t)\xi_2 + y_{T,\varepsilon}(\varepsilon t)\xi_3 + \varepsilon(\dot{y}_{T,\varepsilon}(\varepsilon t)/q)\xi_4$$

in (6.2.4). All $u_c^{T,\varepsilon}(t)$ lie in a large ball Ω. Furthermore, we have

$$U(t, \cdot) = u_c(t) + \varepsilon^2 \Phi_\varepsilon(u_c(t)) = u_c(t) + O(\varepsilon^2)$$

for (6.2.1) on the center manifold considered in (6.2.4). We also note that the $x(t)$-coordinate of $U(t, v)$ in (6.2.1) satisfies (6.1.5). Consequently, if $x^{T,\varepsilon}(\varepsilon t)$ is the x-coordinate of $u_c^{T,\varepsilon}(t) + \varepsilon^2 \Phi_\varepsilon(u_c^{T,\varepsilon}(t))$, then the traveling wave solution of (6.1.2) corresponding to $x_{T,\varepsilon}(t)$, $y_{T,\varepsilon}(t)$ has the form

$$u_n^{T,\varepsilon}(t) = x^{T,\varepsilon}\left(\varepsilon\left(n - \frac{\nu}{\varepsilon}t\right)\right) = x^{T,\varepsilon}(\varepsilon n - \nu t)$$

$$= x_{T,\varepsilon}(\varepsilon n - \nu t) + y_{T,\varepsilon}(\varepsilon n - \nu t) + O(\varepsilon^2).$$

Note $u_n^{T,\varepsilon}(t)$ is $2T/\nu$-periodic in t with the velocity ν and such that

$$u_n^{T,\varepsilon}(t) = \phi\left(\frac{\varepsilon n - \nu t}{\sqrt{1 - \nu^2}}\right) + O(\varepsilon^{1/4})$$

uniformly for $-T \le \varepsilon n - \nu t \le T$ and T satisfying the assumption of Theorem 6.4.3 for a fixed k_0. Finally, we recall (6.1.4). Summarizing we get the main result of this section.

Theorem 6.5.1. *If $h \in C^1$ satisfies the assumption* (A) *then traveling wave solution $u(x,t) = \phi\left(\frac{x-\nu t}{\sqrt{1-\nu^2}}\right)$ for $0 < \nu_1 < \nu < 1$ of* (6.1.3) *can be approximated by periodic traveling wave solutions of* (6.1.2) *with very large periods and with the velocity ν.*

We note that for a C^∞-smooth h, the center manifold graph Φ_ε is C^k-smooth for any fixed $k \in \mathbb{N}$, and then (6.2.5) is also C^k-smooth. Hence the bifurcation function of homoclinics for (6.2.5) is of order $O(\varepsilon^k)$. So it is flat at $\varepsilon = 0$. Since (6.2.5) is not analytical, we do not get further information of this flatness. Hence it seems that the center manifold method is not fruitful for detecting bounded solutions of (6.2.5) near $(\phi, 0)$ on $\mathbb{R}$.

6.6   Traveling Waves in Heteroclinic Cases

Theorem 6.4.3 can not be applied to (6.1.6) since now the limit reduced equation (6.2.6) is a pendulum-like equation

$$(1 - \nu^2)\ddot{x} - \sin x = 0$$

with a heteroclinic connection

$$\widetilde{\Psi}(t) = 4 \arctan \exp\left[t/\sqrt{1 - \nu^2}\right],$$

while we consider in (6.4.1) a homoclinic solution (see assumption (A)). In this part, the *heteroclinic case* is studied for a perturbed Hamiltonian chain of coupled oscillators with an Hamiltonian

$$\mathcal{H} = \sum_{n \in \mathbb{Z}} \left(\frac{1}{2}\dot{u}_n^2 + \frac{1}{2\varepsilon^2}(u_{n+1} - u_n)^2 + H(u_n) + \mu G(u_{n+1} - u_n)\right), \qquad (6.6.1)$$

where $\varepsilon > 0$ is a discretization parameter and μ is a small parameter measuring the relation of intersite and offsite potentials, and $H, G \in C^2(\mathbb{R})$. The corresponding discrete nonlinear Klein-Gordon equation is:

$$\ddot{u}_n - \frac{1}{\varepsilon^2}(u_{n+1} - 2u_n + u_{n-1}) + h(u_n)$$
$$+\mu\Big\{g(u_n - u_{n-1}) - g(u_{n+1} - u_n)\Big\} = 0,$$

(6.6.2)

where $h(x) = H'(x)$ and $g(x) = G'(x)$. The following conditions are supposed.

(B1) $h, g \in C^1(\mathbb{R})$ are odd, h is 2π-periodic and g is globally Lipschitz on $\mathbb{R}$.

(B2) $h(-\pi) = h(\pi) = 0$, $h'(-\pi) = h'(\pi) = a^2 > 0$ and there is a heteroclinic solution Φ of $\ddot{x} - h(x) = 0$ such that $\Phi(t) = 2\pi - \Phi(-t)$ and $\Phi(t) \to 2\pi$ as $t \to +\infty$.

Now the traveling wave equation (6.1.5) takes the form

$$\nu^2 V''(z) - V(z+1) + 2V(z) - V(z-1) + \varepsilon^2 h(V(z))$$
$$+\varepsilon^2\mu\Big\{g(V(z) - V(z-1)) - g(V(z+1) - V(z))\Big\} = 0.$$

(6.6.3)

We need the following definition [182].

Definition 6.6.1. By a *uniform sliding state* of (6.6.3) we mean a smooth function $V(z)$ solving (6.6.3) and satisfying $V(z+T) = V(z) + 2\pi$.

The method of this section is successfully applied in [95] to (6.6.3) and the next analogy of Theorem 6.5.1 is proved.

Theorem 6.6.2. *If h, g satisfy the assumptions* (B1), (B2) *then traveling wave solution $u(x,t) = \Phi\left(\frac{x-\nu t}{\sqrt{1-\nu^2}}\right)$ for $0 < \nu_1 < \nu < 1$ of $u_{tt} - u_{\xi\xi} + h(u) = 0$ can be approximated by both periodic traveling wave solutions and uniform sliding states of (6.6.2) with very large periods and with the velocity ν for $\mu = o(\varepsilon^{1/4})$ small.*

These solutions of (6.6.3) have again tails of periodic waves of small amplitude. This result is consistent with the numerical result of [71, 176] for (6.1.6) mentioned above.

We remind that $\nu = \nu_1$ is the first value from the left of 1 for which the equations of (6.2.3) have a common nonzero solution $\lambda \neq 0$. For $0 < \nu < \nu_1$, we could still use the above method. We know from Lemma 6.2.1 that for any $\nu_{i+1} < \nu < \nu_i$ the linear operator L has the double non semi-simple eigenvalue at 0, and $2i + 1$ pairs of simple imaginary eigenvalues. So after the center manifold reduction, we should get a system like (6.2.5) and we could generalize the bifurcation result of Section 6.4 for such systems. We do not carry out those computations in this book. Related problems are studied in [24, 25] on the existence of periodic solutions of certain singularly perturbed systems of o.d.eqns having symmetry properties with applications to some singular systems of o.d.eqns arising in the study of Hamiltonian systems with a strong restoring force.

We finish this part with a Hamiltonian perturbation of (6.1.8) of the form

$$\ddot{u}_n - \frac{1}{\varepsilon^2}(u_{n+1} - 2u_n + u_{n-1}) + \frac{(1+2\alpha)\sin u}{(1+\alpha(1-\cos u))^2}$$

$$+\mu\Big\{\sin(u_n - u_{n+1}) + \sin(u_n - u_{n-1})\Big\} = 0. \tag{6.6.4}$$

For $\alpha = \mu = 0$ we get the discrete sine-Gordon equation. Changing variables $u_n \longleftrightarrow u_n - \pi$ in (6.6.4), we have

$$\ddot{u}_n - \frac{1}{\varepsilon^2}(u_{n+1} - 2u_n + u_{n-1}) + g_\alpha(u_n)$$

$$+\mu\Big\{\sin(u_n - u_{n+1}) + \sin(u_n - u_{n-1})\Big\} = 0 \tag{6.6.5}$$

for

$$g_\alpha(u) = -\frac{(1+2\alpha)\sin u}{(1+\alpha(1+\cos u))^2}.$$

Clearly $g_\alpha \in C^1(\mathbb{R})$ is odd and 2π-periodic, so (B1) is satisfied. Next $g_\alpha(-\pi) = g_\alpha(\pi) = 0$, $g'_\alpha(-\pi) = g'_\alpha(\pi) = 1 + 2\alpha > 0$ and a heteroclinic solution Ψ_α of $\ddot{x} - g_\alpha(x) = 0$ in (B2) is determined by an implicit equation

$$\sqrt{2\alpha}\arcsin\frac{\sqrt{2\alpha}\sin(\Psi_\alpha(t)/2)}{\sqrt{1+2\alpha}} + \text{arctanh}\frac{\sin(\Psi_\alpha(t)/2)}{\sqrt{1+2\alpha\cos^2(\phi_\alpha(t)/2)}} = \sqrt{1+2\alpha}t.$$

Summarizing, Theorem 6.6.2 can be applied to (6.6.4) uniformly for $\alpha \geq 0$ from bounded intervals.

6.7   Traveling Waves in 2 Dimensions

Finally, further developments on lattice equations are presented in [96] for

$$\ddot{u}_{n,m} = (\Delta u)_{n,m} - f(u_{n,m}) \tag{6.7.1}$$

on the two dimensional integer lattice $(n, m) \in \mathbb{Z}^2$ under conditions that $f \in C^1(\mathbb{R}, \mathbb{R})$ is odd and 2π-periodic. Δ denotes the discrete Laplacian defined as

$$(\Delta u)_{n,m} = u_{n+1,m} + u_{n-1,m} + u_{n,m+1} + u_{n,m-1} - 4u_{n,m}.$$

For $f(u) = \omega\sin u$, we get the 2-dimensional discrete sine-Gordon lattice equation [186] (see Fig. 6.2)

$$\ddot{u}_{n,m} - (\Delta u)_{n,m} + \omega\sin u_{n,m} = 0. \tag{6.7.2}$$

A *traveling wave solution* of (6.7.1) in the direction $e^{i\theta}$ of the form

$$u_{n,m}(t) = U(n\cos\theta + m\sin\theta - \nu t)$$

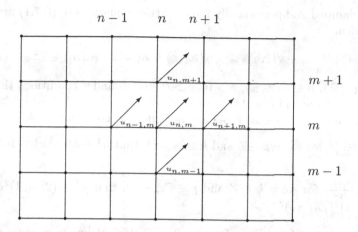

Figure 6.2: The two-dimensional lattice model of rigid rotation molecules with orientation $u_{n,m}$ at site (n, m)

for $U \in C^2(\mathbb{R}, \mathbb{R})$ satisfies the equation

$$\nu^2 U''(z) = U(z + \cos\theta) + U(z - \cos\theta)$$
$$+U(z + \sin\theta) + U(z - \sin\theta) - 4U(z) - f(U(z)) \qquad (6.7.3)$$

with $z = n\cos\theta + m\sin\theta - \nu t$. Topological and variational methods are used in [96] to show periodic traveling waves and *uniform sliding states* of (6.7.1), i.e. solutions of (6.7.3) satisfying either $U(z + T) = U(z)$ or $U(z + T) = U(z) + 2\pi$ for any $z \in \mathbb{R}$ and some T. For instance the following results are proved.

Theorem 6.7.1. *For any $\omega > 16$ and $1.17196 < T < 1.7579$, the 2d discrete sine-Gordon equation (6.7.2) possesses 4 nontrivial/nonconstant traveling wave solutions of the form*

$$u_{n,m}(t) = \pi + U\left(\frac{1}{\sqrt{2}}(n + m) - \frac{1}{2}t\right)$$

for $U(z)$ satisfying either $U(z + T) = U(z) + 2\pi$ with $U(-z) = -U(z)$, or $U(z+T) = -U(z)+2\pi$, or $U(z+T) = -U(z)$, or $U(z+T) = U(z)$ with $U(-z) = -U(z)$, respectively. Moreover, for $\omega = 1$ and $T \in \bigcup_{k=2}^{436}\left(\frac{\pi}{r_0}k, 2\sqrt{3}\pi(k+1)\right)$, (6.7.2) has at least 2 nonzero traveling wave solutions of the form

$$u_{n,m}(t) = U\left(\frac{1}{\sqrt{2}}(n + m) - 2t\right)$$

for $U(z)$ odd and T-periodic, where $r_0 \doteq 0.2880$.

We note that the first interval with $k = 2$ is approximately $(21.8154, 32.6484)$ and the last one with $k = 436$ is approximately $(4755.7599, 4755.7819)$.

In [87], damped and periodically forced lattice equations of (6.7.1) are studied of the form

$$\ddot{u}_{n,m} = -\delta \dot{u}_{n,m} + \chi(\Delta u)_{n,m} - f(u_{n,m}) + h(\mu t), \quad (n,m) \in \mathbb{Z}^2 \qquad (6.7.4)$$

for $f \in C^1(\mathbb{R}, \mathbb{R})$, $h \in C(\mathbb{R}, \mathbb{R})$, $\delta > 0$, $\chi > 0$, $\mu > 0$ under conditions that f is odd and 2π-periodic and $h \neq 0$ is π-antiperiodic, i.e. $h(x + \pi) = -h(x) \ \forall x \in \mathbb{R}$. It is shown among orders that if one of the following conditions holds

(a) $\nu = \mu \frac{2p+1}{2k}$ for some $p \in \mathbb{Z}$ and $k \in \mathbb{N}$ such that $\mu^4 + 4\delta^2 \mu^2 k^2 > 16k^4(L + 8\chi)^2$

(b) $\nu = \mu \frac{2k}{2p+1}$ for some $k \in \mathbb{Z}$ and $p \in \mathbb{Z}_+$ such that $\mu^4 + (2p + 1)^2 \delta^2 \mu^2 > (2p + 1)^4 (L + 8\chi)^2$

where $L := \max_{\mathbb{R}} |f'(x)|$. Then for any $\theta \in \mathbb{R}$, (6.7.4) has a *periodic moving wave* solution of the form

$$u_{n,m}(t) = U(n \cos \theta + m \sin \theta - \nu t, \mu t) \qquad (6.7.5)$$

for some U which is 2π-periodic in the both variables. Now the equation for moving waves is more complicated

$$\nu^2 U_{zz}(z,v) - 2\mu\nu U_{zv}(z,v) + \mu^2 U_{vv}(z,v) + \delta \left(\mu U_v(z,v) - \nu U_z(z,v) \right)$$
$$= \chi \Big(U(z + \cos \theta, v) + U(z - \cos \theta, v) \qquad (6.7.6)$$
$$+ U(z + \sin \theta, v) + U(z - \sin \theta, v) - 4U(z,v) \Big) - f(U(z,v)) + h(v)$$

with $z = n \cos \theta + m \sin \theta - \nu t$ and $v = \mu t$.

Of course, (6.7.1) can be also derived as a spatial discretization of the p.d.eqn

$$u_{tt} - \Delta u + f(u) = 0$$

of the form

$$\ddot{u}_{n,m} - \frac{1}{\varepsilon^2} (\Delta u)_{n,m} + f(u_{n,m}) = 0, \qquad (6.7.7)$$

where $u = u(x, y, t)$, $x, y, t \in \mathbb{R}$ and $\Delta u = u_{xx} + u_{yy}$ is the Laplacian. Then for

$$u_{n,m}(t) = U \left(n \cos \theta + m \sin \theta - \frac{\nu}{\varepsilon} t \right)$$

the traveling wave equation reads

$$\nu^2 U''(z) = U(z + \cos \theta) + U(z - \cos \theta)$$
$$+ U(z + \sin \theta) + U(z - \sin \theta) - 4U(z) - \varepsilon^2 f(U(z)). \qquad (6.7.8)$$

It would be interesting to carry out similar computations for (6.7.8) as it is done above for (6.1.5) considering ν and θ as parameters.

Chapter 7

Periodic Oscillations of Wave Equations

7.1 Periodics of Undamped Beam Equations

7.1.1 Undamped Forced Nonlinear Beam Equations

Let us consider a forced sine-beam p.d.eqn given by

$$u_{tt} + \alpha u_{xxxx} + \sin u + \tau(x) = \mu \sin t, \quad u(x+1,t) = u(x,t), \qquad (7.1.1)$$

where $\alpha > 0$ is a constant, τ is a nonzero continuous 1-periodic function satisfying $\int_0^1 \tau(x)\,dx = 0$, $\mu \in \mathbb{R}$ is a small parameter and α is assumed to be sufficiently large. Letting $\alpha \to \infty$, we also consider a *limit o.d.eqn* of (7.1.1) of the form

$$\ddot{u} + \sin u = \mu \sin t, \qquad (7.1.2)$$

which is a forced pendulum equation. It is well-know that $(-\pi, 0)$ and $(\pi, 0)$ are hyperbolic equilibria of the o.d.eqn

$$\dot{x} = y, \quad \dot{y} = -\sin x \qquad (7.1.3)$$

joined by the upper separatrix $(\gamma(t), \dot{\gamma}(t))$, $\gamma(t) = \pi - 4\arctan(\mathrm{e}^{-t})$. We consider (7.1.2) as an o.d.eqn on the circle $S^{2\pi} := \mathbb{R}/2\pi\mathbb{Z}$. Then (7.1.3) is defined on the cylinder $S^{2\pi} \times \mathbb{R} \ni (x, y)$ and $(-\pi, 0)$, $(\pi, 0)$ are glued to a hyperbolic equilibrium of (7.1.3) joined by the homoclinic orbit $(\gamma(t), \dot{\gamma}(t))$. Furthermore, the Melnikov function $M(\widetilde{\alpha})$ of (7.1.3) has now the form (see computations for (3.1.36))

$$M(\widetilde{\alpha}) = 2\int_{-\infty}^{\infty} \operatorname{sech} t \, \sin(t + \widetilde{\alpha})\,dt = 2\pi \operatorname{sech} \frac{\pi}{2} \sin\widetilde{\alpha}.$$

M. Fečkan, *Topological Degree Approach to Bifurcation Problems*, 199–226.
© Springer Science + Business Media B.V., 2008

Since $\tilde{\alpha} = 0$ is a simple root of $M(\tilde{\alpha})$, Theorem 4.2.2 (see also Remark 5.1.4 and Theorem 5.1.12) is applicable to (7.1.3) with $b = -a = $ small. Consequently, for any $\mu \neq 0$ sufficiently small, (7.1.2) exhibits a chaotic behavior. In particular, it has an infinite number of subharmonic solutions with periods tending to infinity. The purpose of this section is to show that most of them can be traced as $\alpha \to \infty$ for (7.1.1). Homoclinic and heteroclinic bifurcations for p.d.eqns are already studied in [20, 23, 115, 137–139, 170, 197]. But methods of these papers seem to be not applicable to (7.1.1), since it is undamped. We combine a method of bifurcation of periodic solutions from Chapter 3 along with an assumption of incommensurability for eigenvalues of the linear part of (7.1.1) to the time period 2π of (7.1.1) (cf. (7.1.39)). We also derive an estimate on a Lebesgue measure of a set of all parameters satisfying that assumption of incommensurability. Following this, we are able to show only the existence of any finite number of subharmonic solutions. All these results are derived for abstract wave equations on Hilbert spaces modeled by (7.1.1). Related problems are also studied in [30, 150, 198]. Furthermore, we apply our abstract results also to the following p.d.eqns

$$u_{tt} + \alpha u_{xxxx} + \sin u + \tau(x) = \mu \cos t, \quad u(x,t) = u(1-x,t)$$
$$u_{xx}(0,t) = u_{xx}(1,t) = u_{xxx}(0,t) = u_{xxx}(1,t) = 0, \tag{7.1.4}$$

and

$$u_{tt} + \alpha u_{xxxx} - u + u^3 + \tau(x) = \mu \cos t, \quad u(x+1,t) = u(x,t), \tag{7.1.5}$$

where $\alpha > 0$ is a large constant, τ is a nonzero continuous function satisfying $\int_0^1 \tau(x)\,dx = 0$ and $\mu \in \mathbb{R}$ is a small parameter. Now either τ satisfies $\tau(x) = \tau(1-x)$ for (7.1.4) or τ is 1-periodic and C^1-smooth for (7.1.5). The limit o.d.eqn of (7.1.5) is a forced Duffing equation

$$\ddot{u} - u + u^3 = \mu \cos t \tag{7.1.6}$$

which is chaotic (see (3.1.36)). We are motivated to study (7.1.1), (7.1.4) and (7.1.5) by the well-known sine-Gordon and Klein-Gordon partial differential equations [176] when the term αu_{xxxx} is replaced with $-\alpha u_{xx}$.

Finally, similar approach as in this section is used in [79] to the following p.d.eqn

$$u_{tt} + u_{xxxx} + \Gamma u_{xx} + p\left(\int_0^1 u^2(s,t)\,ds, \int_0^1 u_x^2(s,t)\,ds\right) D_{xx}^\xi u = \varepsilon q(x) \cos \frac{2\pi t}{T}$$
$$u(0,\cdot) = u(1,\cdot) = u_{xx}(0,\cdot) = u_{xx}(1,\cdot) = 0,$$

where $D_{xx}u = -u_{xx}$, D_{xx}^ξ is the ξ-power of D_{xx} in $L^2(0,1)$, $0 \leq \xi \leq 1$, $\Gamma \in \mathbb{R}$, $p \in C^2(\mathbb{R} \times \mathbb{R}, \mathbb{R})$, $p(0,0) = 0$, $q \in H^2(0,1) \cap H_0^1(0,1)$, $T > 0$ and $\varepsilon \in \mathbb{R}$ is a small parameter.

7.1.2 Existence Results on Periodics

In this subsection, we introduce abstract wave equations on Hilbert spaces modeled by (7.1.1). Let Y be a Hilbert space with an inner product $\langle \cdot, \cdot \rangle$ and the corresponding norm is denoted by $\| \cdot \|$. Let X be a Banach space $X \subset Y$ and X is dense in Y. Let us consider the equation

$$u_{tt} + \alpha A u = f(u, t), \tag{7.1.7}$$

where $\alpha > 0$ is a parameter and $A : X \to Y$ is a bounded linear operator. We assume

$$0 < d := \dim \mathcal{N} A < \infty.$$

Let $\{u_j\}_{j=-d+1}^{0}$ be an orthonormal basis of $\mathcal{N} A \subset Y$. Let $\{u_j\}_{j=1}^{\infty}$ be eigenvectors of A with corresponding nonzero eigenvalues $\{\lambda_j\}_{j=1}^{\infty}$ such that $\lambda_1 \leq \lambda_2 \leq \lambda_{i_0} < 0 < \lambda_{i_0+1} \leq \cdots \to \infty$. We set $\lambda_0 := 0$. We assume that $\{u_j\}_{j=-d+1}^{\infty}$ represents an orthonormal basis of Y, and the linear span of $\{u_j\}_{j=-d+1}^{\infty}$ is dense in X. Furthermore, we suppose that $f : Y \times S^T \to Y$ is continuous and globally Lipschitz in y with a constant M. Recall $S^T = \mathbb{R}/T\mathbb{Z}$ is the circle.

We are looking for weak T-periodic solutions of (7.1.7) for $\alpha > 0$ large. By a weak T-periodic solution of (7.1.7) we mean $u \in L^\infty(S^T, Y)$ satisfying

$$\int_0^T \langle u(t), v_{tt}(t) + \alpha A v(t) \rangle \, dt = \int_0^T \langle f(u(t), t), v(t) \rangle \, dt \quad \forall v \in C^2(S^T, X).$$

The integrability is considered in the sense of Bochner [102]. Note the Hilbert space $L^2(S^T, Y)$ has an orthonormal basis

$$\left\{ \frac{1}{\sqrt{T}} u_j, \sqrt{\frac{2}{T}} \sin m \frac{2\pi t}{T} u_j, \sqrt{\frac{2}{T}} \cos m \frac{2\pi t}{T} u_j \mid m \in \mathbb{N}, j \geq -d+1 \right\} \subset C^2(S^T, X). \tag{7.1.8}$$

We consider the norm $\||v\|| = \operatorname{ess\ sup}_{S^T} \|v(\cdot)\|$ on $L^\infty(S^T, Y)$. Next, we decompose Y as follows

$$Y = \mathcal{N} A \oplus \mathcal{N} A^\perp$$

and take the orthogonal projections $Q : Y \to \mathcal{N} A^\perp$, $P : Y \to \mathcal{N} A$. Then (7.1.7) has the form

$$\begin{aligned} w_{tt} + \alpha A w &= Q f(w + v, t), \quad w \in \mathcal{N} A^\perp \\ v_{tt} &= P f(w + v, t), \quad v \in \mathcal{N} A. \end{aligned} \tag{7.1.9}$$

Lemma 7.1.1. *Assume that there is a constant $c > 0$ such that*

$$\sqrt{\lambda_j} \left| \sin \frac{\sqrt{\alpha \lambda_j} T}{2} \right| \geq c \quad \forall j > i_0. \tag{7.1.10}$$

Then the equation

$$L_\alpha w = w_{tt} + \alpha A w = h, \quad h \in L^\infty(S^T, \mathcal{N} A^\perp) \tag{7.1.11}$$

has a unique weak T-periodic solution $w = L_\alpha^{-1} h \in L^\infty(S^T, \mathcal{N}A^\perp)$ satisfying

$$|||w||| \leq \beta(\alpha, T, c)|||h||| \tag{7.1.12}$$

with

$$\beta(\alpha, T, c) := \frac{1}{\sqrt{\alpha}} \sqrt{\frac{i_0}{\lambda_{i_0}^2 \alpha} + T^2 \left(\frac{2}{\lambda_{i_0+1}} + \frac{1}{2c^2}\right)}. \tag{7.1.13}$$

Proof. Since $h \in L^\infty(S^T, \mathcal{N}A^\perp)$, we have

$$h(t) = \sum_{j \in \mathbb{N}} h_j(t) u_j, \quad h_j \in L^\infty(S^T, \mathbb{R}), \quad \text{ess sup}_{S^T} \sum_{j \in \mathbb{N}} h_j^2(\cdot) \leq |||h|||^2.$$

From $w \in L^\infty(S^T, \mathcal{N}A^\perp)$, we have $w(t) = \sum_{j \in \mathbb{N}} w_j(t) u_j$. We have to solve

$$\int_0^T \langle w(t), v_{tt}(t) + \alpha A v(t)\rangle \, dt = \int_0^T \langle h(t), v(t)\rangle \, dt \quad \forall v \in C^2(S^T, X). \tag{7.1.14}$$

Hence (7.1.14) gives

$$\ddot{w}_j(t) + \alpha \lambda_j w_j(t) = h_j(t), \quad j \in \mathbb{N}. \tag{7.1.15}$$

Since (7.1.10) holds, for $j > i_0$, (7.1.15) has the solution

$$w_j(t) = \frac{1}{\sqrt{\alpha\lambda_j}} \int_0^t \sin\left(\sqrt{\alpha\lambda_j}(t-s)\right) h_j(s) \, ds$$

$$+ \frac{1}{2\sqrt{\alpha\lambda_j}} \frac{1}{\sin\frac{\sqrt{\alpha\lambda_j}T}{2}} \int_0^T \cos\left(\sqrt{\alpha\lambda_j}\left(s - \frac{T}{2} - t\right)\right) h_j(s) \, ds. \tag{7.1.16}$$

Formula (7.1.16) gives

$$\sum_{j > i_0} w_j^2(t) \leq 2 \sum_{j > i_0} \frac{1}{\alpha\lambda_j} \left(\int_0^t \sin\left(\sqrt{\alpha\lambda_j}(t-s)\right) h_j(s) \, ds\right)^2$$

$$+ \frac{1}{2\alpha} \sum_{j > i_0} \frac{1}{\lambda_j \sin^2 \frac{\sqrt{\alpha\lambda_j}T}{2}} \left(\int_0^T \cos\left(\sqrt{\alpha\lambda_j}\left(s - \frac{T}{2} - t\right)\right) h_j(s) \, ds\right)^2$$

$$\leq \frac{2}{\alpha\lambda_{i_0+1}} \sum_{j > i_0} \int_0^T \sin^2 \sqrt{\alpha\lambda_j}(t-s) \, ds \int_0^T h_j^2(s) \, ds \tag{7.1.17}$$

$$+ \frac{1}{2\alpha c^2} \sum_{j > i_0} \int_0^T \cos^2 \sqrt{\alpha\lambda_j}\left(s - \frac{T}{2} - t\right) \, ds \int_0^T h_j^2(s) \, ds$$

$$\leq \frac{T}{\alpha} \left(\frac{2}{\lambda_{i_0+1}} + \frac{1}{2c^2}\right) \int_0^T \sum_{j > i_0} h_j^2(s) \, ds \leq \frac{T^2}{\alpha} \left(\frac{2}{\lambda_{i_0+1}} + \frac{1}{2c^2}\right) |||h|||^2.$$

For $\lambda_j < 0$, let $|w_j(t_0)|$, $t_0 \in [0, T]$ be the maximum of $|w_j(t)|$ in (7.1.15). Then

$$0 \geq \ddot{w}_j(t_0)w_j(t_0) = -\lambda_j \alpha w_j(t_0)^2 + h_j(t_0)w_j(t_0).$$

Hence $-\lambda_j \alpha w_j(t_0)^2 \leq |h_j(t_0)w_j(t_0)|$ and then

$$\max_{t \in S^T} |w_j(t)| \leq \frac{1}{\alpha|\lambda_j|} |||h||| \leq \frac{1}{\alpha|\lambda_{i_0}|} |||h|||.$$

In this way, we have a function $w \in L^\infty(S^T, \mathcal{N}A^\perp)$ with $w(t) = \sum_{j \in \mathbb{N}} w_j(t)u_j$
satisfying (7.1.12) and (7.1.15). Finally, it is not difficult to see that w is a weak
solution of (7.1.11), i.e. w satisfies (7.1.14) (see more details at the end of the
proof of Lemma 7.2.1). $\qquad\square$

Taking $w \in L^\infty(S^T, \mathcal{N}A)$, we consider the first equation of (7.1.9) in the
form

$$w = L_\alpha^{-1} Q f(w + v, t).$$

Since

$$|||L_\alpha^{-1} Q(f(w_1 + v, t) - f(w_2 + v, t))||| \leq M\beta(\alpha, T, c)|||w_1 - w_2|||,$$

we see that if $M\beta(\alpha, T, c) < 1$ then the first equation of (7.1.9) has, by the
Banach fixed point theorem, a unique solution $w = w(v, \alpha, T, c)$. Summarizing
we obtain the following result [86].

Theorem 7.1.2. *Assume that (7.1.10) holds. If $M\beta(\alpha, T, c) < 1$ then the problem of existence of weak T-periodic solutions of (7.1.7) is reduced to the finite
dimensional equation*

$$\ddot{v} = Pf(w(v, \alpha, T, c) + v, t), \quad v \in L^\infty(S^T, \mathcal{N}A), \tag{7.1.18}$$

*where $w(\cdot, \alpha, T, c) : L^\infty(S^T, \mathcal{N}A) \to L^\infty(S^T, \mathcal{N}A^\perp)$ is Lipschitz continuous with
a constant $\frac{M\beta(\alpha,T,c)}{1-M\beta(\alpha,T,c)}$. Moreover,*

$$|||w(v, \alpha, T, c)||| \leq \frac{\beta(\alpha, T, c)}{1 - M\beta(\alpha, T, c)} (M|||v||| + |||f(0, \cdot)|||).$$

In particular, $|||w(v, \alpha, T, c)||| \to 0$ as $\beta(\alpha, T, c) \to 0$ uniformly for v bounded.

Since by (7.1.13) we see that $\beta(\alpha, T, c) \to 0$ whenever $\alpha \to \infty$, according to
Theorem 7.1.2, the *limit equation* of (7.1.7) as $\alpha \to \infty$ is the o.d.eqn

$$\ddot{v} = Pf(v, t), \quad v \in L^\infty(S^T, \mathcal{N}A). \tag{7.1.19}$$

Of course, any weak solution v of (7.1.19) satisfies $v \in W^{2,\infty}(S^T, \mathcal{N}A)$.

Definition 7.1.3. A result on the existence of a weak T-periodic solution for (7.1.19) is said to be (ε, D)-*stable* for some $\varepsilon > 0$, $D > 0$, if for any perturbation

$$\tilde{f} \in C\left(L^\infty(S^T, \mathcal{N}A), L^\infty(S^T, \mathcal{N}A)\right)$$

of $Pf(v, \cdot)$ with

$$|||\tilde{f}(v) - Pf(v, \cdot)||| \leq \varepsilon$$

on the set $\mathcal{S} := \{v \in L^\infty(S^T, \mathcal{N}A) \mid |||v||| \leq D\}$, there does exist a weak T-periodic solution $v \in \mathcal{S}$ of $\ddot{v} = \tilde{f}(v)$.

Theorem 7.1.4. *Assume that (7.1.10) holds. If $\alpha > 0$ is sufficiently large then a (ε, D)-stable result on the existence of a weak T-periodic solution for (7.1.19) implies also the existence of a weak T-periodic solution for (7.1.7).*

Proof. The result follows immediately from Theorem 7.1.2 and Definition 7.1.3. Indeed, now we have

$$\tilde{f}(v)(t) = Pf(w(v, \alpha, T, c)(t) + v(t), t).$$

Then Theorem 7.1.2 gives

$$|||\tilde{f}(v) - Pf(v(t), t)||| = |||Pf(w(v, \alpha, T, c) + v, \cdot) - Pf(v, \cdot)|||$$

$$\leq M|||w(v, \alpha, T, c)||| \leq \frac{M\beta(\alpha, T, c)}{1 - M\beta(\alpha, T, c)} \left(M|||v||| + |||f(0, \cdot)|||\right) \qquad (7.1.20)$$

$$\leq \frac{M\beta(\alpha, T, c)}{1 - M\beta(\alpha, T, c)} \left(MD + |||f(0, \cdot)|||\right)$$

for any $v \in \mathcal{S}$. Consequently, $|||\tilde{f}(v) - Pf(v(t), t)||| \leq \varepsilon$ for α large. The proof is finished. $\qquad\qquad\square$

Note (ε, D)-stable results for (7.1.19) can be derived by using Mawhin's coincidence degree theory [145].

Now we concentrate on the condition (7.1.10).

Lemma 7.1.5. *Suppose*

$$\sum_{k > i_0} \frac{1}{\sqrt{\lambda_k}} < \infty$$

and take $D \geq 0$. Let $S(c)$ be the set of all $\Omega > 0$ satisfying

$$\sqrt{\lambda_j} \left| \sin \frac{\sqrt{\lambda_j}\Omega}{2} \right| \geq c \quad \forall j > i_0, \qquad (7.1.21)$$

where $c > 0$ is a constant. Then the Lebesgue measure of the complement

$$(\mathbb{R} \setminus S(c)) \cap [D, D+1]$$

satisfies

$$m\left((\mathbb{R} \setminus S(c)) \cap [D, D+1]\right) \leq \sum_{k > i_0} \left(\frac{2c\pi}{\lambda_k} + \frac{c}{\sqrt{\lambda_k}} + \frac{2c^2\pi}{\sqrt{\lambda_k}\lambda_k}\right).$$

Proof. If $\Omega \in (\mathbb{R} \setminus S(c)) \cap [D, D+1]$ then $\sqrt{\lambda_k}\left|\sin\frac{\sqrt{\lambda_k}\Omega}{2}\right| < c$ for some $k > i_0$. Since $|\sin x| \geq \frac{2}{\pi}|x - n\pi|$ for any $|x - n\pi| \leq \pi/2$, $n \in \mathbb{Z}_+$, we obtain

$$\frac{2}{\pi}\left|\frac{\sqrt{\lambda_k}\Omega}{2} - n\pi\right| < c/\sqrt{\lambda_k}$$

for some $n \in \mathbb{Z}_+$. Hence

$$\left|\Omega - \frac{2\pi n}{\sqrt{\lambda_k}}\right| < \frac{\pi c}{\lambda_k}.$$

Consequently, we obtain

$$\frac{D}{2\pi} - \frac{c}{2\lambda_k} \leq \frac{\Omega}{2\pi} - \frac{c}{2\lambda_k} < \frac{n}{\sqrt{\lambda_k}} < \frac{\Omega}{2\pi} + \frac{c}{2\lambda_k} \leq \frac{D+1}{2\pi} + \frac{c}{2\lambda_k},$$

$$\frac{D}{2\pi}\sqrt{\lambda_k} - \frac{c}{2\sqrt{\lambda_k}} < n < \frac{D+1}{2\pi}\sqrt{\lambda_k} + \frac{c}{2\sqrt{\lambda_k}}.$$

We arrive at

$$m\left((\mathbb{R} \setminus S(c)) \cap [D, D+1]\right) \leq \sum_{k > i_0}\left(\frac{c}{\sqrt{\lambda_k}} + \frac{\sqrt{\lambda_k}}{2\pi} + 1\right)\frac{2\pi c}{\lambda_k}$$

$$= \sum_{k > i_0}\frac{2c\pi}{\lambda_k} + \sum_{k > i_0}\frac{c}{\sqrt{\lambda_k}} + \sum_{k > i_0}\frac{2c^2\pi}{\sqrt{\lambda_k}\lambda_k}.$$

The proof is finished. □

Summarizing we have the following result.

Theorem 7.1.6. *Assume that* $\sum_{k>i_0}\frac{1}{\sqrt{\lambda_k}} < \infty$. *Then for any* α *such that*

$$\alpha > \frac{2Mi_0}{|\lambda_{i_0}|} + M^2T^2\left(\frac{4}{\lambda_{i_0+1}} + \frac{1}{c^2}\right) \quad and \quad \sqrt{\alpha}T \in S(c), \tag{7.1.22}$$

assumptions of Theorem 7.1.2 are satisfied. If in addition

$$\alpha > \frac{2i_0}{|\lambda_{i_0}|K} + \frac{T^2}{K^2}\left(\frac{8}{\lambda_{i_0+1}} + \frac{2}{c^2}\right) \quad for \quad K = \frac{\varepsilon}{M(MD + |||f(0,\cdot)||| + \varepsilon)} \tag{7.1.23}$$

then assumptions of Theorem 7.1.4 are also satisfied.

Proof. (7.1.22) implies that (7.1.10) holds along with $\frac{M^2 i_0}{\alpha^2 \lambda_{i_0}^2} \leq \frac{M^2 i_0^2}{\alpha^2 \lambda_{i_0}^2} < 1/4$ and $\frac{M^2 T^2}{\alpha}\left(\frac{2}{\lambda_{i_0+1}} + \frac{1}{2c^2}\right) < 1/2$. Hence $M^2\beta(\alpha, T, c)^2 < \frac{1}{4} + \frac{1}{2} = \frac{3}{4}$ and $M\beta(\alpha, T, c) < \sqrt{3}/2 < 1$, so assumptions of Theorem 7.1.2 are verified.

Similarly, (7.1.23) gives $\frac{i_0}{\alpha^2 \lambda_{i_0}^2} < K^2/4$ and $\frac{T^2}{\alpha}\left(\frac{2}{\lambda_{i_0+1}} + \frac{1}{2c^2}\right) < K^2/4$. Hence $\beta(\alpha, T, c)^2 < K^2/2$, so $\beta(\alpha, T, c) < \sqrt{2}K/2 < K$. Consequently, we obtain $\frac{M\beta(\alpha,T,c)}{1-M\beta(\alpha,T,c)}(MD + |||f(0,\cdot)|||) < \varepsilon$. So by (7.1.20), assumptions of Theorem 7.1.4 are also verified. □

Note (7.1.23) implies the inequality from (7.1.22).

Remark 7.1.7. By Lemma 7.1.5, the smaller c the larger the set $S(c)$. Moreover, $S(c_1) \subset S(c_2)$ whenever $c_1 \geq c_2 > 0$ and (see Theorem 7.2.7)

$$m\left(\cup_{c>0} S(c) \cap [D, D+1]\right) = 1 \quad \forall D \in \mathbb{N}.$$

Finally, let us assume that $f(u,t)$ in (7.1.7) is a gradient operator, i.e. $f(u,t) = \operatorname{grad}_u F(u,t)$ for a $F \in C^1\left(Y \times S^T, \mathbb{R}\right)$.

Theorem 7.1.8. *Assume that* $\sum\limits_{\lambda_k > 0} \frac{1}{\sqrt{\lambda_k}} < \infty$ *and* $f(u,t)$ *in (7.1.7) is a gradient operator in u satisfying*

$$a\|u-v\|^2 \leq \langle f(u,t) - f(v,t), u - v \rangle \leq b\|u-v\|^2 \quad \forall u, \forall v \in Y, \forall t \in S^T$$

for constants $a, b \in \mathbb{R}$. *If* $[a,b] \cap \left\{ -\frac{4\pi^2 n^2}{T^2} \right\}_{n \in \mathbb{Z}_+} = \emptyset$ *then for any α such that*

$$\alpha > \max\left\{ \frac{T^2 a^2}{4c^2}, \frac{T^2 b^2}{4c^2}, \frac{|a|}{|\lambda_{i_0}|}, \frac{|b|}{|\lambda_{i_0}|} \right\} \quad \text{and} \quad \sqrt{\alpha} T \in S(c), \tag{7.1.24}$$

(7.1.7) has a unique weak T-periodic solution u in $L^2(S^T, Y)$.

Proof. Since $|x| \geq |\sin x| \; \forall x \in \mathbb{R}$, condition (7.1.10) gives $\left| \frac{\sqrt{\alpha \lambda_k} T}{2} - \pi n \right| \geq \frac{c}{\sqrt{\lambda_k}}$, $\forall k > i_0$, $\forall n \in \mathbb{Z}_+$ from which we derive

$$\left| \alpha \lambda_k - \frac{4\pi^2 n^2}{T^2} \right| = \left| \sqrt{\alpha \lambda_k} - \frac{2\pi n}{T} \right| \left| \sqrt{\alpha \lambda_k} + \frac{2\pi n}{T} \right| \geq \frac{2c\sqrt{\alpha}}{T}$$

for any $k > i_0$ and $n \in \mathbb{Z}_+$. While for $1 \leq k \leq i_0$, we have $\left| \alpha \lambda_k - \frac{4\pi^2 n^2}{T^2} \right| \geq \alpha |\lambda_{i_0}|$. So the first condition of (7.1.24) gives $\left| \alpha \lambda_k - \frac{4\pi^2 n^2}{T^2} \right| > \max\{|a|, |b|\}$ for any $k \geq 1$ and $n \in \mathbb{Z}_+$. This together with $[a,b] \cap \left\{ -\frac{4\pi^2 n^2}{T^2} \right\}_{n \in \mathbb{Z}_+} = \emptyset$ imply

$$[a,b] \cap \left\{ \alpha \lambda_k - \frac{4\pi^2 n^2}{T^2} \right\}_{n, k \in \mathbb{Z}_+} = \emptyset. \tag{7.1.25}$$

Next we take $H = L^2(S^T, Y)$, $Lu = u_{tt} + \alpha Au$ and $N(u)(t) = f(u(t), t)$. The scalar product on H is the usual one $(u, v) := \int\limits_0^T \langle u(t), v(t) \rangle \, dt$ with the corresponding norm $|\cdot|$. Then a weak T-periodic solution u of (7.1.7) in H is determined by $Lu = N(u)$, which is equivalent to $\widetilde{L} u = \widetilde{N}(u)$ with $\widetilde{L} := L - \varpi \mathbb{I}$ and $\widetilde{N} := N - \varpi \mathbb{I}$ for $\varpi = \frac{a+b}{2}$. Furthermore, $\widetilde{N}(u) = \operatorname{grad}_u \int\limits_0^T \left(F(u(t), t) - \varpi \frac{\|u(t)\|^2}{2} \right) dt$ along with

$$-\varsigma |u-v|^2 \leq (\widetilde{N}(u) - \widetilde{N}(v), u - v) \leq \varsigma |u-v|^2 \quad \forall u, \forall v \in H$$

for $\varsigma = \frac{b-a}{2}$, which by [30, Theorem 2] or [146] gives

$$|\widetilde{N}(u) - \widetilde{N}(v)| \leq \varsigma|u - v| \quad \forall u, \forall v \in H. \tag{7.1.26}$$

From (7.1.8) we see that the spectrum $\sigma(\widetilde{L})$ of $\widetilde{L}$ is

$$\sigma(\widetilde{L}) = \left\{ -\frac{4\pi^2 n^2}{T^2} + \alpha\lambda_k - \varpi \mid n, k \in \mathbb{Z}_+ \right\}.$$

So by (7.1.25) we have $[-\varsigma, \varsigma] \cap \sigma(\widetilde{L}) = \emptyset$. As $\mathbb{R} \setminus \sigma(\widetilde{L})$ is open, we have $[-\widetilde{\varsigma}, \widetilde{\varsigma}] \cap \sigma(\widetilde{L}) = \emptyset$ for a $\widetilde{\varsigma} > \varsigma$. This gives $\|\widetilde{L}^{-1}\| \leq \frac{1}{\widetilde{\varsigma}}$. By (7.1.26) we get that the mapping $\widetilde{L}^{-1}\widetilde{N}$ is a contraction with a constant $\frac{\varsigma}{\widetilde{\varsigma}} < 1$. The Banach fixed point theorem gives a unique solution of $u = \widetilde{L}^{-1}\widetilde{N}(u)$, and so a unique solution of $Lu = N(u)$ in H. $\qquad\square$

7.1.3 Subharmonics from Homoclinics

In this part, we consider a periodically forced abstract wave equation

$$u_{tt} + \alpha Au = g(u) + \mu\phi(t), \tag{7.1.27}$$

where $\mu \in \mathbb{R}$ is a small parameter, $\alpha > 0$ is large and A satisfies assumptions of Subsection 7.1.2. Furthermore, $g : Y \to Y$ is Lipschitz continuous with a constant M and $\phi : S^T \to Y$ is continuous. Concerning the limit o.d.eqn, we assume

(i) $\mathcal{N}A = \mathbb{R}w_0$ with $\|w_0\| = 1$

(ii) Function $\widetilde{g}(x) := \langle g(xw_0), w_0 \rangle$, $x \in \mathbb{R}$ is C^2-smooth

(iii) $\widetilde{g}(0) = 0$ and $\widetilde{g}'(0) > 0$

(iv) There is a nonzero $\gamma \in C^2(\mathbb{R}, \mathbb{R})$ such that $\lim\limits_{t \to \pm\infty} \gamma(t) = 0$ and $\ddot{\gamma} = \widetilde{g}(\gamma)$.

For this case, (7.1.18) and (7.1.19) with $v(t) = x(t)w_0$ have the forms

$$\ddot{x} = \langle g(w(xw_0, \alpha, T, c) + xw_0), w_0 \rangle + \mu\langle \phi(t), w_0 \rangle \tag{7.1.28}$$

and

$$\ddot{x} = \langle g(xw_0), w_0 \rangle + \mu\langle \phi(t), w_0 \rangle. \tag{7.1.29}$$

The assumptions (iii) and (iv) mean that the o.d.eqn

$$\dot{x} = y, \quad \dot{y} = \langle g(xw_0), w_0 \rangle \tag{7.1.30}$$

has a homoclinic orbit $(\gamma, \dot{\gamma})$ to a hyperbolic equilibrium $(0, 0)$. We shall apply a result of Section 3.1 to study bifurcations of periodic solutions of (7.1.28) near γ for $\alpha > 0$ large.

Theorem 7.1.9. *Assume that* $\sum_{k>i_0} \frac{1}{\sqrt{\lambda_k}} < \infty$ *and (i–iv) hold as well. Let $\rho > 5/2$.
If there are constants $a < b$ such that $\mathcal{M}(a)\mathcal{M}(b) < 0$, where*

$$\mathcal{M}(\sigma) = \int_{-\infty}^{\infty} \langle \phi(t+\sigma), w_0 \rangle \dot{\gamma}(t)\, dt\,.$$

Then there are constants $K_1 > 0$, $K_2 > 0$ such that for any $K_1 > |\mu| > 0$, $m \in \mathbb{N}$, $0 < \Gamma \le 1$ satisfying

$$\frac{1}{|\mu|^{1/2}} < m < \frac{|\mu|^{2-\rho}}{T}\,, \qquad \frac{mT}{\Gamma|\mu|^\rho} \in S(c)\,, \qquad (7.1.31)$$

(7.1.27) has a weak mT-periodic solution u_m with $\alpha = \Gamma^{-2}|\mu|^{-2\rho}$ satisfying

$$\max_{-mT/2 \le t \le mT/2} \left| \langle u_m(t), w_0 \rangle - \gamma(t - \delta_m) \right| \le K_2 |\mu| \qquad (7.1.32)$$

for some $\delta_m \in (a, b)$.

Proof. First we note $|||v||| = \|x\|_\infty$ for $v(t) = x(t)w_0$ and $\|x\|_\infty := \operatorname{ess\,sup}_{\mathbb{R}} |x(\cdot)|$. Next, Theorem 7.1.2 implies for $m \in \mathbb{N}$ and $x \in L^\infty\left(S^{mT}, \mathbb{R}\right)$ with $\|x\|_\infty \le \|\gamma\|_\infty + 1$ that

$$\begin{aligned}
&\|\langle g(w(xw_0, \alpha, mT, c) + xw_0), w_0 \rangle - \langle g(xw_0), w_0 \rangle\|_\infty \\
&\le |||g(w(xw_0, \alpha, mT, c) + xw_0) - g(xw_0)||| \le M|||w(xw_0, \alpha, mT, c)||| \\
&\le \frac{M\beta(\alpha, mT, c)}{1 - M\beta(\alpha, mT, c)} \left(M\|x\|_\infty + \|g(0)\| + |\mu||||\phi||| \right) \\
&\le 2M\beta(\alpha, mT, c) \left(M\|\gamma\|_\infty + \|g(0)\| + |\mu||||\phi||| + M \right)
\end{aligned}$$

when $2M\beta(\alpha, mT, c) < 1$, and this holds if (see (7.1.13) and the proof of Theorem 7.1.6)

$$\alpha > \frac{3Mi_0}{|\lambda_{i_0}|} + 4M^2 m^2 T^2 \left(\frac{4}{\lambda_{i_0+1}} + \frac{1}{c^2} \right)\,. \qquad (7.1.33)$$

Taking $\alpha = \Gamma^{-2}|\mu|^{-2\rho}$ for $0 \le \Gamma \le 1$ and $\rho > 5/2$ from (7.1.13), we derive

$$\begin{aligned}
\beta(\alpha, mT, c) &\le \Gamma|\mu|^\rho \sqrt{\frac{i_0 \Gamma^2 |\mu|^{2\rho}}{\lambda_{i_0}^2} + m^2 T^2 \left(\frac{2}{\lambda_{i_0+1}} + \frac{1}{2c^2} \right)} \\
&\le \frac{i_0 \Gamma^2 |\mu|^{2\rho}}{|\lambda_{i_0}|} + \Gamma|\mu|^\rho mT \sqrt{\frac{2}{\lambda_{i_0+1}} + \frac{1}{2c^2}} \qquad (7.1.34) \\
&\le \mu^2 \left(\frac{i_0}{|\lambda_{i_0}|} + \sqrt{\frac{2}{\lambda_{i_0+1}} + \frac{1}{2c^2}} \right)
\end{aligned}$$

provided $|\mu| \leq 1$ and $mT|\mu|^\rho \leq \mu^2$, which follows from (7.1.31) when K_1 is small. To verify (7.1.33), we calculate

$$\left(\frac{3Mi_0}{|\lambda_{i_0}|} + 4M^2 m^2 T^2 \left(\frac{4}{\lambda_{i_0+1}} + \frac{1}{c^2}\right)\right)\alpha^{-1} \leq$$

$$\frac{3Mi_0}{|\lambda_{i_0}|}K_1^{2\rho} + 4M^2\left(\frac{4}{\lambda_{i_0+1}} + \frac{1}{c^2}\right)K_1^4 < 1$$

provided $|\mu| \leq K_1$ and $mT|\mu|^\rho \leq \mu^2$, which follow from (7.1.31). So (7.1.33) is shown for K_1 small. Consequently, when T is replaced with mT, (7.1.28) has the form

$$\ddot{x} = \langle g(xw_0), w_0\rangle + \mu\langle\phi(t), w_0\rangle + O(|\mu|^2) \tag{7.1.35}$$

near γ whenever (7.1.31) is satisfies and K_1 is small. Now, Theorem 3.1.9 implies that if K_1 is sufficiently small, then (7.1.28) has a weak mT-periodic solution for $m > 1/\sqrt{|\mu|}$ satisfying (7.1.32). The proof is finished. □

7.1.4 Periodics from Periodics

Above we suppose that the limit o.d.eqn (7.1.30) has a homoclinic structure and we study bifurcation of periodics from a homoclinic one. Now we investigate bifurcation of periodics from a periodic solution of (7.1.30). Consequently, we assume that (i), (ii) hold and also the following one

(v) There is a nonconstant T-periodic function $\gamma \in C^2(\mathbb{R}, \mathbb{R})$ satisfying the o.d.eqn $\ddot{\gamma} = \tilde{g}(\gamma)$. Moreover, the variational equation $\ddot{z} = \tilde{g}'(\gamma)z$ has a $\dot{\gamma}$ as the only nonzero T-periodic solution up to a scalar multiple.

We shall apply a result of Section 3.3 to study bifurcations of periodic solutions of (7.1.28) near γ for $\alpha > 0$ large.

Theorem 7.1.10. *Assume that* $\sum\limits_{k>i_0} \frac{1}{\sqrt{\lambda_k}} < \infty$ *and (i), (ii), (v) hold as well. If there are constants $a < b$ such that $\mathcal{M}(a)\mathcal{M}(b) < 0$, where*

$$\mathcal{M}(\sigma) = \int\limits_0^T \langle\phi(t+\sigma), w_0\rangle\dot{\gamma}(t)\,dt\,.$$

Then there are constants $K_1 > 0$, $K_2 > 0$ such that for any $K_1 > |\mu| > 0$, $0 < \Gamma \leq 1$ satisfying

$$\frac{T}{\Gamma\mu^2} \in S(c)\,,$$

(7.1.27) has a weak T-periodic solution u with $\alpha = \Gamma^{-2}|\mu|^{-4}$ satisfying

$$\max_{t\in[0,T]} \left|\langle u(t), w_0\rangle - \gamma(t-\delta)\right| \leq K_2|\mu| \tag{7.1.36}$$

for some $\delta \in (a, b)$.

Proof. From the proof of Theorem 7.1.9 with $m = 1$ and for $x \in L^\infty(S^T, \mathbb{R})$ with $\|x\|_\infty \leq \|\gamma\|_\infty + 1$, we have

$$\|\langle g(w(xw_0, \alpha, T, c) + xw_0), w_0\rangle - \langle g(xw_0), w_0\rangle\|_\infty$$
$$\leq 2M\beta(\alpha, T, c)\left(M\|\gamma\|_\infty + \|g(0)\| + |\mu|\|\|\phi\|\| + M\right)$$

when

$$\alpha > \frac{3Mi_0}{|\lambda_{i_0}|} + 4M^2T^2\left(\frac{4}{\lambda_{i_0+1}} + \frac{1}{c^2}\right). \tag{7.1.37}$$

For $\alpha = \Gamma^{-2}|\mu|^{-4}$ from (7.1.34) we know

$$\beta(\alpha, T, c) \leq \mu^2\left(\frac{i_0}{|\lambda_{i_0}|} + T\sqrt{\frac{2}{\lambda_{i_0+1}} + \frac{1}{2c^2}}\right)$$

provided $|\mu| \leq 1$. Next, (7.1.37) is satisfied if

$$|\mu| < \left(\frac{3Mi_0}{|\lambda_{i_0}|} + 4M^2T^2\left(\frac{4}{\lambda_{i_0+1}} + \frac{1}{c^2}\right)\right)^{-1/4}.$$

Consequently, (7.1.28) has the form of (7.1.35) for x near γ and μ is small. Corollary 3.3.5 implies that if K_1 is sufficiently small, then (7.1.28) has a weak T-periodic solution satisfying (7.1.36). The proof is finished. $\qquad\square$

Remark 7.1.11. In Theorems 7.1.9 and 7.1.10 it is enough to suppose that $\forall D \geq 0, \exists M > 0, \forall u_1, \forall u_2 \in Y, \|u_1\| \leq D, \|u_2\| \leq D$ implies $\|g(u_1) - g(u_2)\| \leq M\|u_1 - u_2\|$.

7.1.5 Applications to Forced Nonlinear Beam Equations

In this subsection, we first apply Theorem 7.1.9 to (7.1.5) by putting

$$X = C^5(S^1, \mathbb{R}), \quad Y = W^{1,2}(S^1, \mathbb{R}), \quad Au = u_{xxxx},$$
$$g(u)(x) = u(x) - u(x)^3 - \tau(x), \quad \phi(t) = \cos t, \quad T = 2\pi. \tag{7.1.38}$$

Clearly $\mathcal{N}A = \{\text{constant functions}\} \simeq \mathbb{R}$ and the spectrum of A is $\sigma(A) = \{16\pi^4m^4 \mid m \in \mathbb{Z}_+\}$ with the corresponding eigenfunctions $\{\sin 2\pi mx, \cos 2\pi mx\}$.

Since $\sum_{m=1}^\infty \frac{1}{m^2} < \infty$, Lemma 7.1.5 is applicable and (7.1.10) has now the form

$$4\pi^2j^2\left|\sin\left(4\pi^3j^2\sqrt{\alpha}\right)\right| \geq c, \quad \forall j \in \mathbb{N}. \tag{7.1.39}$$

Note (7.1.39) holds if and only if $2\pi\sqrt{\alpha} \in S(c)$. Applying Lemma 7.1.5 and using formulas

$$\sum_{k\in N}\frac{1}{k^2} = \frac{\pi^2}{6}, \quad \sum_{k\in N}\frac{1}{k^4} = \frac{\pi^4}{90}, \quad \sum_{k\in N}\frac{1}{k^6} = \frac{\pi^6}{945},$$

we derive

$$m\left((\mathbb{R} \setminus S(c)) \cap [D, D+1]\right) \leq \sum_{k \in N} \left(\frac{c}{8\pi^3 k^4} + \frac{c}{4\pi^2 k^2} + \frac{c^2}{32\pi^5 k^6}\right)$$

$$= \frac{c\pi}{720} + \frac{c}{24} + \frac{c^2\pi}{30240}.$$

From $\frac{c\pi}{720} + \frac{c}{24} + \frac{c^2\pi}{30240} < 1$ and $c > 0$ we get

$$0 < c < \frac{3\left(\sqrt{7(6300 + 900\pi + 7\pi^2)} - 210 - 7\pi\right)}{\pi} \doteq 20.7529. \qquad (7.1.40)$$

So for any c satisfying (7.1.40), there is a continuum many $\alpha > 0$ satisfying (7.1.39) in any $\left[\frac{D^2}{4\pi^2}, \frac{(D+1)^2}{4\pi^2}\right]$, $D \geq 0$.

Let $|u| = \sqrt{\int_0^1 u(x)^2 \, dx}$ be the norm on $L^2(S^1, \mathbb{R})$. Then $\|u\| = \sqrt{|u|^2 + |u'|^2}$ is a norm on Y. For any $u \in Y$, $\exists x_0 \in S^1$ such that $|u(x_0)| \leq |u|$ and so $\forall x \in S^1$: $u(x)^2 \leq u(x_0)^2 + 2|u||u'|$ implying $|u| \leq \|u\|_\infty \leq 2\|u\|$. Next, for any $u_{1,2} \in Y$ we derive

$$\|u_1 u_2\|^2 \leq 16\|u_1\|^2 \|u_2\|^2 + 2\left(|u_1' u_2|^2 + |u_1 u_2'|^2\right)$$
$$\leq 16\|u_1\|^2 \|u_2\|^2 + 2\left(4|u_1'|^2 \|u_2\|^2 + 4\|u_1\|^2 |u_2'|^2\right) \leq 36\|u_1\|^2 \|u_2\|^2.$$

Hence $\|u_1 u_2\| \leq 6\|u_1\|\|u_2\|$, which implies

$$\|u_1^3 - u_2^3\| \leq 6\|u_1 - u_2\| \left(\|u_1^2\| + \|u_1 u_2\| + \|u_2^2\|\right)$$
$$\leq 36\|u_1 - u_2\| \left(\|u_1\|^2 + \|u_1\|\|u_2\| + \|u_2\|^2\right).$$

Summarizing, for any $u_{1,2} \in Y$ with $\|u_{1,2}\| \leq D$, we get

$$\|g(u_1) - g(u_2)\| \leq \|u_1 - u_2\| + \|u_1^3 - u_2^3\| \leq \left(1 + 108D^2\right) \|u_1 - u_2\|$$

and Remark 7.1.11 is verified. Moreover, with $w_0(x) = 1$ and $u \in L^\infty(\mathbb{R}, \mathbb{R})$ we derive

$$\langle g(u(t)w_0), w_0 \rangle = \int_0^1 \left(u(t) - u(t)^3 - \tau(x)\right) \, dx = u(t) - u(t)^3,$$

$$\langle \phi(t), w_0 \rangle = \int_0^1 \cos t \, dx = \cos t,$$

where $\langle z(x), u(x) \rangle := \int_0^1 z(x)u(x) \, dx + \int_0^1 z'(x)u'(x) \, dx$. Hence (7.1.29) is (7.1.6), which is the forced Duffing equation. Conditions (i–iv) of Subsection 7.1.3 are clearly satisfied for (7.1.6) with $\gamma(t) = \sqrt{2}\,\mathrm{sech}\,t$. The function $\mathcal{M}(\sigma)$ of Theorem 7.1.9 has now the form

$$\mathcal{M}(\sigma) = \int\limits_{-\infty}^{\infty} \langle \phi(t+\sigma), w_0 \rangle \dot\gamma(t)\, dt = \sqrt{2}\pi\,\mathrm{sech}\,\frac{\pi}{2}\sin\sigma \,.$$

Since $\sigma = 0$ is a simple root of $\mathcal{M}(\sigma)$, Theorem 7.1.9 is applicable to (7.1.5) with $a = -b$ small.

Next, we intend to apply Theorem 7.1.10 to (7.1.5). According to [108, p.198], (7.1.6) with $\mu = 0$ has a family of periodic solutions

$$u_k(t) = \frac{\sqrt{2}}{\sqrt{2-k^2}}\,\mathrm{dn}\left(\frac{t}{\sqrt{2-k^2}}, k\right),$$

where dn is the Jacobi elliptic function and k is the elliptic modulus. The period of these orbits is given by

$$T_k = 2K(k)\sqrt{2-k^2}\,,$$

where $K(k)$ is the complete elliptic integral of the first kind. T_k is monotonically increasing in k with $\lim\limits_{k\to 0} T_k = \sqrt{2}\pi$ and $\lim\limits_{k\to 1} T_k = \infty$. Consequently, there is a unique $k_0 \doteq 0.982635$, $0 < k_0 < 1$ such that $T_{k_0} = 2\pi$. Then assumptions (i), (ii) and (v) of Subsections 7.1.3 and 7.1.4 are satisfied for (7.1.6) with $\gamma(t) = u_{k_0}(t)$. Now we compute

$$\mathcal{M}(\sigma) = \int\limits_{0}^{T_{k_0}} \langle \phi(t+\sigma), w_0 \rangle \dot\gamma(t)\, dt$$

$$= \int\limits_{0}^{T_{k_0}} \dot u_{k_0}(t)\cos(t+\sigma)\, dt = -\sqrt{2}\pi\,\mathrm{sech}\,\frac{\pi K'(k_0)}{K(k_0)}\sin\sigma \,.$$

Since again $\sigma = 0$ is a simple root of $\mathcal{M}(\sigma)$, Theorem 7.1.10 is applicable to (7.1.5) with $a = -b$ small.

The above arguments can be repeated to (7.1.1) when now

$$X = C^4(S^1, \mathbb{R}), \quad Y = L^2(S^1, \mathbb{R}), \quad Au = u_{xxxx}\,,$$
$$g(u)(x) = -\sin u(x) - \tau(x), \quad \phi(t) = \sin t, \quad T = 2\pi \,.$$

Hence (7.1.29) is (7.1.2), which is the forced pendulum equation. From Subsection 7.1.1 we already know that conditions (i–iv) of Subsection 7.1.3 are satisfied, when (7.1.3) is considered on the cylinder $S^{2\pi} \times \mathbb{R}$. Furthermore, the function $\mathcal{M}(\sigma)$ of Theorem 7.1.9 has now the form

$$\mathcal{M}(\sigma) = \int\limits_{-\infty}^{\infty} \langle \phi(t+\sigma), w_0 \rangle \dot\gamma(t)\, dt = 2\pi\,\mathrm{sech}\,\frac{\pi}{2}\sin\sigma \,.$$

Since $\sigma = 0$ is a simple root of $\mathcal{M}(\sigma)$, Theorem 7.1.9 is applicable to (7.1.1) with $b = -a$ small. Similarly, following results of [108, pp. 201–204], Theorem 7.1.10 is applicable to (7.1.1) for subharmonics. Hence like in Chapter 3, there are bifurcations of periodics for (7.1.1) from a heteroclinic cycle of (7.1.2) created by $(\gamma(t), \dot{\gamma}(t))$ and $(\gamma(-t), -\dot{\gamma}(-t))$. But we also get *librational solutions* bifurcating from $(\gamma(t), \dot{\gamma}(t))$, i.e. a weak solution of (7.1.1) satisfying $u(x, t+2m\pi) = u(x,t) + 2\pi$ for some large $m \in \mathbb{N}$.

Finally, we investigate (7.1.4) taking

$$X = \Big\{ u \in C^4([0,1], \mathbb{R}) \mid u(x) = u(1-x), u_{xx}(0) = u_{xx}(1) = 0 ,$$
$$u_{xxx}(0) = u_{xxx}(1) = 0 \Big\},$$
$$Y = \Big\{ u \in L^2([0,1], \mathbb{R}) \mid u(x) = u(1-x) \Big\}, \quad Au = u_{xxxx},$$
$$g(u)(x) = -\sin u(x) - \tau(x), \quad \phi(t) = \cos t, \quad T = 2\pi .$$

First we note that the eigenvalue problem

$$u_{xxxx}(x) = \nu u(x)$$
$$u_{xx}(0) = u_{xx}(1) = u_{xxx}(0) = u_{xxx}(1) = 0 \tag{7.1.41}$$

is known [23] to possess a sequence of eigenvalues $\nu_k = \xi_k^4$, $k = -1, 0, 1, \cdots$ with $\xi_{-1} = \xi_0 = 0$ and

$$\cos \xi_k \cosh \xi_k = 1 . \tag{7.1.42}$$

The corresponding orthonormal system of eigenvectors reads

$$u_{-1}(x) = 1, \quad u_0(x) = \sqrt{3}(2x - 1)$$
$$u_k(x) = \frac{2}{W_k} \Big[\cosh(\xi_k x) + \cos(\xi_k x)$$
$$- \frac{\cosh \xi_k - \cos \xi_k}{\sinh \xi_k - \sin \xi_k} (\sinh(\xi_k x) + \sin(\xi_k x)) \Big]$$

where

$$W_k = \cosh(\xi_k) + \cos(\xi_k) - \frac{\cosh \xi_k - \cos \xi_k}{\sinh \xi_k - \sin \xi_k} (\sinh(\xi_k) + \sin(\xi_k)) . \tag{7.1.43}$$

Then we get $\cos \xi_k = \frac{1}{\cosh \xi_k}$. Numerically we find $\xi_1 \doteq 4.73004075$.

Moreover, $0 < \xi_1 < \xi_2 < \cdots$ and so $\cosh \xi_1 < \cosh \xi_2 < \cdots$. Since $\xi_k \sim \pi(2k+1)/2$ and $\cos(\pi(2k+1)/2) = 0$, we get

$$|\sin \theta_k| \cdot |\xi_k - \pi(2k+1)/2| = |\cos \xi_k - \cos(\pi(2k+1)/2)| = \frac{1}{\cosh \xi_k} \leq 2 e^{-\xi_k}$$

for a $\theta_k \in (\xi_k, \pi(2k+1)/2)$. But we have

$$1 \geq |\sin \xi_k| = \sqrt{1 - \cos^2 \xi_k} \geq \sqrt{1 - \cos^2 \xi_1} \doteq 0.999844212 ,$$

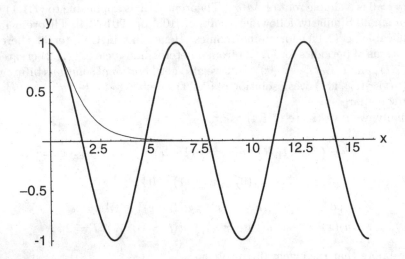

Figure 7.1: The graphs of functions $y = \cos x$ (thick line) and $y = \operatorname{sech} x$

since
$$0 < \cos \xi_k = \frac{1}{\cosh \xi_k} \leq \frac{1}{\cosh \xi_1} = \cos \xi_1 \,.$$

Next, we can easily see (cf. Fig. 7.1) that in fact $(4k - 1)\pi/2 < \xi_{2k-1}, \xi_{2k} < (4k + 1)\pi/2$ and function $\cos x$ is positive on intervals $(\xi_k, \pi(2k + 1)/2)$ for any $k \in \mathbb{N}$. So function $\sin x$ is increasing on these intervals, and it is positive on $[\xi_{2k}, (4k + 1)\pi/2]$ and negative on $[(4k - 1)\pi/2, \xi_{2k-1}]$. From these arguments we deduce
$$|\sin \theta_k| \geq |\sin \xi_k| \geq |\sin \xi_1| \doteq 0.9998444212 \,.$$

This gives
$$\left| \xi_k - \pi(2k + 1)/2 \right| \leq \frac{2}{|\sin \xi_1|} \, e^{-\xi_1} \doteq 0.017654973 \,.$$

So we obtain
$$\xi_k \geq \frac{\pi(2k + 1)}{2} - 0.017654973 \geq \pi k \,.$$

Consequently, we arrive at
$$\left| \xi_k - \pi(2k + 1)/2 \right| \leq \frac{2}{|\sin \xi_1|} \, e^{-\xi_k} \leq \frac{2}{|\sin \xi_1|} \, e^{-\pi k} \leq c \frac{\pi}{4} e^{-\pi k} \qquad (7.1.44)$$

for $c \doteq 2.546875863$. Furthermore, if $u(x)$ solves (7.1.41), then also $u(1 - x)$ is its solution. Moreover, $u_k(x)$, $k \in \mathbb{N}$ is an orthonormal system in Y. This gives $u_k(1-x) = \pm u_k(x)$ for any $k \in \mathbb{N}$. Next, we already know that $\sin \xi_{2k} > 0$. Hence $\sin \xi_{2k} = \sqrt{1 - \cos^2 \xi_{2k}}$. Using also $\cosh \xi_k = \frac{1}{\cos \xi_k}$ and $\sinh \xi_k = \sqrt{\cosh^2 \xi_k - 1}$ form (7.1.43) we derive $W_{2k} = -2$. Similarly, from $\sin \xi_{2k-1} < 0$, $k \in \mathbb{N}$ we derive $\sin \xi_{2k-1} = -\sqrt{1 - \cos^2 \xi_{2k-1}}$ and then $W_{2k-1} = 2$. Using $u_k(0) = \frac{4}{W_k}$ and $u_k(1) = 2$, we see
$$u_{2k}(1 - x) = -u_{2k}(x), \quad u_{2k-1}(1 - x) = u_{2k-1}(x) \quad \forall k \in \mathbb{N} \,.$$

Consequently, the eigenvalues of A for (7.1.4) read $\{\xi_{2k-1}^4 \mid k \in \mathbb{Z}_+\}$ with the corresponding eigenfunctions $\{u_{2k-1}(x) \mid k \in \mathbb{Z}_+\}$. So

$$\mathcal{N}A = \{\text{constant functions}\} \simeq \mathbb{R}$$

and (7.1.29) is again (7.1.2). From (7.1.44) we get $\sum_{k \in \mathbb{N}} \frac{1}{\xi_{2k-1}^2} < \infty$, and Lemma 7.1.5 is applicable. Consequently, the above results for (7.1.1) are extended to (7.1.4). Summarizing we get the following result.

Theorem 7.1.12. *Lemma 7.1.5 and Theorems 7.1.9, 7.1.10 are applicable to (7.1.1), (7.1.4) and (7.1.5), while the corresponding limit o.d.eqns are (7.1.2) and (7.1.6), respectively.*

Theorem 7.1.12 asserts that there are many periodic/subharmonic solutions of (7.1.1), (7.1.4) and (7.1.5) when nonresonant conditions of Lemma 7.1.5 are satisfied.

In [26] a similar problem is studied on the existence of weak $\frac{2T}{\sqrt{\varepsilon}}$–periodic solutions of equation

$$u_{tt} + u_{xxxx} + \varepsilon \mu h(x, \sqrt{\varepsilon}t) = 0 \,,$$
$$u_{xx}(0, \cdot) = u_{xx}(1, \cdot) = 0 \,,$$
$$u_{xxx}(0, \cdot) = -\varepsilon f \left(\int_0^1 u(x, \cdot)\varphi(x)dx \right) , \qquad (7.1.45)$$
$$u_{xxx}(1, \cdot) = \varepsilon g \left(\int_0^1 u(x, \cdot)\varphi(1 - x)dx \right)$$

for $\varepsilon > 0$ small. We assume that $h(x, t)$ is a $2T$–periodic (in t) C^1–function on $[0, 1] \times \mathbb{R}$, $f(x)$, $g(x)$ are sufficiently smooth functions such that $f(0) = g(0) = 0$ and $\varphi(x) = \varphi_a(x) \in L^2(\mathbb{R}, \mathbb{R})$, is a non-negative function whose support $\operatorname{supp} \varphi \subseteq [0, a]$, where a is a fixed positive number such that $0 < a < \frac{1}{3}$, and

$$\int_0^1 \varphi(x)dx = \int_{-\infty}^\infty \varphi(x)dx = 1 \,.$$

Physically conditions

$$u_{xxx}(0, \cdot) = -\varepsilon f \left(\int_0^1 u(x, \cdot)\varphi(x)dx \right), \quad u_{xxx}(1, \cdot) = \varepsilon g \left(\int_0^1 u(x, \cdot)\varphi(1 - x)dx \right)$$

mean that the response at the end points of the beam depends on a small part of the beam near the end points. In [26] under additional assumptions, our main result states that if $h(x, t) = h(x, -t)$, $\varepsilon > 0$ and μ are sufficiently small and the period $2T$ of $h(x, t)$ belongs to a certain non-zero measure subset of the interval $[2\widetilde{T}_0, 2\varepsilon^{-1/4}]$, with $\widetilde{T}_0$ sufficiently large, then (7.1.45) has a weak $\frac{2T}{\sqrt{\varepsilon}}$–periodic solution. When $h(x, t) = 0$, there are several layers of free symmetric weak periodic vibrations of (7.1.45) for any small $\varepsilon > 0$. This achievement is a p.d.eqn analogy of period blow-up results concerning accumulation of periodic solutions to homoclinic orbits in finite dimensional reversible systems studied in Section 5.2.

7.2   Weakly Nonlinear Wave Equations

7.2.1   Excluding Small Divisors

In this section we proceed with investigation on the existence of periodic solutions of undamped nonlinear wave equations. We study a weakly nonlinear equation

$$u_{tt} + Au = \varepsilon f(u,t), \tag{7.2.1}$$

where A, X, Y, f are already defined in the previous Subsection 7.1.2 and $\varepsilon \in \mathbb{R}$ is small. To get bifurcation results, we suppose

$$0 < \dim \mathcal{N}A < \infty.$$

By a *weak T-periodic solution* of (7.2.1) we mean a function $u \in L^2(S^T,Y)$ satisfying

$$\int_0^T \langle u(t), v_{tt}(t) + Av(t) \rangle \, dt = \varepsilon \int_0^T \langle f(u(t),t), v(t) \rangle \, dt, \quad \forall v \in C^2(S^T, X). \tag{7.2.2}$$

The integrability is considered in the sense of Bochner. Generally, problems of those kinds lead to *problems of small divisors*, and for this reason, it is very hard to study such problems [13,37,128] (see also Lemma 7.1.1). This is the case for one-dimensional wave equations when the ratio between the space length and the period T is irrational [30,150]. On the other hand, we can very easy study these problems for specific irrational numbers of the ratio. These irrational numbers can be nicely characterized in notions of the number theory. We present some results in this direction at the end of this section. To be more concrete about a small divisor problem, first we study a linear version of (7.2.1) given by

$$\int_0^T \langle u(t), v_{tt}(t) + Av(t) \rangle \, dt = \int_0^T \langle h(t), v(t) \rangle \, dt, \quad \forall v \in C^2(S^T, X) \tag{7.2.3}$$

for $h \in L^2(S^T, Y)$. Let $P : Y \to \mathcal{N}A$ be the orthogonal projection, and set

$$\widetilde{Q} := \mathbb{I} - \tilde{P}, \quad \tilde{P}h := \frac{1}{T} \int_0^T Ph(t) \, dt.$$

Note, these projections are defined for the following reason

$$\tilde{P}h = 0 \iff \int_0^T \langle h(t), u_p \rangle \, dt = 0, \quad \forall u_p \in \mathcal{N}A.$$

By (7.2.3) we derive

$$\int_0^T \langle h(t), u_p \rangle \, dt = 0, \quad \forall u_p \in \mathcal{N}A, \quad \text{i.e.} \quad h \in \mathcal{R}\widetilde{Q}.$$

Now we prove a non-resonance result for (7.2.3).

Lemma 7.2.1. *Assume the existence of a constant $c > 0$ such that*

$$\left| \alpha^2 - \frac{m^2}{\lambda_i} \right| \geq \frac{c}{\lambda_i} \quad \forall m \in \mathbb{N}, \quad \forall i > i_0, \tag{7.2.4}$$

where $\alpha = T/2\pi$. Then for any $h \in \mathcal{R}\widetilde{Q}$, (7.2.3) has a unique solution $u = Lh \in \mathcal{R}\widetilde{Q}$. Moreover,

$$\|u\|_{L^2(S^T, Y)} \leq \widetilde{c}\|h\|_{L^2(S^T, Y)} \tag{7.2.5}$$

for a constant $\widetilde{c} > 0$.

Proof. By our assumptions, the Hilbert space $\mathcal{R}\widetilde{Q} \subset Y$ has the orthogonal basis

$$\left\{ u_k, \sin m\frac{2\pi t}{T} \cdot u_j, \cos m\frac{2\pi t}{T} \cdot u_j \mid m, k \in \mathbb{N}, j \geq -d + 1 \right\} \subset C^2(S^T, X) \tag{7.2.6}$$

We expand u (formally) and h really in the basis (7.2.6) to get

$$u(t) = \sum_{(m, \lambda_j) \neq (0,0)} \left(u^1_{mj} \sin m\frac{2\pi t}{T} + u^2_{mj} \cos m\frac{2\pi t}{T} \right) u_j$$

$$h(t) = \sum_{(m, \lambda_j) \neq (0,0)} \left(h^1_{mj} \sin m\frac{2\pi t}{T} + h^2_{mj} \cos m\frac{2\pi t}{T} \right) u_j,$$

where we set $\lambda_j := 0$ for $-d + 1 \leq j < 0$. Of course, we take $u^1_{0j} = 0$ and $h^1_{0j} = 0$. If u is a solution of (7.2.3), then we take $v(t) = \sin m\frac{2\pi t}{T} \cdot u_j$ and $v(t) = \cos m\frac{2\pi t}{T} \cdot u_j$, for $(m, \lambda_j) \neq (0,0)$, to get

$$u^i_{mj} = \frac{\alpha^2}{\alpha^2 \lambda_j - m^2} h^i_{mj}, \quad i = 1, 2.$$

Hence if (7.2.3) has a solution $u \in \mathcal{R}\widetilde{Q}$, then it is unique and it should be given by

$$u(t) = \sum_{(m, \lambda_j) \neq (0,0)} \frac{\alpha^2}{\alpha^2 \lambda_j - m^2} \left(h^1_{mj} \sin m\frac{2\pi t}{T} + h^2_{mj} \cos m\frac{2\pi t}{T} \right) u_j. \tag{7.2.7}$$

By (7.2.4) it holds $|u^i_{mj}| \leq \widetilde{c}|h^i_{mj}|$, $i = 1, 2$ for a constant $\widetilde{c} > 0$. Hence

$$\|u\|^2_{L^2(S^T, Y)} = \sum_{\lambda_j \neq 0} T(u^2_{0,j})^2 + \sum_{m \neq 0, j} (T/2)\big((u^1_{mj})^2 + (u^2_{mj})^2\big)$$

$$\leq \widetilde{c}^2 \left(\sum_{\lambda_j \neq 0} T(h^2_{0j})^2 + \sum_{m \neq 0, j} (T/2)\big((h^1_{mj})^2 + (h^2_{mj})^2\big) \right) = \widetilde{c}^2 \|h\|^2_{L^2(S^T, Y)}.$$

This gives $u \in \mathcal{R}\widetilde{Q}$ given by (7.2.7), and we have (7.2.5). So L is continuous.

Now we show that this u satisfies (7.2.3). Our assumptions give that the linear hull L_H of (7.1.8) is dense in $C^2(S^T, X)$: one can prove this by using the $\triangle$-approximation method like in [178], see also Fejér's Theorem [171]. So for any $v \in C^2(S^T, X)$ there is a sequence $v_j \in L_H$ such that $v_j \to v$ in $C^2(S^T, X)$. This gives $v_{jtt} \to v_{tt}$ and $v_j \to v$ in $C(S^T, X)$. Hence $Av_j \to Av$ in $C(S^T, Y)$. The equality (7.2.3) holds for any $v_j \in L_H$, and since $X \subset Y$ continuously, we take the limit $j \to \infty$ in (7.2.3) for $v = v_j$ to get the validity of (7.2.3) for any v. The proof is finished. $\square$

Note (7.2.4) is satisfied when it holds

$$\left| \alpha - \frac{m}{\sqrt{\lambda_i}} \right| \geq \frac{c}{\lambda_i} \quad \forall m \in \mathbb{N}, \quad \forall i > i_0 \tag{7.2.8}$$

for a constant $c > 0$. Indeed, from (7.2.8) we derive

$$\left| \alpha^2 - \frac{m^2}{\lambda_i} \right| = \left| \alpha - \frac{m}{\sqrt{\lambda_i}} \right| \left| \alpha + \frac{m}{\sqrt{\lambda_i}} \right| \geq \alpha \frac{c}{\lambda_i}.$$

Considering a one-dimensional operator

$$Au = -u_{xx}, \quad u(0) = u(\pi) = 0, \quad u \in C^2\left([0, \pi]\right)$$

determined by the one-dimensional wave operator

$$u_{tt} - u_{xx}, \quad u(\cdot, 0) = u(\cdot, \pi) = 0, \tag{7.2.9}$$

we get $\lambda_i = i^2$, $i \in \mathbb{N}$. Then (7.2.4) has the form

$$\inf_{i, m \in \mathbb{N}} |i^2 \alpha^2 - m^2| > 0, \tag{7.2.10}$$

while (7.2.8) has the form

$$\left| \alpha - \frac{m}{i} \right| \geq \frac{c}{i^2} \quad \forall m, i \in \mathbb{N}. \tag{7.2.11}$$

Next, the real number α can be uniquely expressed in the form

$$\alpha = a_0 + \theta_1$$

with a_0 integer and $0 \leq \theta_1 < 1$. If $\theta_1 \neq 0$ then there is a unique $\alpha_1 > 1$ with

$$\alpha = a_0 + \frac{1}{\alpha_1}.$$

If α_1 is not an integer number then it has a unique representation

$$\alpha_1 = a_1 + \frac{1}{\alpha_2}$$

with a_1 integer and $\alpha_2 > 1$. This procedure terminates only if α is rational. For irrational α we get its *continued fraction expansion*

$$\alpha = a_0 + \cfrac{1}{a_1 + \cfrac{1}{a_2 + \cdots}} = [a_0, a_1, a_2, \cdots] .$$

The integers $a_0, a_1, \cdots$ are the *partial quotients* of α. We set

$$p_n/q_n := [a_0, a_1, a_2, \cdots, a_{n-1}, a_n]$$

with $(p_n, q_n) = 1$ and $q_n > 0$. Here as usually (p, q) is the largest common divisor of integer numbers p and q. The following interesting results are well-known [30, 51, 126, 150].

Proposition 7.2.2. *If α is irrational then:*

(i) *The integers p_n, q_n recursively satisfy relations*

$$p_0 = a_0, \quad q_0 = 1, \quad p_1 = a_0 a_1 + 1, \quad q_1 = a_1 ,$$
$$p_n = a_n p_{n-1} + p_{n-2}, \quad q_n = a_n q_{n-1} + q_{n-2}, \quad n \geq 2 .$$

(ii) *The rational number $\frac{p_n}{q_n}$ is the best rational approximation in the sense that there is no rational number $\frac{p}{q}$ with $0 < q < q_n$ and*

$$\left| \alpha - \frac{p}{q} \right| < \left| \alpha - \frac{p_n}{q_n} \right| .$$

(iii) *If*

$$\left| \alpha - \frac{p}{q} \right| < \frac{1}{2q^2}$$

for some $p \in \mathbb{Z}$ and $q \in \mathbb{N}$ then $\frac{p}{q} = \frac{p_n}{q_n}$ for some $n \in \mathbb{Z}_+$.

(iv) *For $n = 0, 1, 2, \cdots$ one has*

$$\frac{1}{q_n^2(a_{n+1} + 2)} \leq \left| \alpha - \frac{p_n}{q_n} \right| \leq \frac{1}{q_n^2 a_{n+1}} .$$

Definition 7.2.3. An irrational number α has a *bounded continued fraction expansion* $[a_0, a_1, \cdots]$ if $\max_{i \in \mathbb{N}} a_i < \infty$.

Proposition 7.2.4. *An irrational number α has a bounded continued fraction expansion if and only if α satisfies either (7.2.10) or (7.2.11) for some $c > 0$*

So (7.2.10) and (7.2.11) are equivalent. From (iii) and (iv) of Proposition 7.2.2 we get the following improvement of Propositions 7.2.4.

Proposition 7.2.5. *If $\alpha = [a_0, a_1, \cdots]$ with $a_k \leq M$ for some $M > 0$ and all $k \geq 1$, then it holds*

$$\left| \alpha - \frac{p}{q} \right| \geq \frac{1}{M+2} \frac{1}{q^2}$$

for any $p, q \in \mathbb{N}$.

Unfortunately, the author does not know generally such a nice criterion like in Proposition 7.2.4 for α in (7.2.8) with general eigenvalues λ_i. On the other hand when α has no a bounded continued fraction expansion, then

$$\inf_{i,m \in \mathbb{N}} |i^2 \alpha^2 - m^2| = 0 \,,$$

and we encounter to a small divisor problem in (7.2.7). This is mentioned at the beginning of this subsection.

7.2.2   Lebesgue Measures of Nonresonances

We start with the following well-known result [51, 126].

Proposition 7.2.6. *The Lebesgue measure of all positive irrational numbers with bounded continued fraction expansions is zero.*

So for almost all T there is a problem of small divisors for the one-dimensional wave operator (7.2.9) with T-periodic time conditions $u(x, t + T) = u(x, t)$. On the other hand, like in Lemma 7.1.5 we have the following result.

Theorem 7.2.7. *Assume*

$$\sum_{i > i_0} \frac{1}{\sqrt{\lambda_i}} < \infty \,.$$

Then the Lebesgue measure of the set of all positive α not satisfying (7.2.4) is zero.

Proof. If (7.2.4) is false for some $\alpha \in (K, K+1)$, $K > 0$, then for any $d > 0$ there exist $m \in \mathbb{N}$ and $i > i_0$ such that

$$\left| \alpha^2 - \frac{m^2}{\lambda_i} \right| \leq \frac{d}{\lambda_i} \,.$$

This implies

$$\left| \alpha - \frac{m}{\sqrt{\lambda_i}} \right| \leq \frac{d}{K \lambda_i} \,.$$

From $\alpha \in (K, K+1)$, we have $\frac{m^2}{\lambda_i} < (K+2)^2$ for any $0 < d < K\lambda_{i_0+1}$. Thus

$$m \leq (K+2)\sqrt{\lambda_i} \,.$$

Denote by $\mathcal{M}$ the set of all $\alpha \in (K, K+1)$ for which (7.2.4) does not hold. Then the Lebesgue measure $\mu(\mathcal{M})$ of $\mathcal{M}$ satisfies

$$\mu(\mathcal{M}) \leq \sum_{i > i_0} \frac{2d(K+2)}{\lambda_i K} \sqrt{\lambda_i} = \frac{2d(K+2)}{K} \sum_{i > i_0} \frac{1}{\sqrt{\lambda_i}}.$$

Since d can be arbitrarily small, it holds $\mu(\mathcal{M}) = 0$. The proof is finished. $\square$

Since $\sum_{i=1}^{\infty} \frac{1}{i} = \infty$, this theorem is not applicable (and it can not be by Proposition 7.2.6) for the one-dimensional wave operator (7.2.9). But taking a one-dimensional beam operator $u_{tt} + u_{xxxx}$ for which

$$Au = u_{xxxx}, \quad u(0) = u(\pi) = 0, \quad u_{xx}(0) = u_{xx}(\pi) = 0, \quad u \in C^4\left([0,\pi]\right),$$

then we have $\lambda_i = i^4, \forall i \in \mathbb{N}$, so $\sum_{i=1}^{\infty} \frac{1}{i^2} = \frac{\pi^2}{6}$ and Theorem 7.2.7 is applicable to the beam operator.

7.2.3 Forced Periodic Solutions

Now we are ready to study a weakly nonlinear problem (7.2.2) from which we see that any weak T-periodic solution of (7.2.1) satisfies

$$\int_0^T \langle f(u(t), t), u_p \rangle \, dt = 0, \quad \forall u_p \in \mathcal{N}A. \tag{7.2.12}$$

Then (7.2.1) has the form

$$\begin{aligned} w_{tt} + Aw &= \varepsilon \widetilde{Q} f(w + u_p, \cdot), \quad w \in \mathcal{R}\widetilde{Q} \\ 0 &= \tilde{P} f(w + u_p, \cdot), \quad u_p \in \mathcal{N}A. \end{aligned} \tag{7.2.13}$$

Note $u = w + u_p$ in (7.2.1). Using Lemma 7.2.1, the first equation of (7.2.13) has the form

$$w = F(w, u_p, \varepsilon) := \varepsilon L \widetilde{Q} f(w + u_p, \cdot). \tag{7.2.14}$$

Since $f \in C\left(Y \times S^T, Y\right)$ is globally Lipschitz in y with a constant M and $\widetilde{Q} : L^2\left(Y \times S^T, Y\right) \to L^2\left(Y \times S^T, Y\right)$ is orthogonal, from Lemma 7.2.1 we derive

$$\begin{aligned} &\|F(w_1, u_{p_1}, \varepsilon) - F(w_2, u_{p_2}, \varepsilon)\|_{L^2\left(S^T, Y\right)} \\ &\leq |\varepsilon| \|L\| \|\widetilde{Q}\| \| \left(f\left(w_1 + u_{p_1}, \cdot\right) - f\left(w_2 + u_{p_2}, \cdot\right)\right)\|_{L^2\left(S^T, Y\right)} \\ &\leq |\varepsilon| \tilde{c} M \left(\|w_1 - w_2\|_{L^2\left(S^T, Y\right)} + \|u_{p_1} - u_{p_2}\|_{L^2\left(S^T, Y\right)} \right) \\ &\leq |\varepsilon| \tilde{c} M \left(\|w_1 - w_2\|_{L^2\left(S^T, Y\right)} + \sqrt{T} \|u_{p_1} - u_{p_2}\| \right) \end{aligned} \tag{7.2.15}$$

for any $w_{1,2} \in \mathcal{R}\widetilde{Q}$ and $u_{p_{1,2}} \in \mathcal{N}A$. Moreover, we have

$$||F(0,0,\varepsilon)||_{L^2(S^T,Y)} = |\varepsilon| \, ||L\widetilde{Q}f(0,\cdot)||_{L^2(S^T,Y)} \leq |\varepsilon| |\widetilde{c}| \, ||f(0,\cdot)||_{L^2(S^T,Y)} \, .$$

$$(7.2.16)$$

Using the Banach fixed point theorem, for ε small, i.e. $|\varepsilon|\widetilde{c}M < 1$, we are able to solve (7.2.14) in $w := w(\varepsilon, u_p)$. From (7.2.15) and (7.2.16) we derive

$$||w(\varepsilon, u_{p_1}) - w(\varepsilon, u_{p_2})||_{L^2(S^T,Y)} \leq \frac{|\varepsilon| \widetilde{c} M \sqrt{T}}{1 - \varepsilon \widetilde{c} M} ||u_{p_1} - u_{p_2}||$$

$$||w(\varepsilon, u_p)||_{L^2(S^T,Y)} \leq \frac{|\varepsilon| \widetilde{c}}{1 - \varepsilon \widetilde{c} M} \left(M\sqrt{T} ||u_p|| + ||f(0,\cdot)||_{L^2(S^T,Y)} \right)$$

$$(7.2.17)$$

for any $u_p, u_{p_{1,2}} \in \mathcal{N}A$. By inserting this solution $w(\varepsilon, u_p)$ into the second equation of (7.2.13) we arrive at the *bifurcation equation*

$$B(\varepsilon, u_p) := \widetilde{P}f\big(w(\varepsilon, u_p) + u_p, \cdot\big) = 0 \, .$$

Note $B \in C\left(\left(-\frac{1}{\widetilde{c}M}, \frac{1}{\widetilde{c}M} \right) \times \mathcal{N}A, \mathcal{N}A \right)$, and

$$B(0, u_p) = \frac{1}{T} \int_0^T Pf(u_p, t) \, dt, \quad B(0, \cdot) : \mathcal{N}A \to \mathcal{N}A \, .$$

Summing up we obtain the next result [78].

Theorem 7.2.8. *Let (7.2.4) be satisfied. Assume the existence of an open and bounded subset $\Omega \subset \mathcal{N}A$ such that $0 \notin B(0, \partial\Omega)$ and $\deg(B(0, \cdot), \Omega, 0) \neq 0$. Then (7.2.1) has a weak T-periodic solution for any ε small.*

Proof. From (7.2.17) we derive

$$||B(\varepsilon, u_p) - B(0, u_p)|| = \left\| \frac{1}{T} \int_0^T P\Big(f\big(w(\varepsilon, u_p)(t) + u_p, t\big) - f(u_p, t) \Big) \, dt \right\|$$

$$\leq \frac{M}{T} \int_0^T ||w(\varepsilon, u_p)(t)|| \, dt \leq \frac{M}{\sqrt{T}} ||w(\varepsilon, u_p)||_{L^2(S^T,Y)}$$

$$\leq \frac{|\varepsilon| \widetilde{c} M}{(1 - \varepsilon \widetilde{c} M)\sqrt{T}} \left(M\sqrt{T} ||u_p|| + ||f(0,\cdot)||_{L^2(S^T,Y)} \right)$$

for any $u_p \in \mathcal{N}A$. Then from $0 \notin B(0, \partial\Omega)$ we get $0 \notin B(\varepsilon, \partial\Omega)$ for ε small. Consequently, we obtain

$$\deg(B(\varepsilon, \cdot), \Omega, 0) = \deg(B(0, \cdot), \Omega, 0) \neq 0$$

for ε small. Hence $B(\varepsilon, u_p) = 0$ has a solution in Ω. The proof is finished. $\square$

Example 7.2.9. Consider

$$u_{tt} - u_{xx} - n^2 u = \varepsilon f(u, t)$$
$$u(t + T, \cdot) = u(t, \cdot), \quad u(t, 0) = u(t, \pi) = 0 \quad \forall t \in \mathbb{R}, \tag{7.2.18}$$

where $n \in \mathbb{N}$, $f : \mathbb{R} \times S^T \to \mathbb{R}$ is continuous and globally Lipschitz in u. Now, we take

$$X = \{u \in W^{2,2}([0,\pi], \mathbb{R}) \mid u(0) = u(\pi) = 0\}, \quad Y = L^2([0,\pi], \mathbb{R})$$
$$Au = -u_{xx} - n^2 u, \quad \mathcal{N}A = \{\sin nx\}, \quad \lambda_i = i^2 - n^2,$$

and the condition (7.2.4) reads as follows

$$\left| \frac{i^2 - n^2}{m^2} - \frac{1}{\alpha^2} \right| \geq \frac{c}{\alpha^2 m^2}, \quad \forall m, \forall i \in \mathbb{N}, \quad i > n, \tag{7.2.19}$$

for a constant $c > 0$. Note (7.2.19) is equivalent to

$$\inf_{i, m \in \mathbb{N}, i > n} |i^2 - n^2 - \omega^2 m^2| > 0 \tag{7.2.20}$$

for $\omega = 1/\alpha$.

Theorem 7.2.10. *Equation (7.2.18) has a weak solution, provided (7.2.20) holds and there are $z_1, z_2 \in \mathbb{R}$ such that*

$$\int_0^T \int_0^\pi f(z_1 \sin nx, t) \sin nx \, dx \, dt \int_0^T \int_0^\pi f(z_2 \sin nx, t) \sin nx \, dx \, dt < 0.$$

Proof. We see that for this case

$$Pu = \frac{2}{\pi} \int_0^\pi u(x) \sin nx \, dx \cdot \sin nx, \quad u_p = z \sin nx$$

$$B(0, u_p) = \frac{2}{T\pi} \int_0^T \int_0^\pi f(z \sin nx, t) \sin nx \, dx \, dt \cdot \sin nx.$$

Applying Theorem 7.2.8, the proof is finished. $\qquad\square$

Further results on periodic solutions for abstract wave equation are presented in [40, 80].

7.2.4 Theory of Numbers and Nonresonances

This section is devoted to results concerning the condition (7.2.20). We already know from Proposition 7.2.4 that (7.2.20) holds for $n = 0$ if and only if ω has a bounded continued fraction expansion. Note that ω has a bounded continued fraction expansion if and only if $\alpha = 1/\omega$ has the same property. We intend to derive similar results for $n \in \mathbb{N}$. This situation is different from $n = 0$. First we study the case when ω is rational.

Theorem 7.2.11. *Let* $\omega = \frac{p}{q}$, $p, q \in \mathbb{N}$, $(p, q) = 1$. *Then* (7.2.20) *holds if and only if any* $n_2 \in \mathbb{N}$ *with* $n_2 \mid \frac{n}{(p,n)}$ *satisfies*

(i) *If* n_2 *is odd then* $p/(p, n)$ *does not divide* $(a^2 - b^2)/2$ *for any* $a > b$, $a, b \in \mathbb{N}$ *such that* $n_2 = ab$ *and* $(a, b) = 1$.

(ii) *If* n_2 *is even then* $p/(p, n)$ *does not divide* $a^2 - b^2$ *for any* $a > b$, $a, b \in \mathbb{N}$ *such that* $n_2 = 2ab$ *and* $(a, b) = 1$.

Here as usually $a \mid b$ *means that* a *is a divisor of* b.

Proof. The condition (7.2.20) does not hold if and only if there are $i, m \in \mathbb{N}$, $i > n$ such that

$$q^2 i^2 = p^2 m^2 + q^2 n^2. \tag{7.2.21}$$

Hence $q \mid pm$ implies $q \mid m$, i.e. $m = rq$, $r \in \mathbb{N}$ and (7.2.21) gives

$$i^2 = p^2 r^2 + n^2. \tag{7.2.22}$$

After dividing (7.2.22) by $(p, n)^2$, we get

$$i_1^2 = p_1^2 r^2 + n_1^2, \quad p_1 = p/(p, n), \quad n_1 = n/(p, n), \quad i_1 = i/(p, n).$$

Similarly we have

$$i_2^2 = p_1^2 r_1^2 + n_2^2, \quad r_1 = r/(r, n_1), \quad n_2 = n_1/(r, n_1), \quad i_2 = i_1/(r, n_1).$$

Note $(n_2, p_1 r_1) = 1$. We have two possibilities:

1. If n_2 is odd, then $p_1 r_1$ is even and we get

$$\frac{i_2 + n_2}{2} \frac{i_2 - n_2}{2} = \left(\frac{p_1 r_1}{2}\right)^2.$$

Since $(p_1 r_1, n_2) = 1$, we get $\left(\frac{i_2 + n_2}{2}, \frac{i_2 - n_2}{2}\right) = 1$ and so

$$i_2 + n_2 = 2A^2, \quad i_2 - n_2 = 2B^2, \quad p_1 r_1 = 2AB$$
$$i_2 = A^2 + B^2, \quad n_2 = A^2 - B^2 = (A - B)(A + B)$$

for some $A, B \in \mathbb{N}$ with $A > B$ and $(A, B) = 1$. So we derive

$$A = \frac{a + b}{2}, \quad B = \frac{a - b}{2}, \quad n_2 = ab, \quad p_1 r_1 = \frac{a^2 - b^2}{2}$$

for some $a, b \in \mathbb{N}$ with $a > b$ and $(a, b) = 1$. Hence $p_1 = p/(p, n) \mid (a^2 - b^2)/2$. This contradiction with (i) justifies (7.2.20).

2. If n_2 is even, then $p_1 r_1$ is odd and we get

$$\frac{i_2 + p_1 r_1}{2} \frac{i_2 - p_1 r_1}{2} = \left(\frac{n_2}{2}\right)^2.$$

Similarly like above we get

$$i_2 = A^2 + B^2, \quad p_1 r_1 = A^2 - B^2, \quad n_2 = 2AB$$

for some $A, B \in \mathbb{N}$ with $A > B$ and $(A, B) = 1$. This contradiction with (ii) justifies (7.2.20). The proof is finished. $\qquad\square$

Corollary 7.2.12. *Let* $\omega = p/q$, $p, q \in \mathbb{N}$, $(p, q) = 1$. *Then*
  (a) condition (7.2.20) holds if $n \mid p$
  (b) condition (7.2.20) holds for $n = 2$, *and for a prime number* $n > 2$ *if and only if either* $(n, p) > 1$, *or* $(n, p) = 1$ *and* p *does not divide* $(n^2 - 1)/2$

In the rest, we study (7.2.20) for more general ω.

Theorem 7.2.13. *If* $\omega = \sqrt{p/q}$ *is irrational for* $p, q \in \mathbb{N}$, $(p, q) = 1$, *then* (7.2.20) *does not hold for any* $n \in \mathbb{N}$.

Proof. Since $\sqrt{pq}$ is irrational, the Pelle equation $i^2 = pqm^2 + 1$ has a natural number solution i_0 and m_0. Then $i = i_0 n$, $m = m_0 qn$ satisfy $i^2 = \omega^2 m^2 + n^2$. The proof is finished. $\qquad\square$

On the other hand, we have the following positive result.

Theorem 7.2.14. *Condition* (7.2.20) *is satisfied for any irrational number* $\omega > 0$ *with* $\omega = [a_0, a_1, \cdots]$ *such that* $a_k \leq M$ *for some* $M > 0$ *and all* $k \geq 1$, *and* $\omega > (M + 2)n^2$.

Proof. From Proposition 7.2.5, we get

$$\left| \omega - \frac{p}{q} \right| \geq \frac{1}{M+2} \frac{1}{q^2}$$

for any $p, q \in \mathbb{N}$. Hence

$$\left| \frac{i^2}{m^2} - \omega^2 \right| = \left| \frac{i}{m} - \omega \right| \left| \frac{i}{m} + \omega \right| \geq \frac{\omega}{M+2} \frac{1}{m^2}$$

for any $i, m \in \mathbb{N}$. Consequently, for $i > n$, $m \in \mathbb{N}$, we obtain

$$|i^2 - \omega^2 m^2 - n^2| \geq |i^2 - \omega^2 m^2| - n^2 \geq \frac{\omega}{M+2} - n^2 > 0.$$

The proof is finished. $\qquad\square$

For instance, we have the following consequence.

Corollary 7.2.15. *If* $\omega = \frac{2M-1+\sqrt{5}}{2}$ *for some* $M \in \mathbb{N}$ *then* (7.2.20) *holds when*

$$2M - 1 + \sqrt{5} > 6n^2. \qquad (7.2.23)$$

Proof. From $\omega = \frac{2M-1+\sqrt{5}}{2}$ for some $M \in \mathbb{N}$ we get $\omega = [M, 1, 1, \cdots]$, and then (7.2.23) implies $\omega > (M + 2)n^2$. The proof is finished. $\qquad\square$

Furthermore, following [51, 58], we can get more characterizations of ω to satisfy (7.2.20).

Definition 7.2.16. The set of values taken by

$$\mu(\alpha)^{-1} = \liminf_{q \to +\infty, p, q \in \mathbb{Z}} |q(q\alpha - p)|$$

as α varies is called the *Lagrange spectrum*.

We know that if $\alpha = [a_0, a_1, a_2, \cdots]$ then

$$\mu(\alpha) = \limsup_{k \to +\infty} \left([a_{k+1}, a_{k+2}, \cdots] + [0, a_k, a_{k-1}, \cdots, a_1] \right).$$

Moreover, $\mu(\alpha) < +\infty$ if and only if α is irrational and all a_i are uniformly bounded. Next, $\mu(\alpha) \geq \sqrt{5}$ for any α and the Lagrange spectrum on the interval $[\sqrt{5}, 3)$ is the set $\{\sqrt{9 - 4m^{-2}}\}$, where m is a positive integer number such that

$$m^2 + m_1^2 + m_2^2 = 3mm_1m_2, \quad m_{1,2} \leq m$$

holds for some positive integers m_1 and m_2. Then $\widetilde{\omega}$ is a root of the Markoff form F_m such that $\mu(\widetilde{\omega}) = \sqrt{9 - 4m^{-2}}$. We also note that according to the Hall theorem, the Lagrange spectrum contains every number greater or equal to $\sqrt{21}$.

Definition 7.2.17. Two real numbers θ and θ' are *equivalent* if

$$\theta = \frac{r\theta' + s}{t\theta' + u}$$

for some integer numbers r, s, t, u satisfying $ru - ts = \pm 1$.

Theorem 7.2.18. *Let $\omega' > 0$ be irrational with a bounded continued fraction expansion, i.e $\mu(\omega') < \infty$. Then (7.2.20) holds for any $\omega > 0$ equivalent to ω' satisfying $\omega > \mu(\omega')n^2$.*

Proof. We know that $\mu(\omega) = \mu(\omega') < \infty$. Next, there are $M_0, M_1 \in \mathbb{N}$, $n < M_0$ such that if either $M_0 < i \in \mathbb{N}$ or $M_1 < m \in \mathbb{N}$ then

$$\left| i^2 - \omega^2 m^2 - n^2 \right| \geq |i - \omega m||i + \omega m| - n^2 \geq \frac{1}{2} \left[\frac{\omega}{\mu(\omega)} - n^2 \right] > 0. \quad (7.2.24)$$

On the other hand, if it could be $i_0^2 - \omega^2 m_0^2 - n^2 = 0$ for some $n < i_0 \leq M_0$, $m_0 \leq M_1$, $i_0, m_0 \in \mathbb{N}$, then $\sqrt{i_0^2 - n^2}$ is irrational and so there are $x_0, y_0 \in \mathbb{N}$ satisfying the Pelle equation $x_0^2 - \left(i_0^2 - n^2 \right) m_0^2 y_0^2 = 1$. Then the iteration

$$i_{k+1} = x_0 i_k + \left(i_0^2 - n^2 \right) y_0 m_k, \quad m_{k+1} = y_0 m_0^2 i_k + x_0 m_k, \quad k \in \mathbb{Z}_+$$

would also satisfy $i_k^2 - \omega^2 m_k^2 - n^2 = 0$ for all $k \in \mathbb{N}$. This would contradict to (7.2.24) for k large. Hence $i^2 - \omega^2 m^2 - n^2 \neq 0$ for any $n, m \in \mathbb{N}$, $n < i \leq M_0, m \leq M_1$. The proof is finished. $\qquad \square$

For instance, if we take $u = t = 1, r = s + 1, s \in \mathbb{Z}$, then $\omega = s + \frac{\omega'}{\omega' + 1}$ is equivalent to ω', and Theorem 7.2.18 is applicable for any $s \in \mathbb{N}$ such that

$$s > n^2 \mu(\omega') - \frac{\omega'}{\omega' + 1}.$$

This is related to (7.2.23).

Chapter 8

Topological Degree for Wave Equations

8.1 Discontinuous Undamped Wave Equations

In this chapter, we study bifurcation of weak 2π-periodic solutions with large amplitudes to the discontinuous semilinear wave equation

$$u_{tt} - u_{xx} - \eta u - g(u) - f(x,t,u) = h(x,t),$$
$$u(0,\cdot) = u(\pi,\cdot) = 0,$$

$$(8.1.1)$$

where $f : \Omega \times \mathbb{R} \to \mathbb{R}$, $\Omega = (0,\pi) \times (0,2\pi)$ is continuous and nondecreasing in u, $g : \mathbb{R} \to \mathbb{R}$ is bounded nondecreasing, $h \in L^2(\Omega)$ and $\eta > 0$ is a parameter. Moreover, we suppose

$$|f(x,t,u)| \leq c_0|u|^\alpha + h_0(x,t) \quad \forall u \in \mathbb{R}, \, \forall (x,t) \in \Omega \qquad (8.1.2)$$

for constants $c_0 > 0$, $1 > \alpha \geq 0$ and $h_0 \in L^2(\Omega)$. Continuous undamped wave equations are studied in [33,34], based on [35] where a construction of a topological degree is introduced for a class of monotone single-valued mappings. Related problems are earlier studied in [43, 147–149]. The purpose of this chapter is to extend that method to monotone multi-valued mappings modeled by (8.1.1). So our method constitutes a combination of a multivalued Browder-Skrypnik degree [46,179] with Mawhin's coincidence index [145]. We note that in Sections 7.1 and 7.2 we avoid many resonances for the linear operator of wave or beam equations assuming one of conditions (7.1.10) and (7.2.4). On the other hand, since the linear boundary value problem $u_{tt} - u_{xx} = 0$, $u(0,\cdot) = u(\pi,\cdot) = 0$ has an infinitely many 2π-periodic solutions $\sin nt \sin nx$, $\cos nt \sin nx$, $n \in \mathbb{N}$, the linear part of (8.1.1) has an infinitely dimensional kernel. So here we study a complementary case to Sections 7.1, 7.2 and moreover, we investigate discontinuous equations. Other topological degrees for multi-valued mappings have been introduced in [36,53,54,125]. The dual variational principle is applied to solve elliptic problems with discontinuous nonlinearities in [3].

M. Fečkan, *Topological Degree Approach to Bifurcation Problems*, 227–241.
© Springer Science + Business Media B.V., 2008

8.2   Standard Classes of Multi-Mappings

Now we recall some known definitions for multi-valued mappings [125] defined on a real separable Hilbert space H with an inner product $(\cdot, \cdot)$ and with the corresponding norm $|\cdot|$. A multi-valued mapping $F : H \to 2^H \setminus \{\emptyset\}$ is

- **Monotone** (denote $F \in (mMON)$), if

$$(f_u^* - f_v^*, u - v) \geq 0$$

  for all $u, v \in H$ and all selections $f_u^* \in F(u)$, $f_v^* \in F(v)$

- **Quasimonotone** ($F \in (mQM)$), if for any sequence $\{u_n\}_{n\in\mathbb{N}}$ in H with $u_n \rightharpoonup u$ and for all selections $f_n^* \in F(u_n)$ we have

$$\liminf_{n\to\infty}(f_n^*, u_n - u) \geq 0$$

- **Of class** (mS_+) ($F \in (mS_+)$), if for any sequence $\{u_n\}_{n\in\mathbb{N}}$ in H with $u_n \rightharpoonup u$, the existence of selections $f_n^* \in F(u_n)$ with $\limsup_{n\to\infty}(f_n^*, u_n - u) \leq 0$ implies $u_n \to u$

- **Compact** ($F \in (mCOMP)$), if for any bounded sequence $\{u_n\}_{n\in\mathbb{N}}$ in H and for any $f_n^* \in F(u_n)$ the sequence $\{f_n^*\}_{n\in\mathbb{N}}$ has a convergent subsequence

- **Bounded**, if for any bounded set $B \subset H$ the set $\bigcup_{u\in B} F(u)$ is bounded

- **Convex-valued**, if $F(u)$ is a non-empty convex set in H for any $u \in H$

- **Weakly upper semicontinuous** (F is w-usc), if for any sequence $\{u_n\}_{n\in\mathbb{N}} \in H$, $u_n \to u \in H$, the existence of selections $f_n^* \in F(u_n)$ with $f_n^* \rightharpoonup f^* \in H$ implies $f^* \in F(u)$

Remark 8.2.1. We always assume that all mappings used are bounded, w-usc and convex-valued. When a mapping is defined only on a subset of H, the above definitions can be modified in an obvious way.

Proposition 8.2.2. *The following inclusion hold for the above defined classes:*

$$(mS_+) \cup (mMON) \cup (mCOMP) \subset (mQM).$$

Proposition 8.2.3. *If $F \in (mQM)$ and $G \in (mS_+)$ then $F + G \in (mS_+)$.*

Since proofs of Propositions 8.2.2, 8.2.3 are simple, we omit them.

8.3 *M*-Regular Multi-Functions

We show concrete mappings motivated by (8.1.1) which fit into the framework of Section 8.2.

Definition 8.3.1. A function $p : \Omega \times \mathbb{R} \to \mathbb{R}$ is called

(a) *superpositionally measurable* if $p(x, t, u(x, t))$ is measurable for any Lebesgue measurable function $u : \Omega \to \mathbb{R}$

(b) *lower semicontinuous* (for short - lsc) in u if $\forall (x, t, c) \in \Omega \times \mathbb{R}$ the set $\{u \in \mathbb{R} \mid p(x, t, u) > c\}$ is open

(c) *upper semicontinuous* (for short - usc) in u if $\forall (x, t, c) \in \Omega \times \mathbb{R}$ the set $\{u \in \mathbb{R} \mid p(x, t, u) < c\}$ is open [52, 171]

Definition 8.3.2. A multi-function $S : \Omega \times \mathbb{R} \to 2^{\mathbb{R}}$ is called *measurable-bounded* if there exist two superpositionally measurable functions $q_-(x, t, u)$ and $q_+(x, t, u)$ such that

$$q_-(x, t, u) \le q_+(x, t, u) \quad \text{and} \quad S(x, t, u) = [q_-(x, t, u), q_+(x, t, u)]$$

for any $(x, t, u) \in \Omega \times \mathbb{R}$ where the function $q_-(x, t, u)$ is lsc in u, the function $q_+(x, t, u)$ is usc in u and there exist positive constants d_1, d_2 and positive $c_1, c_2 \in L^2(\Omega)$ such that

$$|q_-(x, t, u)| \le c_1(x, t) + d_1|u| \quad \text{and} \quad |q_+(x, t, u)| \le c_2(x, t) + d_2|u|$$

for any $(x, t, u) \in \Omega \times \mathbb{R}$. We denote by (mMB) the set of all measurable-bounded multi-functions.

By using a multi-function $S \in (mMB)$, for any $u \in L^2(\Omega)$ we put

$$N(u) := \left\{ v \in L^2(\Omega) \mid v(x, t) \in S(x, t, u(x, t)) \right\}$$
$$= \left\{ v \in L^2(\Omega) \mid q_-(x, t, u(x, t)) \le v(x, t) \le q_+(x, t, u(x, t)) \right\}$$

and call it *an M-regular multi-function*. We denote the set of all such multi-functions by (mMr). Note $q_\pm(x, t, u(x, t)) \in N(u)$, so $N(u)$ is nonempty. Now we are ready to show the following result.

Lemma 8.3.3. *If* $N \in (mMr)$ *then* $N : L^2(\Omega) \to 2^{L^2(\Omega)}$ *is w-usc.*

Proof. Let $N \in (mMr)$ and $u_n \to u$ in $L^2(\Omega)$. We have to show that if a sequence $\{w_n^*\}_{n \in \mathbb{N}}$ satisfies $w_n^* \in N(u_n)$ and $w_n^* \rightharpoonup w^*$ in $L^2(\Omega)$ then $w^* \in N(u)$. First, since $u_n \to u$ in $L^2(\Omega)$, we can assume by passing to a subsequence that $u_n(x, t) \to u(x, t)$ almost everywhere in (x, t) [171]. Next, we have

$$q_-(x, t, u_n(x, t)) \le w_n^*(x, t) \tag{8.3.1}$$

for every $n \in \mathbb{N}$. Since $w_n^* \rightharpoonup w^*$, using the Mazur Theorem 2.1.2 we can choose a sequence $\{v_n\}_{n \in \mathbb{N}}$, $v_n \in \mathrm{con}\left[\{w_n^*, w_{n+1}^*, \dots\}\right]$ such that $v_n \to w^*$ in $L^2(\Omega)$. Thus we have

$$v_n = \sum_{k=n}^{m_n} \lambda_{n,k} w_k^*; \quad 0 \le \lambda_{n,k} \le 1; \quad \sum_{k=n}^{m_n} \lambda_{n,k} = 1$$

$$\text{for } n \le m_n \in \mathbb{N}, \, n \le k \le m_n.$$

Since $v_n \to w^*$ in $L^2(\Omega)$, we can again assume that $v_n(x,t) \to w^*(x,t)$ almost everywhere in (x,t). From (8.3.1) we have

$$\sum_{k=n}^{m_n} \lambda_{n,k} q_-(x,t,u_k(x,t)) \le \sum_{k=n}^{m_n} \lambda_{n,k} w_k^*(x,t) = v_n(x,t).$$

Let $(x_0, t_0) \in \Omega$ be such an element that $u_n(x_0, t_0) \to u(x_0, t_0)$ and $v_n(x_0, t_0) \to w^*(x_0, t_0)$. The mapping $s \to q_-(x_0, t_0, s)$ is lsc and so for every $\varepsilon > 0$ there exists a positive integer n_0 such that for every $k \ge n_0$ we have

$$q_-(x_0, t_0, u(x_0, t_0)) - \varepsilon \le q_-(x_0, t_0, u_k(x_0, t_0)).$$

Summing this inequality for $k = n, n+1, \dots, m_n$ with weights $\lambda_{n,k}$ we get

$$q_-(x_0, t_0, u(x_0, t_0)) - \varepsilon \le \sum_{k=n}^{m_n} \lambda_{n,k} q_-(x_0, t_0, u_k(x_0, t_0)) \le v_n(x_0, t_0)$$

for all $n \ge n_0$. Hence by the convergence $v_n(x_0, t_0) \to w^*(x_0, t_0)$ we have

$$q_-(x_0, t_0, u(x_0, t_0)) - \varepsilon \le w^*(x_0, t_0) \quad \text{for every } \varepsilon > 0.$$

Finally, we get

$$q_-(x_0, t_0, u(x_0, t_0)) \le w^*(x_0, t_0)$$

as we need. Similar argument leads to

$$w^*(x_0, t_0) \le q_+(x_0, t_0, u(x_0, t_0)).$$

Thus $w^* \in N(u)$. $\square$

8.4   Classes of Admissible Mappings

We introduce certain multi-mappings to solve our example (8.1.1) in latter subsections. Let G be a bounded open subset in H, M a closed subspace of H and let Q and P be the orthogonal projections to M and $M^\perp$, respectively. Let $C \in L(\mathcal{R}Q, H)$ be compact. The family

$$\mathcal{F}_G^C := \left\{ F : \overline{G} \to 2^H \mid F = Q - (QCQ - P)f \text{ for some } f \in (mS_+) \right\}$$

is called *the class of admissible mappings*.

Let $L : H \subset D(L) \to H$ be a closed densely defined linear operator with $\mathcal{R}L = (\mathcal{N}L)^{\perp}$. Let $L_0 := L/\mathcal{R}L$ and assume that the right inverse $L_0^{-1} : \mathcal{R}L \to \mathcal{R}L$ is compact. We choose $M = \mathcal{R}L$ and $M^{\perp} = \mathcal{N}L$. Let $N : H \to 2^H$. Then similarly to [35], we consider the mapping

$$F = Q - (QL_0^{-1}Q - P)N.$$

Clearly, $F \in \mathcal{F}_G^C$ for $N \in (mS_+)$ with $C = L_0^{-1}$.

Lemma 8.4.1. *Let F and N be defined as above. Then*

$$0 \in Lu - N(u) \quad with \quad u \in D(L) \cap \overline{G} \tag{8.4.1}$$

if and only if

$$0 \in F(u) \quad with \quad u \in \overline{G}.$$

Proof. $(8.4.1) \Leftrightarrow \exists u \in D(L) \cap \overline{G}, \exists f^* \in N(u): Lu = f^* \Leftrightarrow Qu = QL_0^{-1}Qf^*$, $Pf^* = 0$, $u \in \overline{G} \Leftrightarrow 0 = Qu - (QL_0^{-1}Q - P)f^*$, $u \in \overline{G} \Leftrightarrow 0 \in F(u)$, $u \in \overline{G}$. $\square$

8.5 Semilinear Wave Equations

We show how the previous results can be applied to the semilinear wave equation (8.1.1). We state the precise setting of (8.1.1) by putting

$$q_-(x,t,u) := g_-(u) + f(x,t,u), \quad q_+(x,t,u) := g_+(u) + f(x,t,u),$$
$$g_+(u) := \lim_{s \to u_+} g(u), \quad g_-(u) := \lim_{s \to u_-} g(u).$$

We note that $g_{\pm}$ are Borel measurable, g_+ is usc and g_- is lsc. By Lemma 8.3.3, the Nemytskij operator $N : H \to 2^H$, $H = L^2(\Omega)$ defined by

$$N(u) := \{v \in L^2(\Omega) \mid q_-(x,t,u(x,t)) \leq v(x,t) \leq q_+(x,t,u(x,t))\}$$

is bounded and w-usc. Clearly $N \in (mMON)$ and hence by Propositions 8.2.2 and 8.2.3, $\eta \mathbb{I} + N \in (mS_+)$ for any $\eta > 0$.

Let C^2 be the set of twice continuously differentiable functions $v : [0,\pi] \times \mathbb{R} \to \mathbb{R}$ satisfying $v(0, \cdot) = v(\pi, \cdot) = 0$ and 2π-periodic in $t \in \mathbb{R}$.

A weak 2π-periodic solution of (8.1.1) for $h \in H$ is any $u \in H$ satisfying

$$(u, v_{tt} - v_{xx}) - \eta(u, v) - (u^*, v) = (h, v) \tag{8.5.1}$$

for some $u^* \in N(u)$ and for all $v \in C^2$. Here $(\cdot, \cdot)$ is the usual integral scalar product on $L^2(\Omega)$, i.e. $(u, v) = \int_\Omega u(x,t)\overline{v(x,t)}\, dx\, dt$. Let $\varphi_{m,n}(x,t) = \pi^{-1} e^{imt} \sin nx$ for all $m \in \mathbb{Z}, n \in \mathbb{N}$. Each $u \in L^2(\Omega)$ has a representation

$$u = \sum_{m \in \mathbb{Z}, n \in \mathbb{N}} u_{m,n} \varphi_{m,n},$$

where $u_{m,n} = (u, \varphi_{m,n})$ and $\overline{u}_{m,n} = u_{-m,n}$, since u is a real function. The abstract realization of the wave operator $\frac{\partial^2}{\partial t^2} - \frac{\partial^2}{\partial x^2}$ in $L^2(\Omega)$ is the linear operator $L : D(L) \to L^2(\Omega)$ defined by

$$Lu = \sum_{m \in \mathbb{Z}, n \in \mathbb{N}} (n^2 - m^2) u_{m,n} \varphi_{m,n} \,,$$

where

$$D(L) = \left\{ u \in L^2(\Omega) \mid \sum_{m \in \mathbb{Z}, n \in \mathbb{N}} (n^2 - m^2)^2 |u_{m,n}|^2 < \infty \right\}.$$

It is easy to show that $u \in L^2(\Omega)$ is a weak solution of (8.5.1) if and only if

$$h \in Lu - \eta u - N(u)$$

with $u \in D(L)$. Moreover, L is densely defined, self-adjoint, closed, $\mathcal{R}L = (\mathcal{N}L)^\perp$ and L has a pure point spectrum of eigenvalues

$$\sigma(L) = \left\{ n^2 - m^2 \mid m \in \mathbb{Z}, n \in \mathbb{N} \right\}$$

with the corresponding eigenfunctions $\varphi_{m,n}$. Clearly $\sigma(L)$ is unbounded both from above and from below, any eigenvalue $\lambda \neq 0$ has a finite multiplicity, but $\mathcal{N}L$ is infinite dimensional. The right inverse $L_0^{-1} : \mathcal{R}L \to \mathcal{R}L$ is compact. Hence (8.1.1) fulfils all basic conditions for the inclusion (8.4.1) when $N(u)$ is replaced with $\eta \mathbb{I} + N(u) + h$.

8.6   Construction of Topological Degree

We construct a topological degree function for $\mathcal{F}_G^C$. The method is based on the construction of continuous one-parametric generalized Galerkin projections which we use for the derivation of a one-parametric family of multi-valued mappings possessing the Leray-Schauder degree. The basic Lemma 8.6.4 below is the stabilization of this degree for large parameters. In this way, we can define a topological degree for our multi-valued mappings. First, we define a class of admissible homotopies.

Definition 8.6.1. A mapping: $(t, u) \to f_t(u)$ from $[0, 1] \times \overline{G}$ to 2^H is a *(multi-) homotopy of the class* (mS_+), if for any sequences $\{u_n\}_{n \in \mathbb{N}}$ in $\overline{G}$, $\{t_n\}_{n \in \mathbb{N}}$ in $[0, 1]$, $f_n^* \in f_{t_n}(u_n)$ with $u_n \rightharpoonup u$, $t_n \to t$ and $\limsup_{n \to \infty} (f_n^*, u_n - u) \leq 0$ we have $u_n \to u$.

We denote

$$\mathcal{H}_G^C = \{ F_t \mid F_t = Q - (QCQ - P) f_t \}$$

where f_t is a homotopy of the class (mS_+). The set $\mathcal{H}_G^C$ is called *the class of admissible homotopies*. Obviously $F_t = (1-t)F_1 + tF_2 \in \mathcal{H}_G^C$, $0 \leq t \leq 1$ for any $F_1, F_2 \in \mathcal{F}_G^C$. Finally, we recall Remark 8.2.1.

Since H is separable there exists a sequence $\{N_n\}_{n\in\mathbb{N}}$ of finite dimensional subspaces of $M^\perp$ with $N_n \subset N_{n+1}$ for all n, and $\cup_{n=1}^\infty N_n$ is dense in $M^\perp$. We denote by P_n the orthogonal projection from H to N_n. We extend this to generalized Galerkin approximations defined by

$$P_\lambda = (\lambda - n)P_{n+1} + (n+1-\lambda)P_n \quad \text{for any } \lambda \in [n, n+1].$$

We have the following obvious result.

Proposition 8.6.2. *The generalized Galerkin approximations satisfy*

(i) $(P_\lambda u, v) = (u, P_\lambda v)$ *for every* $\lambda \geq 1$, $u, v \in H$

(ii) $\|P_\lambda\| \leq 1$ *for all* $\lambda \geq 1$

(iii) $P_\lambda v \to Pv$ *for every* $v \in H$ *as* $\lambda \to \infty$

(iv) $P_n P_\lambda = P_n$ *for every* $\lambda \geq n \in \mathbb{N}$

(v) $(z, P_\lambda z) \geq 0$ *for every* $z \in H$ *and* $\lambda \geq 1$

For each $F = Q - (QCQ - P)f \in \mathcal{F}_G^C$, we define the approximations $\{F_\lambda \mid \lambda \geq 1\}$ by

$$F_\lambda = \mathbb{I} - (QCQ - \lambda P_\lambda)f.$$

We note that $(QCQ - \lambda P_\lambda)f$ is compact, convex-valued and usc for each $\lambda \geq 1$.

Similarly, for each admissible homotopy $F_t = Q - (QCQ - P)f_t$, $0 \leq t \leq 1$, we have

$$(F_t)_\lambda = \mathbb{I} - (QCQ - \lambda P_\lambda)f_t,$$

which is obviously a homotopy of the Leray-Schauder type for any $\lambda \geq 1$.

Proposition 8.6.3. *Let* $\{u_k\}_{n\in\mathbb{N}} \subset H$ *and let* $\{P_\lambda\}_{\lambda \geq 1}$ *be the projections defined as above. Then*

(a) *if* $u_k \rightharpoonup u$ *and* $\lambda_k \to \infty$ *then* $P_{\lambda_k} u_k \rightharpoonup Pu$

(b) *if* $u_k \to u$ *and* $\lambda_k \to \infty$ *then* $P_{\lambda_k} u_k \to Pu$

Now we can formulate the basic lemma.

Lemma 8.6.4. *Let* $F_t = Q - (QCQ - P)f_t$ *be an admissible homotopy and let* A *be a closed subset in* $\overline{G}$. *If* $0 \notin F_t(A)$ *for all* $t \in [0, 1]$, *then there exists* $\lambda_0 > 0$ *such that*

$$0 \notin (F_t)_\lambda(A) \qquad \text{for all } t \in [0, 1] \text{ and } \lambda \geq \lambda_0.$$

Proof. Assume by contrary that there exist sequences $\{u_k\}_{k\in\mathbb{N}}$ in A and $\{\lambda_k\}_{k\in\mathbb{N}}$ in $[1, \infty)$, $\lambda_k \to \infty$, and $\{t_k\}_{k\in\mathbb{N}}$ in $[0, 1]$ such that $0 \in (F_{t_k})_{\lambda_k}(u_k)$. This is equivalent to the existence of selections $g_k^* \in f_{t_k}(u_k)$ and $c_k = -CQg_k^*$ for which

$$u_k + Qc_k^* + \lambda_k P_{\lambda_k} g_k^* = 0.$$

Writing this equation in both subspaces M and $M^\perp$ we get

$$Qu_k + Qc_k^* = 0,\qquad\qquad (8.6.1)$$

$$Pu_k + \lambda_k P_{\lambda_k} g_k^* = 0.\qquad\qquad (8.6.2)$$

The sequence $\{u_k\}_{k\in\mathbb{N}}$ is bounded therefore we can (taking a subsequence, if necessary) assume that $u_k \rightharpoonup u$ for some $u \in H$. Since the sequence g_k^* is bounded, we suppose $g_k^* \rightharpoonup g^*$, $g^* \in H$ and the compactness of C gives $c_k^* \to z^*$, $z^* = -CQg^*$.

Hence we have $Qc_k^* \to Qz^*$ and by (8.6.1), $Qu_k \to Qu$. By (8.6.2),

$$\frac{1}{\lambda_k} Pu_k + P_{\lambda_k} g_k^* = 0.$$

The set $\{u_k \mid k = 1, 2, \dots\}$ is bounded, thus $\frac{1}{\lambda_k} Pu_k \to 0$ for $k \to \infty$. This leads to $P_{\lambda_k} g_k^* \to 0$. On the other hand, for $g_k^* \rightharpoonup g^*$ we conclude $P_{\lambda_k} g_k^* \rightharpoonup Pg^*$, which yields $Pg^* = 0$. Hence we have $Pg_k^* \rightharpoonup Pg^* = 0$ followed by $\lim_{k\to\infty}(g_k^*, Pu) = 0$.

We continue with calculating of $\limsup_{k\to\infty}(g_k^*, u_k - u)$:

$$\limsup_{k\to\infty}(g_k^*, u_k - u) = \limsup_{k\to\infty}(g_k^*, Pu_k - Pu) = \limsup_{k\to\infty}(g_k^*, Pu_k).\qquad (8.6.3)$$

From (8.6.2) we obtain $Pu_k = -\lambda_k P_{\lambda_k} g_k^*$. Inserting it in to the last term of (8.6.3) we get

$$\limsup_{k\to\infty}(g_k^*, u_k - u) = -\liminf_{k\to\infty}\lambda_k(g_k^*, P_{\lambda_k} g_k^*).$$

By (v) of Proposition 8.6.2, $\lambda_k(g_k^*, P_{\lambda_k} g_k^*) \geq 0$ and it immediately follows that

$$\limsup_{k\to\infty}(g_k^*, u_k - u) \leq 0.$$

Hence, by the definition of the homotopy f_t of the class (mS_+) we have $u_k \to u$, $u \in A$. We have $t_n \to t$, $g_k^* \rightharpoonup g^*$ and $u_k \to u$. Since $(t, u) \to f_t(u)$ is w-usc, we get $g^* \in f_t(u)$. From $Qu_k \to Qu$, $Qc_k^* \to Qz^*$ using (8.6.1), we obtain $Qu + Qz^* = 0$. Since $Pg^* = 0$ we have

$$0 = Qu - (QCQ - P)g^* \in F_t(u),$$

a contradiction. The proof is complete. $\square$

We have the following important consequence for the stabilization of a degree.

Lemma 8.6.5. *Let $F \in \mathcal{F}_G^C$ and $0 \notin F(\partial G)$. Then there exist $\lambda_1 \in [1, \infty)$ such that*

$$d_{LS}(F_\lambda, G, 0) = constant \quad \text{for all } \lambda \geq \lambda_1,$$

where d_{LS} is the Leray-Schauder topological degree.

Proof. Choosing a constant homotopy $F_t = F$ and $A = \partial G$, we obtain that there exists $\lambda_1 \geq 1$ such that $0 \notin F_\lambda(\partial G)$ for all $\lambda \geq \lambda_1$. $\qquad \square$

Due to Lemma 8.6.5 we can define a degree function for the class $\mathcal{F}_G^C$. We put

$$d(F, G, 0) := \lim_{\lambda \to \infty} d_{LS}(F_\lambda, G, 0) \qquad (8.6.4)$$

for any given $F \in \mathcal{F}_G^C$ with $0 \notin F(\partial G)$. In the next theorem we show that the degree function defined by (8.6.4) has all the usual properties.

Theorem 8.6.6. *Let H be a real separable Hilbert space, G be a bounded open subset of H, M be a closed subspace of H, Q be the orthogonal projection onto M, $C \in L(\mathcal{R}Q, H)$ be compact, $\mathcal{F}_G^C$ be the class of admissible mappings and $\mathcal{H}_G^C$ be the class of admissible homotopies defined above. Then there exists a classical $\mathbb{Z}$-defined topological degree function d on $\mathcal{F}_G^C$ satisfying the following properties:*

(i) If $0 \notin F(\partial G)$ and $d(F, G, 0) \neq 0$ then $0 \in F(G)$.

(ii) $d(F, G, 0) = d(F, G_1, 0) + d(F, G_2, 0)$ (thus $F \in \mathcal{F}_{G_1}^C$ and $F \in \mathcal{F}_{G_2}^C$), whenever G_1 and G_2 are disjoint open subsets of G such that $0 \notin F(\overline{G} \setminus (G_1 \cup G_2))$).

(iii) $d(F_t, G, 0)$ is independent of $t \in [0, 1]$ if $F_t \in \mathcal{H}_G^C$ and $0 \notin F_t(\partial G)$ for all $t \in [0, 1]$.

Proof. If $0 \notin F(\partial G)$ and $d(F, G, 0) \neq 0$ then by Lemma 8.6.4, $0 \notin F_\lambda(\partial G)$ and $d_{LS}(F_\lambda, G, 0) \neq 0$ for λ large. Hence $0 \in F_\lambda(G)$. Then Lemma 8.6.4 gives $0 \in F(G)$. So (i) is shown.

If $0 \notin F(\overline{G} \setminus (G_1 \cup G_2))$ then by Lemma 8.6.4, $0 \notin F_\lambda(\overline{G} \setminus (G_1 \cup G_2))$ for λ large. Then $d_{LS}(F_\lambda, G, 0) = d_{LS}(F_\lambda, G_1, 0) + d_{LS}(F_\lambda, G_2, 0)$. So (ii) is shown.

If $0 \notin F_t(\partial G)$ for all $t \in [0, 1]$, then by Lemma 8.6.4, $0 \notin (F_t)_\lambda(\partial G)$ for all $t \in [0, 1]$ and λ large. Then $d_{LS}((F_t)_\lambda, G, 0)$ is independent of $t \in [0, 1]$. So (iii) is shown. The proof is finished. $\qquad \square$

More general mappings are considered in [36, 92] but we do not present it here, since we are focussing on extension of some known results on local bifurcations [32–35].

8.7 Local Bifurcations

We consider the inclusion

$$0 \in Lu - \eta u - N(u), \qquad (8.7.1)$$

where L is given in Section 8.4, $\eta > 0$ is a parameter and $N \in (mQMN)$ is bounded, w-usc, convex-valued. Hence $\eta \mathbb{I} + N \in (mS_+)$. Then we set

$$\deg(L - \eta \mathbb{I} - N, G, 0) := d\left(Q - (QL_0^{-1}Q - P)(\eta \mathbb{I} + N), G, 0\right)$$

for an open bounded subset $G \subset H$ such that $0 \notin (L - \eta\mathbb{I} - N)(\partial G \cap D(L))$.

We also suppose that L is *self-adjoint*. Then $L_0^{-1} : \mathcal{R}L \to \mathcal{R}L$ is a compact, self-adjoint operator, so the spectrum $\sigma(L)$ of L is $\sigma(L) = \{\lambda_j\}_{j\in\mathbb{Z}}$, with $\lambda_0 = 0$, and the multiplicity m_j of each $\lambda_j \neq 0$ is finite. Concerning $N : H \to 2^H$ we in addition assume

(A1) $\sup \{|v| \mid v \in N(u)\} = o(|u|)$ as $u \to 0$.

In this and the next sections, we intend to extend some results of [32] to (8.7.1).

Lemma 8.7.1. *If $\eta \notin \sigma(L)$, $\eta > 0$, then*

$$d(L - \eta\mathbb{I}, B_r, 0) = (-1)^\chi$$

with $\chi = \sum\limits_{0 < \lambda_j < \eta} m_j$ and B_r is the ball of H centered at 0 with radius $r > 0$.

Proof. Now we have $F = Q(\mathbb{I} - \eta L_0^{-1}Q) + \eta P$ and $F_\lambda = Q(\mathbb{I} - \eta L_0^{-1}Q) + P + \eta\lambda P_\lambda$. Let H_{-1}, H_0, H_1 be spanned by eigenvectors of L with the corresponding eigenvalues $\lambda_j < 0$, $0 < \lambda_j < \eta$, $\lambda_j > \eta$, respectively. Clearly $\dim H_0 = \chi$ and $H = H_{-1} \oplus \mathcal{N}L \oplus H_0 \oplus H_1$. So if $u = u_{-1} + w + u_0 + u_1$, $u_j \in H_j$, $w \in \mathcal{N}L$ then

$$F_\lambda(u) = u_{-1} - \eta L_0^{-1}u_{-1} + w + \eta\lambda P_\lambda w + u_0 - \eta L_0^{-1}u_0 + u_1 - \eta L_0^{-1}u_1 .$$

We note

$$(P_\lambda w, w) \geq 0 \quad \forall w \in \mathcal{N}L$$
$$(L_0^{-1}u_{-1}, u_{-1}) \leq 0 \quad \forall u_{-1} \in H_{-1},$$
$$(L_0^{-1}u_1, u_1) < \eta^{-1}|u_1|^2 \quad \forall 0 \neq u_1 \in H_1$$

and $A_0 = (\mathbb{I} - \eta L_0^{-1})/H_0 : H_0 \to H_0$ is an invertible diagonal linear operator with negative eigenvalues. Then we can easily verify that $0 \notin F_{\lambda t}(u)\forall u \neq 0$ and $\forall t \in [0, 1]$, where

$$F_{\lambda t}(u) = u_{-1} - \eta t L_0^{-1}u_{-1} + w + t\eta\lambda P_\lambda w + u_0 - \eta L_0^{-1}u_0 + u_1 - \eta t L_0^{-1}u_1 .$$

Hence $\deg(L - \eta\mathbb{I}, B_r, 0) = d_{LS}(F_{\lambda 1}(u), B_r, 0) = d_{LS}(F_{\lambda 0}(u), B_r, 0) = \det A_0 = (-1)^\chi$ for λ large. The proof is finished. $\qquad\square$

Now we present the following extension of a Krasnoselski result [60].

Theorem 8.7.2. *Let (A1) hold. If $\eta_0 \in \sigma(L)$, $\eta_0 > 0$ with odd multiplicity $m(\eta_0)$, then $(\eta_0, 0)$ is a bifurcation point of (8.7.1), i.e. there is a sequence $\{(u_i, \eta_i)\}_{i=1}^\infty$ such that $u_i \in D(L) \setminus \{0\}$, $\eta_i \to \eta_0$ and $u_i \to 0$ as $i \to \infty$, and $u = u_i$, $\eta = \eta_i$ satisfy (8.7.1).*

Proof. Let $\delta > 0$ be so small that $\eta_0 > \delta$ and $(\eta_0 - \delta, \eta_0 + \delta) \cap \sigma(L) = \{\eta_0\}$. Then there is an $\tau_1 > 0$ such that

$$|Lu - (\eta_0 \pm \delta)u| \geq \tau_1|u| \quad \forall u \in D(L) .$$

According to (A1) there is an $r_0 > 0$ such that $\forall r$, $0 < r \leq r_0$, $\forall t \in [0, 1]$, $\forall u \in D(L)$, $|u| = r$ it holds

$$0 \notin Lu - (\eta_0 \pm \delta)u - tN(u).$$

Then we get

$$\deg\left(Lu - (\eta_0 \pm \delta)u - N(u), B_r, 0\right) = \deg\left(Lu - (\eta_0 \pm \delta)u, B_r, 0\right)$$

$$= (-1)^{\sum_{0 < \lambda_j < \eta_0 \pm \delta} m_j}.$$

Since $m(\eta_0)$ is odd, we obtain

$$\deg\left(Lu - (\eta_0 - \delta)u - N(u), B_r, 0\right) \neq \deg\left(Lu - (\eta_0 + \delta)u - N(u), B_r, 0\right)$$
$$(8.7.2)$$

Then (8.7.2) implies the existence of $u \in D(L)$, $|u| = r$ and $\eta \in (\eta_0 - \delta, \eta_0 + \delta)$ solving (8.7.1). Since δ, r can be arbitrarily small, we get a sequence $\{(\eta_i, u_i)\}_{i=1}^{\infty}$ such that $\eta_i \to \eta_0$, $u_i \to 0$, $u_i \in D(L)$ and $0 \neq u_i$ solves (8.7.1) with η_i. The proof is finished. □

8.8 Bifurcations from Infinity

Now we study (8.7.1) under assumption

(A2) $\sup\{|v| \mid v \in N(u)\} = o(|u|)$ as $|u| \to \infty$.

Theorem 8.8.1. *Let (A2) hold. If $\eta_0 \in \sigma(L)$, $\eta_0 > 0$ with odd multiplicity $m(\eta_0)$, then (η_0, ∞) is a bifurcation point of (8.7.1), i.e. there is a sequence $\{(u_i, \eta_i)\}_{i=1}^{\infty}$ such that $u_i \in D(L)$, $\eta_i \to \eta_0$ and $|u_i| \to \infty$ as $i \to \infty$, and $u = u_i$, $\eta = \eta_i$ satisfy (8.7.1).*

Proof. Let $\delta > 0$ be so small that $\eta_0 > \delta$ and $(\eta_0 - \delta, \eta_0 + \delta) \cap \sigma(L) = \{\eta_0\}$. Then there is an $\tau_1 > 0$ such that

$$|Lu - (\eta_0 \pm \delta)u| \geq \tau_1|u| \quad \forall u \in D(L).$$

According to (A2) there is an $r_0 > 0$ such that $\forall r$, $r \geq r_0$, $\forall t \in [0, 1]$, $\forall u \in D(L)$, $|u| = r$ it holds

$$0 \notin Lu - (\eta_0 \pm \delta)u - tN(u).$$

Like in the proof of Theorem 8.7.2, then there are $u \in D(L)$, $|u| = r$ and $\eta \in (\eta_0 - \delta, \eta_0 + \delta)$ solving (8.7.1). Since δ can be arbitrarily small and then r can be arbitrarily large, we get a sequence $\{(\eta_i, u_i)\}_{i=1}^{\infty}$ such that $\eta_i \to \eta_0$, $|u_i| \to \infty$, $u_i \in D(L)$ and u_i solves (8.7.1) with η_i. The proof is finished. □

8.9   Bifurcations for Semilinear Wave Equations

We apply the previous bifurcation result to the semilinear wave equation (8.1.1). First we verify that (A2) is satisfied under (8.1.2). Note now

$$N(u) = \{v \in L^2(\Omega) \mid v(x,t) \in [\widetilde{q}_-(x,t,u(x,t)), \widetilde{q}_+(x,t,u(x,t))]\}$$

for $\widetilde{q}_\pm(x,t,u) := q_\pm(x,t,u) + h(x,t)$ and $q_\pm(x,t,u)$ are defined in Subsection 8.5. Let $K_1 := \sup_{u \in \mathbb{R}} |g(u)|$. If $v \in N(u)$ then the Jensen inequality [171] together with condition (8.1.2) imply

$$|v|^2 = \int_\Omega v(x,t)^2 \, dx \, dt \le 3 \int_\Omega \left(K_1^2 + c_0^2 |u(x,t)|^{2\alpha} + \widetilde{h}_0(x,t)^2 \right) \, dx \, dt$$

$$\le 6K_1^2 \pi^2 + 3|\widetilde{h}_0|^2 + 6\pi^2 c_0^2 \left(\int_\Omega u(x,t)^2 \, dx \, dt \right)^\alpha = 6K_1^2 \pi^2 + 3|\widetilde{h}_0|^2 + 6\pi^2 c_0^2 |u|^{2\alpha}$$

for $\widetilde{h}_0(x,t) := h_0(x,t) + |h(x,t)|$. Hence

$$|v| \le 3K_1 \pi + 2|\widetilde{h}_0| + 3\pi c_0 |u|^\alpha .$$

Since $0 \le \alpha < 1$, we see that (A2) holds for (8.5.1). It remains to calculate the multiplicity of a positive eigenvalue $\eta_0 \in \sigma(L)$. So we must find the number of solutions of $n^2 - m^2 = \eta_0$, $n \in \mathbb{N}$, $m \in \mathbb{Z}$. If n, m, $m \ne 0$ solves it then $n, -m$ is its another solution. So $m(\eta_0)$ is odd if and only if $\eta_0 = n^2$ for some $n \in \mathbb{N}$. Summarizing, we can apply Theorem 8.8.1 to get the following result.

Proposition 8.9.1. *Under condition (8.1.2), each $\eta_0 = n^2$, $n \in \mathbb{N}$ is a bifurcation point of (8.1.1) at infinity for weak 2π-periodic solutions.*

Similarly we have the following result.

Proposition 8.9.2. *Under condition (8.1.2), each $\eta_0 = n^4$, $n \in \mathbb{N}$ is a bifurcation point of the problem*

$$\begin{aligned} u_{tt} + u_{xxxx} - \eta u - g(u) - f(x,t,u) &= h(x,t), \\ u(0,\cdot) = u(\pi,\cdot) = 0, \quad u_{xx}(0,\cdot) &= u_{xx}(\pi,\cdot) = 0, \end{aligned} \tag{8.9.1}$$

at infinity for weak 2π-periodic solutions.

Proof. Now

$$Lu = u_{tt} + u_{xxxx}$$

with $\sigma(L) = \{n^4 - m^2 \mid m \in \mathbb{Z}, n \in \mathbb{N}\}$. So $0 < \eta_0 \in \sigma(L)$ has an odd multiplicity if and only if $\eta_0 = n^4$ for some $n \in \mathbb{N}$. The proof is finished. □

8.10   Chaos for Discontinuous Beam Equations

Combining methods of Chapters 3 and 8 together with paper [27], it would be possible to show chaotic solutions to a weakly discontinuous system modeling a compressed beam with small damping and subjected to a small periodic forcing described by the following p.d.eqn

$$u_{tt} + \varepsilon \delta u_t + u_{xxxx} + P_0 u_{xx} - \kappa u_{xx} \left(\int_0^\pi u_x^2(\xi, t) d\xi \right) + \varepsilon \operatorname{sgn} u = \varepsilon h(x, t),$$

$$u(0, t) = u(\pi, t) = 0 = u_{xx}(0, t) = u_{xx}(\pi, t),$$

$$(8.10.1)$$

where $u(x, t) \in \mathbb{R}$ is the transverse deflection of the axis of the beam; $P_0 > 0$ is an external load, $\kappa > 0$ is a ratio indicating the external rigidity and $\delta > 0$ is the damping, ε is a small parameter, the function $h(x, t)$ is continuous and periodic in t representing the periodic forcing that is distributed along the whole beam.

The first work on oscillations of an elastic beam subject to an axial compression was done in [115]. More recent works on the full equation are presented in [38, 91, 170]. An undamped buckled beam is investigated in [26, 197] (see (7.1.45)). All these results are about beam equations with continuous terms like the following one:

$$u_{tt} + \mu_1 u_t + u_{xxxx} + P_0 u_{xx} - u_{xx} \left(\int_0^\pi u_x^2(\xi, t) d\xi \right) = \mu_2 \cos \omega_0 t,$$

$$(8.10.2)$$

$$u(0, t) = u(\pi, t) = 0 = u_{xx}(0, t) = u_{xx}(\pi, t),$$

where P_0, ω_0 are constants and μ_1, μ_2 are small parameters. This is a model for oscillations of an elastic beam with a compressive axial load P_0 (see Fig. 8.1).

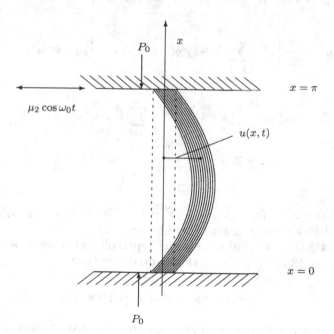

Figure 8.1: The forced buckled beam (8.10.2)

When P_0 is sufficiently large, (8.10.2) can exhibit chaotic behavior for certain small parameters μ_1, μ_2. For instance, we have the following result [91, 115]:

Theorem 8.10.1. *If* $1 < P_0 < 4$ *and* $\omega_0 \neq n^2 (n^2 - P_0)$, $\forall n \in \mathbb{N}$, $n \geq 2$. *Then* (8.10.2) *possesses a Smale horseshoe with the associated chaotic dynamics for any* $\mu_1 \neq 0$ *and* $\mu_2 \neq 0$ *small satisfying*

$$\left| \frac{\mu_1}{\mu_2} \right| < \frac{3\sqrt{\pi}\omega_0}{a^3} \text{ sech } \frac{\pi\omega_0}{2a}, \tag{8.10.3}$$

where $a := \sqrt{P_0 - 1}$.

We expect a similar result also for (8.10.1). Finally, we note that conditions (1.2.10) and (8.10.3) are related to each order as follows. In (8.10.2) substitute $u(x,t) = \sum_{k=1}^{\infty} u_k(t) \sin kx$, multiply by $\sin nx$ and integrate from 0 to π. This yields the infinite set of ordinary differential equations

$$\ddot{u}_n = n^2(P_0 - n^2)u_n - \frac{\pi}{2}n^2 \left[\sum_{k=1}^{\infty} k^2 u_k^2 \right] u_n - \mu_1 \dot{u}_n + 2\mu_2 \left[\frac{1 - (-1)^n}{\pi n} \right] \cos \omega_0 t,$$

$$n = 1, 2, \dots.$$

For $1 < P_0 < 4$, then only the equation with $n = 1$ is hyperbolic while the system of remaining equations has a center. To emphasize this let us define $p = u_1$ and $q_n = u_{n+1}$, $n = 1, 2, \dots$. The preceding equations now take the form

$$\ddot{p} = a^2 p - \frac{\pi}{2} \left[p^2 + \sum_{k=1}^{\infty} (k+1)^2 q_k^2 \right] p - \mu_1 \dot{p} + \frac{4}{\pi} \mu_2 \cos \omega_0 t, \tag{8.10.4}$$

and

$$\ddot{q}_n = -\omega_n^2 q_n - \frac{\pi}{2}(n+1)^2 \left[p^2 + \sum_{k=1}^{\infty} (k+1)^2 q_k^2 \right] q_n$$

$$-\mu_1 \dot{q}_n + 2\mu_2 \left[\frac{1 - (-1)^{n+1}}{\pi(n+1)} \right] \cos \omega_0 t, \tag{8.10.5}$$

$$n = 1, 2, \dots$$

where we define $\omega_n^2 = (n+1)^2 \left[(n+1)^2 - P_0 \right]$. Conditions $\omega_0 \neq \omega_n$, $\forall n \in \mathbb{N}$ of Theorem 8.10.1 are non-resonance ones for (8.10.5).

In (8.10.4–8.10.5) we project onto the hyperbolic subspace by setting $q_n = 0$ for all $n \in \mathbb{N}$ in (8.10.4) to obtain the reduced equation

$$\ddot{p} = a^2 p - \frac{\pi}{2} p^3 - \mu_1 \dot{p} + \frac{4}{\pi} \mu_2 \cos \omega_0 t. \tag{8.10.6}$$

So (8.10.6) is derived from (8.10.2) when only the first (hyperbolic) mode of vibration is considered. We see that this is the forced damped Duffing equation with negative stiffness and with similar form like (1.2.1) for which standard theory, mentioned also in this book, yields chaotic dynamics under condition (8.10.3) (see [91] for further details). As a matter of fact, taking transformations

$$p(t) = \frac{2a}{\sqrt{\pi}} x(at), \quad \mu_1 = a\widetilde{\mu}_1, \quad \mu_2 = \frac{\sqrt{\pi}a^3}{2}\widetilde{\mu}_2, \quad \omega_0 = a\omega$$

in (8.10.6) we derive

$$\ddot{x} = x - 2x^3 - \widetilde{\mu}_1\dot{x} + \widetilde{\mu}_2 \cos\omega t, \tag{8.10.7}$$

which is (1.2.1). Then condition (1.2.10) is just

$$|\widetilde{\mu}_1| < |\widetilde{\mu}_2|\frac{3\pi\omega}{2} \operatorname{sech} \frac{\pi\omega}{2}. \tag{8.10.8}$$

But (8.10.8) is precisely (8.10.3) in the original parameters μ_1, μ_2 and ω_0.

Bibliography

[1] R. ABRAHAM, J.E. MARSDEN AND T. RATIU: *Manifolds, Tensor Analysis and Applications*, Addison-Wesley, Reading MA, 1983.

[2] A.A. AIGNER, A.R. CHAMPNEYS AND V.M. ROTHOS: *A new barrier to the existence of moving kinks in Frenkel-Kontorova lattices*, Physica D **186** (2003), 148–170.

[3] A. AMBROSETTI AND M. BADIALE: *The dual variational principle and elliptic problems with discontinuous nonlinearities*, J. Math. Anal. Appl. **140** (1989), 363–373.

[4] C.J. AMICK AND K. KIRCHGÄSSNER: *A theory of solitary water-waves in the presence of surface tension*, Arch. Ration. Mech. Anal. **105** (1989), 1–49.

[5] J. ANDRES AND L. GÓRNIEWICZ: *Topological Fixed Point Principles for Boundary Value Problems*, Kluwer, Dordrecht, The Netherlands 2003.

[6] A.A. ANDRONOW, A.A. WITT AND S.E. CHAIKIN: *Theorie der Schwingungen I*, Akademie Verlag, Berlin, 1965.

[7] S. AUBRY AND R.S. MACKAY: *Proof of existence of breathers for time-reversible or Hamiltonian networks of weakly coupled oscillators*, Nonlinearity **6** (1994), 1623–1643.

[8] J. AWREJCEWICZ, M. FEČKAN AND P. OLEJNIK: *Bifurcations of planar sliding homoclinics*, Math. Probl. Eng. **2006** (2006), 1–13.

[9] J. AWREJCEWICZ, M. FEČKAN AND P. OLEJNIK: *On continuous approximation of discontinuous systems*, Nonlinear Anal.-Theor. **62** (2005), 1317–1331.

[10] J. AWREJCEWICZ AND M.M. HOLICKE: *Melnikov's method and stick-slip chaotic oscillations in very weakly forced mechanical systems*, Int. J. Bifur. Chaos **9** (1999), 505–518.

[11] J. AWREJCEWICZ AND M.M. HOLICKE: *Smooth and Nonsmooth High Dimensional Chaos and the Melnikov-Type Methods*, World Scientific Publishing, Singapore, 2007.

[12] J. AWREJCEWICZ AND C.H. LAMARQUE: *Bifurcation and Chaos in Nonsmooth Mechanical Systems*, World Scientific Publishing, Singapore, 2003.

[13] D. BAMBUSI: *Lyapunov center theorems for some nonlinear PDEs: a simple proof*, Ann. Scuola Norm. Sup. Pisa Ser. 4, **29** (2000), 823–837.

[14] D. BAMBUSI AND D. VELLA: *Quasiperiodic breathers in Hamiltonian lattices with symmetries*, Discr. Cont. Dyn. Syst. B **2** (2002), 389–399.

[15] F. BATTELLI: *Exponentially small bifurcation functions in singular systems of O.D.E.*, Diff. Int. Eq. **9** (1996), 1165–1181.

[16] F. BATTELLI AND M. FEČKAN: *Subharmonic solutions in singular systems*, J. Differ. Equations **132** (1996), 21–45.

[17] F. BATTELLI AND M. FEČKAN: *Heteroclinic period blow–up in certain symmetric ordinary differential equations*, Zeit. Ang. Math. Phys. (ZAMP) **47** (1996), 385–399.

[18] F. BATTELLI AND M. FEČKAN: *Chaos arising near a topologically transversal homoclinic set*, Top. Meth. Nonl. Anal. **20** (2002), 195–215.

[19] F. BATTELLI AND M. FEČKAN: *Some remarks on the Melnikov function*, Electr. J. Differ. Equations **2002, 13** (2002), 1–29.

[20] F. BATTELLI AND M. FEČKAN: *Chaos in the beam equation*, J. Differ. Equations **209** (2005), 172–227.

[21] F. BATTELLI AND M. FEČKAN: *Global center manifolds in singular systems*, NoDEA: Nonl. Diff. Eq. Appl. **3** (1996), 19–34.

[22] F. BATTELLI AND M. FEČKAN: *Homoclinic trajectories in discontinuous systems*, J. Dynam. Diff. Eq. **20** (2008), 337–376.

[23] F. BATTELLI AND M. FEČKAN: *Homoclinic orbits of slowly periodically forced and weakly damped beams resting on weakly elastic bearings*, Adv. Diff. Equations **8** (2003), 1043–1080.

[24] F. BATTELLI AND M. FEČKAN: *Periodic solutions of symmetric elliptic singular systems*, Adv. Nonl. Studies **5** (2005), 163–196.

[25] F. BATTELLI AND M. FEČKAN: *Periodic solutions of symmetric elliptic singular systems: the higher codimension case*, Adv. Nonl. Studies **6** (2006), 109–132.

[26] F. BATTELLI, M. FEČKAN AND M. FRANCA: *Periodic solutions of a periodically forced and undamped beam resting on weakly elastic bearings*, Zeit. Ang. Math. Phys. (ZAMP) **59** (2008), 212–243.

[27] F. BATTELLI, M. FEČKAN AND M. FRANCA: *On the chaotic behaviour of a compressed beam*, Dynamics of PDE **4** (2007), 55–86.

[28] F. BATTELLI AND C. LAZZARI: *Bounded solutions to singularly perturbed systems of O.D.E.*, J. Differ. Equations **100** (1992), 49–81.

[29] F. BATTELLI AND C. LAZZARI: *Heteroclinic orbits in systems with slowly varying coefficients*, J. Differ. Equations **105** (1993), 1–29.

[30] K. BEN-NAOUM AND J. MAWHIN: *The periodic-Dirichlet problem for some semilinear wave equations*, J. Differ. Equations **96** (1992), 340–354.

[31] M.S. BERGER: *Nonlinearity and Functional Analysis*, Academic, New York, 1977.

[32] J. BERKOVITS: *Some bifurcation results for a class of semilinear equations via topological degree method*, Bull. Soc. Math. Belg. **44** (1992), 237–247.

[33] J. BERKOVITS: *Local bifurcation results for systems of semilinear equations*, J. Differ. Equations **133** (1997), 245–254.

[34] J. BERKOVITS: *On the bifurcation of large amplitude solutions for a system of wave and beam equations*, Nonlinear Anal.-Theor. **52** (2003), 343–354.

[35] J. BERKOVITS AND V. MUSTONEN: *An extension of Leray-Schauder degree and applications to nonlinear wave equations*, Diff. Int. Eqns. **3** (1990), 945–963.

[36] J. BERKOVITS AND M. TIENARI: *Topological degree theory for some classes of multis with applications to hyperbolic and elliptic problems involving discontinuous nonlinearities*, Dyn. Sys. Appl. **5** (1996), 1–18.

[37] M. BERTI AND PH. BOLE: *Multiplicity of periodic solutions of nonlinear wave equations*, Nonlinear Anal.-Theor. **56** (2004), 1011–1046.

[38] M. BERTI AND C. CARMINATI: *Chaotic dynamics for perturbations of infinite dimensional Hamiltonian systems*, Nonlinear Anal.-Theor. **48** (2002), 481–504.

[39] P.A. BLIMAN AND A.M. KRASNOSEL'SKII: *Periodic solutions of linear systems coupled with relay*, Nonlinear Anal.-Theor. **30** (1997), 687–696.

[40] A.A. BOICHUK, I.A. KOROSTIL AND M. FEČKAN: *Bifurcation conditions for a solution of an abstract wave equation*, Diff. Equat. **43** (2007), 495–502.

[41] E. BOSETTO AND E. SERRA: *A variational approach to chaotic dynamics in periodically forced nonlinear oscillators*, Ann. Inst. H. Poincaré, Anal. Nonl. **17** (2000), 673–709.

[42] E.M. BRAVERMAN, S.M. MEERKOV AND E.S. PYATNITSKII: *Conditions for applicability of the method of harmonic balance for systems with hysteresis nonlinearity (in the case of filter hypothesis)*, Automat. Rem. Control **37** (1976), 1640–1650.

[43] H. BREZIS: *Periodic solutions of nonlinear vibrating strings and duality principles*, Bull. Amer. Math. Soc. **8** (1983), 409–426.

[44] H.W. BROER, I. HOVEIJN AND M. VAN NOORT: *A reversible bifurcation analysis of the inverted pendulum*, Physica D **112** (1998), 50–63.

[45] B. BROGLIATO: *Nonsmooth Impact Mechanics: Models, Dynamics, and Control*, Lecture Notes in Control and Information Sciences 220, Springer, Berlin, 1996.

[46] F.E. BROWDER: *Degree theory for nonlinear mappings*, Proc. Sympos. Pure Math. 45, Part 1, AMS, Providence, R.I. (1986), 203–226.

[47] R.F. BROWN: *A Topological Introduction to Nonlinear Analysis*, Birhkhäuser, Boston, MA, 1993.

[48] A. BUICA AND J. LLIBRE: *Averaging methods for finding periodic orbits via Brouwer degree*, Bull. Sci. Math. **128** (2004), 7–22.

[49] K. BURNS AND H. WEISS: *A geometric criterion for positive topological entropy*, Commun. Math. Phys. **172** (1995), 95–118.

[50] N.V. BUTENIN, Y.I. NEJMARK AND N.A. FUFAEV: *An Introduction to the Theory of Nonlinear Oscillations*, Nauka, Moscow, 1987 (in Russian).

[51] J.W.S. CASSELS: *An Introduction to Diophantine Approximation*, Cambridge University, Press Cambridge, 1957.

[52] K.C. CHANG: *Free boundary problems and the set-valued mappings*, J. Differ. Equations **49** (1983), 1–28.

[53] Y. CHEN AND D. O'REGAN: *Coincidence degree theory for mappings of class $L - (S_+)$*, Appl. Analysis **85** (2006), 963–970.

[54] Y. CHEN AND D. O'REGAN: *Generalized degree theory for semilinear operator equations*, Glasgow Math. J. **48** (2006), 65–73.

[55] C. CHICONE: *Lyapunov–Schmidt reduction and Melnikov integrals for bifurcation of periodic solutions in coupled oscillators*, J. Differ. Equations **112** (1994), 407–447.

[56] S.N. CHOW AND J.K. HALE: *Methods of Bifurcation Theory*, Springer, New York, 1982.

[57] L.O. CHUA, M. KOMURO AND T. MATSUMOTO: *The double scroll family*, IEEE Trans. CAS **33** (1986), 1072–1118.

[58] TH.W. CUSICK AND M.E. FLAHIVE: *The Markoff and Lagrange Spectra*, Mathematical Surveys and Monographs **30**, American Mathematical Society, Providence, RI, 1989.

[59] K. DEIMLING: *Multivalued Differential Equations*, W. De Gruyter, Berlin, 1992.

[60] K. DEIMLING *Nonlinear Functional Analysis*, Springer, Berlin, 1985.

[61] K. DEIMLING: *Multivalued differential equations and dry friction problems*, In: Proc. Conf. Differential and Delay Equations, Ames, Iowa 1991, A.M. Fink, R.K. Miller and W. Kliemann eds., World Scientific Publishing. Singapore 1992, 99–106.

[62] K. DEIMLING AND P. SZILÁGYI: *Periodic solutions of dry friction problems*, Z. Angew. Math. Phys. (ZAMP) **45** (1994), 53–60.

[63] K. DEIMLING, G. HETZER AND W. SHEN: *Almost periodicity enforced by Coulomb friction*, Adv. Diff. Eq. **1** (1996), 265–281.

[64] J.P. DEN HARTOG: *Mechanische Schwingungen*, 2nd ed., Springer, Berlin, 1952.

[65] R. DEVANEY: *Blue sky catastrophes in reversible and Hamiltonian systems*, Indiana Univ. Math. J. **26** (1977), 247–263.

[66] R. DEVANEY: *Reversible diffeomorphisms and flows*, Tran. Amer. Math. Soc. **218** (1976), 89–113.

[67] R. DEVANEY: *Homoclinic bifurcations and the area-conserving Hénon mapping*, J. Differ. Equations **51** (1984), 254–266.

[68] R. DEVANEY: *An Introduction to Chaotic Dynamical Systems*, Benjamin/Cummings, Menlo Park, CA, 1986.

[69] A. DONTCHEV, T.Z. DONTCHEV AND I. SLAVOV: *A Tikhonov-type theorem for singularly perturbed differential inclusions*, Nonlinear. Anal.-Theor. **26** (1996), 1547–1554.

[70] W. ECKHAUS: *Singular perturbations of homoclinic orbits in* $\mathbb{R}^4$, SIAM J. Math. Anal. **23** (1992), 1269–1290.

[71] J.C. EILBECK AND R. FLESCH: *Calculation of families of solitary waves on discrete lattices*, Phys. Lett. A **149** (1990), 200–202.

[72] M. FEČKAN: *Melnikov functions for singularly perturbed ordinary differential equations*, Nonlinear. Anal.-Theor. **19** (1992), 393–401.

[73] M. FEČKAN: *Bifurcation from homoclinic to periodic solutions in ordinary differential equations with multivalued perturbations*, J. Differ. Equations **130** (1996), 415–450.

[74] M. FEČKAN: *Bifurcation from homoclinic to periodic solutions in singularly perturbed differential inclusions*, Proc. Royal Soc. Edinburgh **127A** (1997), 727–753.

[75] M. FEČKAN: *Bifurcation of periodic solutions in differential inclusions*, Appl. Math. **42** (1997), 369–393.

[76] M. FEČKAN: *Chaotic solutions in differential inclusions: chaos in dry friction problems*, Tran. Amer. Math. Soc. **351** (1999), 2861–2873.

[77] M. FEČKAN: *Singularly perturbed boundary value problems*, Diff. Int. Equations **7** (1994), 109–120.

[78] M. FEČKAN: *Periodic solutions of certain abstract wave equations*, Proc. Amer. Math. Soc. **123** (1995), 465–470.

[79] M. FEČKAN: *Bifurcation of periodics and subharmonics in abstract nonlinear wave equations*, J. Differ. Equations **153** (1999), 41–60.

[80] M. FEČKAN: *Forced vibrations of abstract wave equations*, Funkc. Ekvacioj **45** (2002), 209–222.

[81] M. FEČKAN: *Chaos in nonautonomous differential inclusions*, Int. J. Bifur. Chaos **15** (2005), 1919–1930.

[82] M. FEČKAN: *Blue sky catastrophes in weakly coupled chains of reversible oscillators*, Discr. Cont. Dyn. Syst. **3** (2003), 193–200.

[83] M. FEČKAN: *Dynamics of nonlinear diatomic lattices*, Miskolc Math. Notes **4** (2003), 111–125.

[84] M. FEČKAN: *Periodic solutions in systems at resonances with small relay hysteresis*, Math. Slovaca **49** (1999), 41–52.

[85] M. FEČKAN: *Topologically transversal reversible homoclinic sets*, Proc. Amer. Math. Soc. **130** (2002), 3369–3377.

[86] M. FEČKAN: *Periodic oscillations of abstract wave equations*, J. Dynam. Diff. Eq. **10** (1998), 605–617.

[87] M. FEČKAN: *Periodic moving waves on 2D lattices with nearest neighbor interactions*, Ukrainian Math. J. **60** (2008), 127–139.

[88] M. FEČKAN: *Transversal bounded solutions for difference equations*, J. Diff. Eq. Appl. **8** (2002), 33–51.

[89] M. FEČKAN: *Multiple solutions of nonlinear equations via Nielsen fixed-point theory: A Survey*, In: Nonlinear Analysis in Geometry and Topology, Th. M. Rassias ed., Hadronic Press, Palm Harbor, FL 2000, 77–97.

[90] M. FEČKAN AND J. GRUENDLER: *Bifurcation from homoclinic to periodic solutions in singular ordinary differential equations*, J. Math. Anal. Appl. **246** (2000), 245–264.

[91] M. FEČKAN AND J. GRUENDLER: *The existence of chaos in infinite dimensional non-resonant systems*, preprint 2008.

[92] M. FEČKAN AND R. KOLLÁR: *Discontinuous wave equations and a topological degree for some classes of multi-valued mappings*, Appl. Math. **44** (1999), 15–32.

[93] M. FEČKAN, R. MA AND B. THOMPSON: *Weakly coupled oscillators and topological degree*, Bull. Sci. Math. **131** (2007), 559–571.

[94] M. FEČKAN AND V. ROTHOS: *Bifurcations of periodics from homoclinics in singular O.D.E.: Applications to discretizations of travelling waves of p.d.e.*, Comm. Pure Appl. Anal. **1** (2002), 475–483.

[95] M. FEČKAN AND V. ROTHOS: *Kink-like periodic travelling waves for lattice equations with on-site and inter-site potentials*, Dynamics of PDE **2** (2005), 357–370.

[96] M. FEČKAN AND V. ROTHOS: *Travelling waves in Hamiltonian systems on 2D lattice with nearest neighbour interactions*, Nonlinearity **20** (2007), 319–341.

[97] N. FENICHEL: *Geometric singular perturbation theory for ordinary differential equations*, J. Differ. Equations **31** (1979), 53–98.

[98] A. FIDLIN: *Nonlinear Oscillations in Mechanical Engineering*, Springer, Berlin, 2006.

[99] S. FLACH AND C.R. WILLIS: *Discrete breathers*, Phys. Rep. **295** (1998), 181–264.

[100] L. FLATTO AND N. LEVINSON: *Periodic solutions to singularly perturbed systems*, J. Ration. Mech. Anal. **4** (1955), 943–950.

[101] I. FONSECA AND W. GANGBO: *Degree Theory in Analysis and Applications*, Oxford Lecture Series in Mathematics and its Applications **2**, Clarendon Press, Oxford (1995).

[102] H. GAJEWSKI, K. GRÖGER AND K. ZACHARIAS: *Nichtlineare Operatorgleichungen und Operatordifferentialgleichungen*, Akademie-Verlag, Berlin, 1974.

[103] L. GÓRNIEWICZ: *Topological Fixed Point Theory for Multivalued Mappings*, Kluwer, Dordrecht, The Netherlands, 1999.

[104] G. GRAMMEL: *Singularly perturbed differential inclusions: An averaging approach*, Set-Valued Anal. **4** (1996), 361–374.

[105] A. GRANAS, R. GUENTHER AND J. LEE: *Nonlinear boundary value problems for ordinary differential equations*, Dissertationes Math. **244**, 1985.

[106] J. GRUENDLER: *Homoclinic solutions for autonomous ordinary differential equations with nonautonomous perturbations*, J. Differ. Equations **122** (1995), 1–26.

[107] J. GRUENDLER: *The existence of transverse homoclinic solutions for higher order equations*, J. Differ. Equations **130** (1996), 307–320.

[108] J. GUCKENHEIMER AND P. HOLMES: *Nonlinear Oscillations, Dynamical Systems, and Bifurcations of Vector Fields*, Springer, New York, 1983.

[109] J.M. HAMMERSLEY AND G. MAZZARINO: *Computational aspects of some autonomous differential equations*, Proc. Royal Soc. London Ser. A **424** (1989), 19–37.

[110] P. HARTMAN: *Ordinary Differential Equations*, Wiley, New York, 1964.

[111] HARTONO AND A.H.P. VAN DER BURGH: *Higher-order averaging: periodic solutions, linear systems and an application*, Nonlinear. Anal.-Theor. **52** (2003), 1727–1744.

[112] M. HASKINS AND J.M. SPEIGHT: *Breather initial profiles in chains of weakly coupled anharmonic oscillators*, Phys. Lett. A **299** (2002), 549–557.

[113] M. HASKINS AND J.M. SPEIGHT: *Breathers in the weakly coupled topological discrete sine-Gordon system*, Nonlinearity **11** (1998), 1651–1671.

[114] M.W. HIRSCH: *Differential Topology*, Springer, New York (1976).

[115] P. HOLMES AND J. MARSDEN: *A partial differential equation with infinitely many periodic orbits: chaotic oscillations of a forced beam*, Arch. Rational Mech. Anal. **76** (1981), 135–165.

[116] F.C. HOPPENSTEADT: *Singular perturbations on the infinite interval*, Trans. Amer. Math. Soc. **123** (1966), 521–535.

[117] F.C. HOPPENSTEADT: *Properties of solutions of ordinary differential equations with small parameters*, Comm. Pure Appl. Math. **24** (1971), 807–840.

[118] F.A. HOWES: *Differential inequalities of higher order and the asymptotic solution of nonlinear boundary value problems*, SIAM J. Math. Anal. **13** (1982), 61–80.

[119] F.A. HOWES: *The asymptotic solution of a class of third-order boundary value problems arising in the theory of thin film flows*, SIAM J. Appl. Math. **43** (1983), 993–1004.

[120] G. IOOSS: *Travelling waves in the Fermi-Pasta-Ulam lattice*, Nonlinearity **13** (2000), 849–866.

[121] G. IOOSS AND M. ADELMEYER: *Topics in Bifurcation Theory and Applications*, World Scientific Publishing, Singapore, 1992.

[122] G. IOOSS AND K. KIRCHGÄSSNER: *Traveling waves in a chain of coupled nonlinear oscillators*, Comm. Math. Phys. **211** (2000), 439–464.

[123] H. KAUDERER: *Nichtlineare Mechanik*, Springer, Berlin, 1958.

[124] H. KIELHÖFER: *Bifurcation Theory, An introduction with Applications to PDEs*, Springer, New-York, 2004.

[125] A. KITTILÄ: *On the topological degree for a class of mappings of monotone type and applications to strongly nonlinear elliptic problems*, Ann. Acad. Sci. Fenn. Ser. A I Math. Disser. **91**, 1994.

[126] A.YA. KHINCHIN: *Continued Fractions*, 3rd ed., Moscow, 1961 (in Russian).

[127] S.G. KRANTZ AND H.R. PARKS: *The Implicit Function Theorem, History, Theory, and Applications*, Birkhäuser, Boston, MA, 2003.

[128] S.B. KUKSIN: *Nearly Integrable Infinite-Dimensional Hamiltonian Systems*, LNM 1556, Springer, Berlin, 1993.

[129] M. KUNZE AND T. KÜPPER: *Qualitative bifurcation analysis of a non-smooth friction-oscillator model*, Z. Angew. Meth. Phys. (ZAMP) **48** (1997), 87–101.

[130] M. KUNZE AND T. KÜPPER: *Non-smooth dynamical systems: an overview*, In: Ergodic Theory, Analysis and Efficient Simulation of Dynamical Systems, B. Fiedler ed., Springer, Berlin, 2001, 431–452.

[131] Yu.A. KUZNETSOV, S. RINALDI AND A. GRAGNANI: *One-parametric bifurcations in planar Filippov systems*, Int. J. Bifur. Chaos **13** (2003), 2157–2188.

[132] D.F. LAWDEN: *Elliptic Functions and Applications*, Applied Mathematical Sciences **80**, Springer, New York, 1989.

[133] C. LAZZARI: *Symmetries and exponentional smallness of bifurcation functions of a class of singular reversible systems*, Nonlinear Anal.-Theor. Meth. Appl. **33** (1998), 759–772.

[134] R.I. LEINE, D.H. VAN CAMPEN AND B. L. VAN DE VRANDE: *Bifurcations in nonlinear discontinuous systems*, Nonlinear. Dynam. **23** (2000), 105–164.

[135] M. LEVI: *Geometry and physics of averaging with applications*, Physica D **132** (1999), 150–164.

[136] N. LEVINSON: *A boundary value problem for a singularly perturbed differential equation*, Duke Math. J. **25** (1958), 331–342.

[137] Y. LI: *Persistent homoclinic orbits for nonlinear Schrödinger equation under singular perturbation*, Dynamics of PDE **1** (2004), 87–123.

[138] Y. LI AND D. McLAUGHLIN: *Morse and Melnikov functions for NLS Pde's.*, Comm. Math. Phys. **162** (1994), 175–214.

[139] Y. LI, D. McLAUGHLIN, J. SHATAH AND S. WIGGINS: *Persistent homoclinic orbits for perturbed nonlinear Schrödinger equation*, Comm. Pure Appl. Math. **49** (1996), 1175–1255.

[140] X.-B. LIN: *Using Melnikov's method to solve Silnikov's problems*, Proc. Roy. Soc. Edinburgh **116A** (1990), 295–325.

[141] J. LLIBRE, J.S. PÉREZ DEL RIO AND J.A. RODRIGUEZ: *Averaging analysis of a perturbed quadratic center*, Nonlinear Anal.-Theor. **46** (2001), 45–51.

[142] R.S. MACKAY AND S. AUBRY: *Proof of existence of breathers for time-reversible or Hamiltonian networks of weakly coupled oscillators*, Nonlinearity **7** (1994), 1623–1643.

[143] J.W. MACKI, P. NISTRI AND P. ZECCA: *Mathematical models for hysteresis*, SIAM Review **35** (1993), 94–123.

[144] J.W. MACKI, P. NISTRI AND P. ZECCA: *Periodic oscillations in systems with hysteresis*, Rocky Mt. J. Math. **22** (1992), 669–681.

[145] J. MAWHIN: *Topological Degree Methods in Nonlinear Boundary Value Problems*, CBMS Regional Conference Series in Mathematics **40**, American Mathematical Society, Providence, RI, 1979.

[146] J. MAWHIN: *Contractive mappings and periodically perturbed conservative systems*, Arch. Math. **12** (1976), 67–74.

[147] J. MAWHIN: *Solutions périodiques d'équations aux dérivés partielles hyperboliques nonlinéaires*, In: Mélanges Théodore Vogel, Rybak, Janssens and Jessel eds., Presses Univ. Bruxelles, 1978, 301–319.

[148] J. MAWHIN: *Nonlinear functional analysis and periodic solutions of semilinear wave equations*, In: Nonlinear Phenomena in Mathematical Sciences, V. Lakshmikantham ed., Academic, New York, 1982, 671–681.

[149] J. MAWHIN AND M. WILLEM: *Operators of monotone type and alternative problems with infinite dimensional kernel*, In: Recent Advances in Differential Equations, Trieste 1978, R. Conti ed., Academic, New York/London, 1981, 295–307.

[150] P.J. McKENNA: *On solutions of a nonlinear wave equation when the ratio of the period to the length of the interval is irrational*, Proc. Amer. Math. Soc. **93** (1985), 59–64.

[151] M. MEDVEĎ: *Fundamentals of Dynamical Systems and Bifurcation Theory*, Adam Hilger, Bristol, 1992.

[152] K.R. MEYER AND G. R. SELL: *Melnikov transforms, Bernoulli bundles, and almost periodic perturbations*, Trans. Amer. Math. Soc. **314** (1989), 63–105.

[153] R.K. MILLER AND A.N. MICHEL: *Sinusoidal input–periodic response in nonlinear differential equations containing discontinuous elements*, In: Proc. Conf. on Integral and Functional Differential Equations, T.L. Herdman, S.M. Rankin and H.W. Stech eds., Marcel Dekker, New York, 1981, 109–117.

[154] K. MISCHAIKOW AND M. MROZEK: *Isolating neighbourhoods and chaos*, Jap. J. Ind. Appl. Math. **12** (1995), 205–236.

[155] K. NIPP: *Smooth attractive invariant manifolds of singularly perturbed ODE's*, SAM-ETH preprint **92–13** (1992)

[156] J.P. PALIS AND W. DE MELO: *Geometric Theory of Dynamical Systems, An Introduction*, Springer, New York, 1982.

[157] K.J. PALMER: *Exponential dichotomies and transversal homoclinic points* J. Differ. Equations **55** (1984), 225–256.

[158] K.J. PALMER: *Exponential dichotomies, the shadowing lemma and transversal homoclinic points*, Dynamics Reported **1** (1988), 265–306.

[159] K.J. PALMER: *Shadowing in Dynamical Systems, Theory and Applications*, Kluwer, Dordrecht, 2000.

[160] K.J. PALMER AND D. STOFFER: *Chaos in almost periodic systems*, Zeit. Ang. Math. Phys. (ZAMP) **40** (1989), 592–602.

[161] M. PEYRARD, ST. PNEVMATIKOS AND N. FLYTZANIS: *Dynamics of two-component solitary waves in hydrogen-bounded chains*, Phys. Rev. A **36** (1987), 903–914.

[162] ST. PNEVMATIKOS, N. FLYTZANIS AND M. REMOISSENET: *Soliton dynamics of nonlinear diatomic lattices*, Phys. Rev. B **33** (1986), 2308–2321.

[163] K. POPP: *Some model problems showing stick–slip motion and chaos*, In: ASME WAM, Proc. Symp. Friction–Induced Vibration, Chatter, Squeal and Chaos, R.A. Ibrahim and A. Soom eds., **49**, ASME New York, 1992, 1–12.

[164] K. POPP, N. HINRICHS AND M. OESTREICH: *Dynamical behaviour of a friction oscillator with simultaneous self and external excitation* In: Sadhana: Academy Proceedings in Engineering Sciences **20**, Part 2-4, Indian Academy of Sciences, Bangalore, India, 1995, 627–654.

[165] K. POPP AND P. STELTER: *Stick–slip vibrations and chaos*, Philos. Trans. R. Soc. London A **332** (1990), 89–105.

[166] T. PRUSZKO: *Some applications of the topological degree theory to multi–valued boundary value problems*, Dissertationes Math. **229**, 1984.

[167] T. PRUSZKO: *Topological degree methods in multi–valued boundary value problems*, Nonlinear Anal.-Theor. **5** (1981), 959–973.

[168] R. REISSIG: *Erzwungene Schwingungen mit zäher Dämpfung und starker Gleitreibung II*, Math. Nachr. **12** (1954), 119–128.

[169] R. REISSIG: *Über die Stabilität gedämpfter erzwungener Bewegungen mit linearer Rückstellkraft*, Math. Nachr. **13** (1955), 231–245.

[170] H.M. RODRIGUES AN M. SILVEIRA: *Properties of bounded solutions of linear and nonlinear evolution equations: homoclinics of a beam equation*, J. Differ. Equations **70** (1987), 403–440.

[171] W. RUDIN: *Real and Complex Analysis*, McGraw-Hill, New York, 1974.

[172] R. RUMPEL: *Singularly perturbed relay control systems*, preprint 1996.

[173] J.A. SANDERS AND F. VERHULST: *Averaging Methods in Nonlinear Dynamical Systems*, Springer, New York, 1985.

[174] A.M. SAMOILENKO AND YU.V. TEPLINSKIJ: *Countable Systems of Differential Equations*, Walter de Gruyter Inc., Utrecht, 2003.

[175] A.V. SAVIN, Y. ZOLOTARYUK AND J.C. EILBECK: *Moving kinks and nanopterons in the nonlinear Klein-Gordon lattices*, Physica D **138** (2000), 267–281.

[176] A. SCOTT: *Nonlinear Sciences: Emergence and Dynamics of Coherent Structures, 2nd ed.*, Oxford University Press, Oxford, 2003.

[177] J.A. SEPULCHRE AND R.S. MACKAY: *Localized oscillations in conservative or dissipative networks of weakly coupled autonomous oscillators,* Nonlinearity **10** (1997), 679–713.

[178] G.J. SHILOV: *Mathematical Analysis*, Nauka, Moscow, 1969 (in Russian).

[179] I.V. SKRYPNIK: *Methods of Investigation of Nonlinear Elliptic Boundary Value Problems*, Nauka, Moscow, 1990 (in Russian).

[180] R. SRZEDNICKI AND K. WÓJCIK: *A geometric method for detecting chaotic dynamics*, J. Differ. Equations **135** (1997), 66–82.

[181] D. STOFFER: *Transversal homoclinic points and hyperbolic sets for nonautonomous maps I, II*, Zeit. Ang. Math. Phys. (ZAMP) **39** (1988), 518–549, 783–812.

[182] T. STRUNZ AND F.J. ELMER: *Driven Frenkel-Kontorova model. I. Uniform sliding states and dynamical domains of different particle densities*, Phys. Rev. E **58** (1998), 1601–1611.

[183] V.V. STRYGIN AND V.A. SOBOLEV: *Separation of Motions by the Method of Integral Manifolds*, Nauka, Moscow, 1988 (in Russian).

[184] P. SZMOLYAN: *Heteroclinic orbits in singularly perturbed differential equations*, IMA preprint **576**.

[185] P. SZMOLYAN: *Transversal heteroclinic and homoclinic orbits in singular perturbation problems*, J. Differ. Equations **92** (1991), 252–281.

[186] J.M. TAMGA, M. REMOISSENET AND J. POUGET: *Breathing solitary waves in a sine-Gordon two dimensional lattice*, Phys. Rev. Lett. **75** (1995), 357–361.

[187] A.N. TIKHONOV: *Systems of differential equations containing small parameters multiplying some of the derivatives*, Mat. Sb. **31** (1952), 575–586.

[188] A. VANDERBAUWHEDE: *Heteroclinic cycles and periodic orbits in reversible systems*, In: Ordinary and Delay Differential Equations, J. Wiener and J.K. Hale eds., Pitman Research Notes in Mathematics Series **272**, Pitman, 1992, 250–253.

[189] A. VANDERBAUWHEDE AND B. FIEDLER: *Homoclinic period blow-up in reversible and conservative systems*, Z. Angew. Math. Phys. (ZAMP) **43** (1992), 292–318.

[190] A. VANDERBAUWHEDE AND G. IOOSS: *Center manifold theory in infinite dimensions*, Dynamics Reported **1** (1992), 125–163.

[191] A. VANDERBAUWHEDE AND S.A. VAN GILS: *Center manifolds and contractions on a scale of Banach spaces*, J. Funct. Anal. **72** (1987), 209–224.

[192] S. VARIGONDA AND T.T. GEORGIOU: *Dynamics of relay relaxation oscillators*, IEE Trans. Aut. Contr. **46** (2001), 65–77.

[193] A.B. VASIL'EVA and B.F. BUTUZOV: *Asymptotic Expansions of the Solutions of Singularly Perturbed Equations*, Nauka, Moscow, 1973.

[194] Z. WEILI: *Singular perturbations of boundary value problems for a class of third order nonlinear ordinary differential equations*, J. Differ. Equations **88** (1990), 265–278.

[195] S. WIGGINS: *Chaotic Transport in Dynamical Systems*, Springer, New York, 1992.

[196] Z. XIA: *Homoclinic points and interactions of Lagrangian submanifolds*, Discr. Cont. Dyn. Systems **6** (2000), 243–253.

[197] K. YAGASAKI: *Homoclinic and heteroclinic behaviour in an infinite-degree-of-freedom Hamiltonian system: chaotic free vibrations of an undamped, buckled beam*, Phys. Lett. A **285** (2001), 55–62.

[198] M. YAMAGUCHI: *Existence of periodic solutions of second order nonlinear evolution equations and applications*, Funkc. Ekvacioj **38** (1995), 519–538.

[199] K. YOSIDA: *Functional Analysis*, Springer, Berlin, 1965.

Index